O linary Lo

HIGH VOLTAGE DEVICES AND CIRCUITS IN STANDARD CMOS TECHNOLOGIES

HIGH VOLTAGE DEVICES AND CIRCUITS IN STANDARD CMOS TECHNOLOGIES

by

Hussein Ballan
Swiss Federal Institute of Technology, Lausanne

and

Michel Declercq
Swiss Federal Institute of Technology, Lausanne

KLUWER ACADEMIC PUBLISHERS
DORDRECHT / BOSTON / LONDON

A C.I.P. Catalogue record for this book is available from the Library of Congress.

ISBN 0-7923-8234-X

Published by Kluwer Academic Publishers,
P.O. Box 17, 3300 AA Dordrecht, The Netherlands.

Sold and distributed in the North, Central and South America
by Kluwer Academic Publishers,
101 Philip Drive, Norwell, MA 02061, U.S.A.

In all other countries, sold and distributed
by Kluwer Academic Publishers,
P.O. Box 322, 3300 AH Dordrecht, The Netherlands.

Printed on acid-free paper

Printed in the Netherlands.

CONTENTS

CONTENTS v

FOREWORD ix

1 INTRODUCTION 1

2 SUPPLY VOLTAGE LIMITS IN STANDARD CMOS TECHNOLOGIES 5

2.1 Channel hot carrier effects 6
2.1.1 Substrate current model 8
2.1.2 Gate current model 12
2.1.3 MOSFET degradation 16
2.1.4 Static device lifetime model 20
2.1.5 Dynamic device lifetime model 21

2.2 Breakdown voltage 23
2.2.1 Avalanche breakdown 23
2.2.2 Surface breakdown 35
2.2.3 Snapback breakdown 39
2.2.4 Gate-oxide breakdown 42

2.3 Drain Induced Barrier-Lowering (DIBL) 43

2.4 References 48

3 MOSFET HIGH-VOLTAGE TECHNOLOGIES **51**

3.1 Isolated high-voltage MOSFETs 53
3.1.1 Vertical buried layer DMOSFET 54
3.1.2 Lateral DMOSFETs 56

3.2 PICs and HVICs in HV device technologies 68
3.2.1 P-N junction isolation 68
3.2.2 Dielectric isolation 69

3.3 PICs and HVICs in LV CMOS technologies 71

3.4 References 75

4 DESIGN OF HIGH-VOLTAGE DEVICES USING THE SVX TECHNIQUE77

4.1 Main steps of a standard n-WELL CMOS process 78
4.1.1 The n-WELL implant and drive-in (Mask M26) 78
4.1.2 Active areas definition (Mask M10) 78
4.1.3 Isolation between transistors (Mask M11) 79
4.1.4 Field oxide growth (LOCOS) 79
4.1.5 Polysilicon gate deposition and gate oxide growth (Mask M40) 79
4.1.6 Source and drain implant (Masks M25 and M24) 79
4.1.7 CVD SiO_2 and metal1 deposition (Masks M50 and M60) 80
4.1.8 Metal2 deposition and passivation (Masks M51 and M61) 80

4.2 Design of the SVX devices 85
4.2.1 High-Voltage nMOS transistor (HVNMOS) 85
4.2.2 High-voltage pMOS transistor (HVPMOS) 91

4.3 Scaling of the high-voltage devices 97
4.3.1 High-voltage nMOS transistor (HVNMOS) 97
4.3.2 High-voltage pMOS transistor (HVPMOS) 104

4.4 Computer simulations of the high-voltage devices with a specific scaling of L_c, L_g and L_{dd} 106
4.4.1 High-voltage nMOS transistor 107
4.4.2 High-voltage pMOS transistor 110

4.5 References 116

5 MEASUREMENT AND MODELLING OF THE HIGH-VOLTAGE DEVICES **117**

5.1 Measurement results of the high-voltage devices 118
5.1.1 Static characteristics of the HVNMOS transistor 118
5.1.2 Dynamic characteristics of the HVNMOS transistor 126

5.1.3 Static characteristics of the HVPMOS transistor 132
5.1.4 Dynamic characteristics of the HVPMOS transistor 138

5.2 Device modelling for circuit simulators 143
5.2.1 Theoretical considerations of the macro-models 145
5.2.2 Simulations versus measurements 157

5.3 Layout precautions against parasitic effects 166
5.3.1 Layout rules and precautions at a device level 167
5.3.2 Layout rules and precautions at a circuit level 172

5.4 References 178

6 HIGH-VOLTAGE ANALOG AND DIGITAL OUTPUT INTERFACES 179

6.1 High-voltage digital output interfaces 180
6.1.1 Basic static level shifter cell 182
6.1.2 Improved static level shifter cell 185
6.1.3 low standby current static level shifter cell 188
6.1.4 Transients considerations 192

6.2 High-voltage analog output interfaces 207
6.2.1 Basic operational transconductance amplifier (OTA) 207
6.2.2 Class-AB output stage operational amplifier 210
6.2.3 Transients considerations 219

6.3 References 226

7 12V DELTA-SIGMA CLASS-D AUDIO AMPLIFIER 227

7.1 General system considerations 228

7.2 Modulation stage 231
7.2.1 Theoretical considerations 231
7.2.2 Switched capacitor implementation 236

7.3 Output stage 242
7.3.1 High-voltage power devices 242
7.3.2 Output drivers 246
7.3.3 Protection features 248

7.4 Measured performances 255
7.4.1 Signal-to-noise ratio 255
7.4.2 Total harmonic distortion 258
7.4.3 Intermodulation distortion 259
7.4.4 Amplifier efficiency 260

7.5 System improvements 261

7.6 References 265

8 CONCLUSIONS 267

APPENDIX A 271

APPENDIX B 277

INDEX 283

FOREWORD

Standard voltages used in today's IC's may vary from about 1.3V to more than 100V, depending on the technology and the application. High-voltage is therefore a relative notion.

This book is mainly focused on standard CMOS technologies, where high-voltage (HV) is defined as any voltage higher than the nominal (low) voltage, i.e. 5V, 3.3V, or even lower. In this standard CMOS environment, IC designers are more and more frequently confronted with HV problems, particularly at the I/O level of the circuit.

In a first group of applications, a large range of industrial or consumer circuits either require HV driving capabilities, or are supposed to work in a high-voltage environment. This includes ultrasonic drivers, flat panel displays, robotics, automotive, etc. On the other hand, in the emerging field of integrated microsystems, MEMS actuators mainly make use of electrostatic forces involving voltages in the typical range of 30 to 60V. Last but not least, with the advent of deep sub-micron and/or low-power technologies, the operating voltage tends towards levels ranging from 1V to 2.5V, while the interface needs to be compatible with higher voltages, such as 5V.

For all these categories of applications, it is usually preferable to perform most of the signal processing at low voltage, while the resulting output is risen to a higher voltage level. Solving this problem requires some special actions at three levels : technology, circuit design and layout. The purpose of this book is to address these topics. Theoretical background is supported by practical information and design examples.

1 INTRODUCTION

The increasing integration density of VLSI (*Very-Large-Scale-Integrated*) circuits and the low-power requirements of complex signal processing applications, force the continuous reduction of the power supply voltages in modern ICs (*Integrated Circuits*). While the low-voltage CMOS (*Complementary Metal Oxide Semiconductor*) technologies implementing these ICs are optimized for speed, minimum power consumption and maximum integration density, they cannot meet the requirements of system applications where high-voltage capabilities are needed.

When electronic signal processing involves high-voltage and/or high-current, different solutions are selected according to the power level of the application.

High-power systems with a rating above 50 or 100 watts, such as AC motor controls and factory automation, usually make use of power modules containing discrete devices, HVICs (*high-voltage integrated circuits*) and standard signal processing ICs. Medium-power systems, involving power driving capability up to about 30 watts, can be fully integrated on the basis of a specially dedicated technology. Audio power amplifiers or some automotive applications enter in this category.

Finally there is a bunch of low-power systems that involve some high-voltage signal processing. Typical applications are, for instance, drivers for ultrasonic transducers, display drivers, telecommunication circuits, EEPROM circuits, integrated micro-systems (MEMS), some automotive electronics, small DC motor control, ink-jet printers, switching regulators, medical instrumentation and many others. Moreover, 5V I/O interfaces can be considered as "high-voltage" parts for deep sub-micron technologies. This range of applications is specifically addressed by this book.

For most of these applications, it is usually preferable to perform most of the signal processing at low voltage, while the resulting output is risen to a higher voltage level. Solving this problem requires some special actions both at the technology level and at the circuit design level. These two topics will be adressed in the following chapters.

At the technology level, two different approaches can be selected. The most straightforward, but usually not the cheapest one, is to use a CMOS technology offering a HV option (with HV DMOS devices) at the cost of one or two supplementary masks and ion implantation steps. While such an approach optimizes area efficiency, it may result in a significant cost increase.

A more original approach, named SVX (Smart Voltage Extension), consists in creating high-voltage devices in a standard, unmodified low-voltage CMOS technology without any process change or addition. In this option, which is thoroughly addressed in the book, existing technological layers are combined in an unconventional way, locally using special layout rules for creating "free" high-voltage devices usually capable to sustain about 10 times the nominal voltage of the technology.

The motivations for extending a low-voltage CMOS technology with high-voltage features are related to the significant advantages offered by this approach in comparison to dedicated technologies. From an economical point of view, the access to the technology is easy and the production cost of the resulting HVICs are as low as a standard low-voltage CMOS circuit. Moreover, since no modification of the process parameters is required to combine both features on the same substrate, existing standard libraries can be used for the low-voltage part of the circuit. CAD (*Computer Aided Design*) tools remain unchanged since special layout rules and models are added for the high-voltage part of the circuit. From a technical point of view, high-voltage devices with thin gate oxides exhibit standard threshold voltage values as well as high current gains. Thus, considering the same current ratings, the surface density that can be achieved with such devices is maximized.

At the circuit and layout levels, using high-voltage devices and handling high voltages on a chip require some very special care. As a matter of fact, a high-voltage transistor is designed to sustain high-voltage between its source and drain terminals, while the gate-to-source voltage drop is limited to much lower values. This requires special circuit design techniques for respecting these limits both in steady state and in transient conditions. On the other hand, special design techniques, such as guard rings and shields, must be applied as necessary for avoiding latch-up or parasitic channel problems.

The structure of this book is organized as follows:

Chapter 2 presents the different undesirable effects that can take place in a low-voltage transistor when one of its terminals is pushed beyond the voltage limit set by the technology. It includes destructive as well as degrading effects. This chapter summarizes all the basic rules required to implement reliable high-voltage devices.

Chapter 3 presents the existing device-level solutions which avoid the different undesirable effects described in chapter 2. It also overviews the existing technological solutions combining high-voltage and low-voltage features on the same substrate. More attention is focused on technologies starting from a low-voltage CMOS process and where the required technological additions or modifications are performed to implement the high-voltage devices.

Chapter 4 details the SVX approach, starting with the layout techniques used to implement the high-voltage devices without modifying the processing steps of a standard n-well CMOS technology. The scaling of the transistors are discussed with the support of two dimensional computer simulations. Reliability aspects are also discussed in details.

Chapter 5 presents the measured static and dynamic characteristics of the high-voltage devices. On the basis of these results, a first order macro-model is proposed for both the n-channel and p-channel HV transistors. The last section of this chapter deals with the layout rules and precautions that must be used to avoid undesirable parasitic effects when high-voltage signals are handled on the chip.

The validation of the SVX technique is achieved by the design of high-voltage integrated circuits. To reach this objective, circuit design techniques, such as level shifters, are first developed to overcome the limitation of the reduced gate-to-source voltage swing. Then, the design and optimization of high-voltage analog and digital output interfaces are performed. These interfaces are currently used in actual industrial HVICs designed with the SVX technique. Finally, a medium-power system application is developed to point out the power limits of such an approach, and to demonstrate its flexibility in combining low and high voltage circuits on the same substrate.

Chapter 6 is dedicated to the design of high-voltage integrated circuits. To reach this objective, circuit design techniques, such as level shifters, are first developed to overcome the limitation of the reduced gate-to-source voltage swing.. The sensitivity of the digital output interfaces to the high-voltage power supply and output stage transients is analyzed. The concept of analog level shifting is also introduced with the example of two high-voltage analog output interfaces. Their sensitivities to the same transients as for the digital interfaces are also considered.

Chapter 7 presents a medium-power system application which consists of a 12V Δ-Σ class-D audio amplifier. It helps to delimit the power limits of the SVX technique and to point out its flexibility in combining complex low-voltage analog and digital circuits with high-voltage power output devices.

Chapter 8 provides the conclusions and future trends in HV.

Annexe A contains the code of the macro-models as well as the resulting SPICE parameters of the high-voltage devices. It summarizes the high-voltage layout rules as well.

Annexe B contains the code of the of the behavioral analog model of the Δ-Σ modulator. The schematics of the general purpose blocks of the modulator are attached as well.

2 SUPPLY VOLTAGE LIMITS IN STANDARD CMOS TECHNOLOGIES

Abstract:
Various undesirable effects can take place in a low-voltage transistor if one of its terminals is pushed beyond the voltage limit set by the technology. The scope of this chapter is the investigation and modelling of these effects. A particular interest is first focused on the degradation of the device characteristics resulting from channel hot-carrier effects. *Then, destructive mechanisms such as,* avalanche breakdown, surface breakdown, snapback breakdown, punchthrough breakdown and gate oxide breakdown *are analysed. These theoretical considerations are the basics required to implement the design of reliable high-voltage devices.*

As was discussed in the previous chapter, high-voltage integrated circuits are usually used to interface the signal processing units, which are running under low-voltage power supply, and the external high-voltage application. Typical low-voltage and high-voltage values range between 2.9V..5V and 12V..1000V respectively. The ideal technical and economical case would combine the low-voltage units and the high-voltage ones on the same chip, without any loss in the performances of each block. Unfortunately, it is not possible in all cases for different reasons.

On the one hand, standard low-voltage CMOS technologies are developed for optimum electric characteristics, minimum parasitic components and maximum integration density which does not optimise them for high-voltage use. Indeed, optimising transconductance g_m, threshold voltage V_t and parasitic components suggests an oxide thickness and doping levels that increase hot-carrier effects and lower breakdown voltages. Moreover, and even in low-voltage applications, the continuous down scaling of MOSFET devices for VLSI raises a serious problem of device degradation due to the hot-carrier effects as well as punchthrough current due to the drain induced barrier lowering. These non desirable effects have been the main motivation in scaling down the power supply in recent submicron technologies.

On the other hand, high-voltage technologies are not dedicated for complex signal processing for the opposite reasons as the previous case. For instance, gate oxide thickness and doping levels are such that to achieve an equivalent g_m huge devices are required, which significantly decreases the integration density. Some particular technologies, however, perform the combination of high and low-voltage devices on the same substrate at the cost of an increased complexity of the process. These technologies will be discussed in detail in the next chapter.

In order to emphasise the difficulties in combining both types of devices on the same substrate, the following sections discuss the limiting effects in standard technologies resulting from an increase of the supply voltage beyond a critical value.

2.1 Channel hot carrier effects

In thermal equilibrium, minority carriers in silicon such as electrons (holes) are close to the conduction band (valence band) with a thermal energy $E\text{-}E_c \approx kT$ ($E_v\text{-}E \approx kT$), where T is the absolute temperature of the device and k the boltzman constant. Under the influence of a small electric field these electrons are not any more in thermal equilibrium with the lattice and they drift with a velocity proportional to the electric field $v=\mu_n E$. The proportionality factor μ_n is the mobility of the electrons. The small increase in energy, in comparison with the initial thermal energy, is such that they are still very close to the conduction band $E\text{-}E_c \approx kT$, but they move along the direction of the electric field. As soon as they enter a reverse biased junction they keep moving along the direction of the electric field, but they start to acquire a continuous increase in kinetic energy until they collide with the lattice and lose their energy by phonon emission. This situation is represented schematically in Figure 2.1. A phonon is the elementary amount of energy which quantifies the vibration waveforms of the lattice atoms that are in continuous thermal vibration around their equilibrium position.

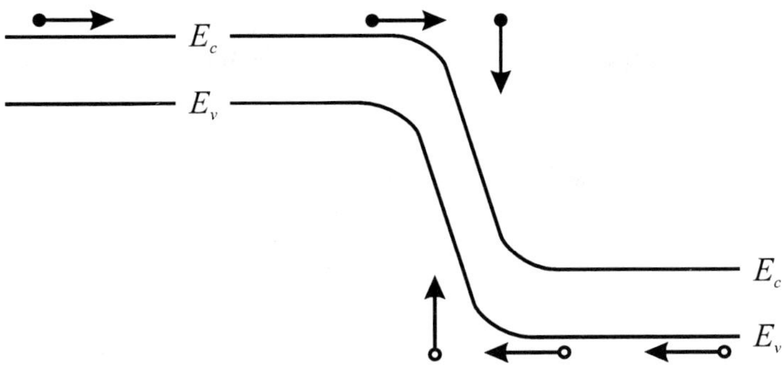

Figure 2.1: Energy band diagram showing electrons close to E_c for low electric fields, above E_c for high electric fields

At the instant of collision the electron energy is $E\text{-}E_c=kT_e > kT$, where T_e is its effective temperature. T_e can be several orders of magnitude grater than the ambient temperature even if the device is at room temperature. The same behaviour is valid for holes as well. Such energetic electrons and holes are called *hot-carriers.*

The mechanism of hot-carriers is typically present in a MOSFET structure when, considering an nMOS transistor (nMOST) for example, the electrons drift in the channel and enter the high field region of the drain junction. These heated electrons are known as *channel hot-electrons* (CHE). Under particular bias conditions, they can acquire enough kinetic energy to cause impact ionisation or some of them can even overcome the Si-SiO_2 interface barrier and enter into the gate oxide. Different effects arise from these CHE. They are shown in the cross-section of the nMOST shown in Figure 2.2.

When a primary CHE acquires enough energy ($\varphi_I \geq 1.3$eV) to cause an impact ionisation, it generates a secondary electron-hole pair. While the primary and secondary electrons constitute the drain-source current I_{ds}, the secondary holes drift through the substrate until they reach the substrate contact and constitute the substrate current I_{sub} [1]. Thus, I_{sub} could be a good monitor to predict CHE generation. The model of I_{sub} is developed in section 2.1.1. If I_{sub} is large enough, it can create a local ohmic voltage drop in the substrate and forward bias the source-to-substrate junction resulting in a snapback breakdown due to the open base transistor discussed in section 2.2.2. In CMOS technologies the same mechanism can trigger latchup as well.

When the energy of the hot-carriers is greater than a critical value ($\varphi_b \geq 3.2$eV for electrons and $\varphi_b \geq 4.9$eV for holes) they can overcome the energy barrier of the Si-SiO_2 interface and are injected into the gate oxide. The majority of these carriers is collected by the gate electrode and constitute the gate current I_g. A fraction N_{ot} is either trapped in the oxide by electrically active defects or create fast interface states N_{it} at the Si-SiO_2 interface. In both cases, device degradation (threshold voltage shift and decrease of the transconductance and subthreshold slope) is the main issue of this channel-hot-carriers injection. The Gate current I_g is a good monitor for the channel-hot-carriers injection as it is very sensitive to the trapping effects [2]. The model of I_g is developed in section 2.1.2.

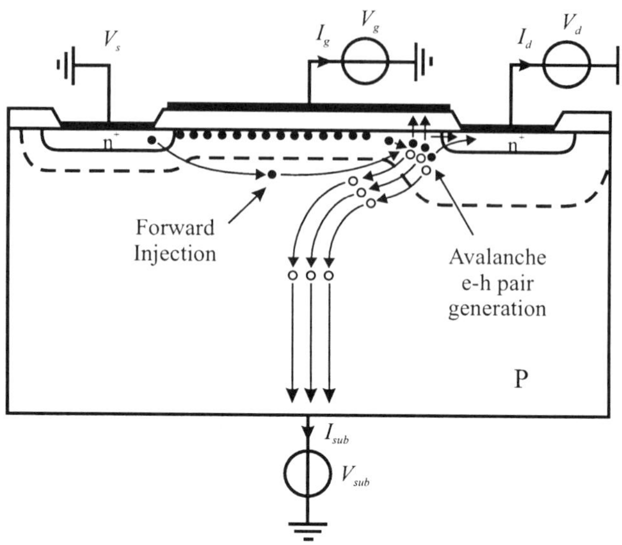

Figure 2.2: Cross-section of an nMOST in saturation with the different CHE effects

At this point, It should be noted that hot-carrier-effects are more serious in the case of nMOSTs than the case of pMOSTs due to the smaller energy barrier and longer mean-free-path of electrons. Higher electron energy results from a longer mean-free-path leading to a higher impact ionisation rate. However, hot-carrier-effects become increasingly important for half-micron pMOS devices and below.

Both, I_g and I_{sub} are the basic monitors to predict device lifetime or degradation. While I_g is a direct measurement of the injected carriers and is very sensitive to the trapping mechanism, I_{sub} is an indirect way of measuring hot-carriers injection. The reason is that I_{sub} results from impact ionisation whether the carriers have enough energy to overcome the Si-SiO_2 barrier or only to create impact ionisation. However, as the common driving force of all hot-electron effects is the maximum electric field E_m at the drain end of the channel there is a correlation between I_{sub} and I_g [3]. Moreover, the difference of magnitude between I_{sub} and I_g makes I_{sub} easier to measure.

2.1.1 Substrate current model

A model for the substrate current was developed by [1] on the basis of a pseudo 2D-model of the electric field E near the drain end. Figure 2.3 represents a cross-section of an nMOST in the saturation region and the distribution of the electric field.

Neglecting the mobile carriers in the space charge region ABCD, the electric field at any point y' along the surface and particularly at the drain end of the channel is given by

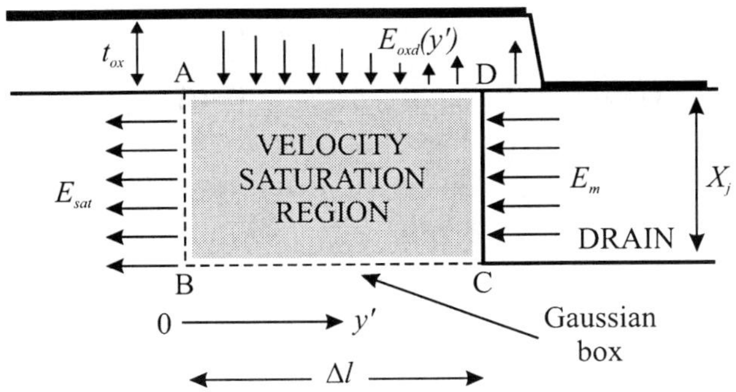

Figure 2.3: Cross-section of an nMOST in the saturation region and its electric field distribution. From P. K. KO, “approaches to scaling”, chap. 1, in *Advanced MOS Device Physics*, N.G. Einspruch and G. Gildenblatt, Eds., Vol. 18 VLSI Electronics. Copyright ©1989 by Academic Press. Reprinted with permission.

$$E_m = \sqrt{\frac{(V_{ds} - V_{dsat})^2}{l^2} + E_{sat}^2} \tag{2.1}$$

where

$$l^2 = \frac{\varepsilon_{si}}{\varepsilon_{ox}} t_{ox} X_j \tag{2.2}$$

X_j is the drain junction depth and t_{ox} the gate oxide thickness. Later in [4], a simplified version of E_m was derived from Equation (2.1) and is given in Equation (2.3). It was based on the following observation: for a V_{ds} approximately 2V larger than V_{dsat}, $(V_{ds}-V_{dsat})/l >> E_{sat}$ such that E_m is roughly proportional to $V_{ds}-V_{dsat}$. Hence, E_m can be expressed as

$$E_m = \frac{V_{ds} - V_{dsat}}{l} = \frac{V_{ds} - V_{dsat}}{0.22 t_{ox}^{1/3} X_j^{1/2}} \tag{2.3}$$

where l, t_{ox} and X_j are in centimetres, and l is interpreted as the effective length of the velocity-saturation region. A good model of the saturation voltage is given by

$$V_{dsat} = \frac{(V_g - V_t) L E_{sat}}{V_g - V_t + L E_{sat}} \tag{2.4}$$

where E_{sat} is the critical field for velocity saturation of the carriers (~4x10^{4}V/cm) and L is the effective channel length. Considering that the substrate current is given by $I_{sub}=(M-1)I_d$ where $(M-1)$ is the multiplication factor, I_{sub} can be written as [1]

$$I_{sub} = \int_0^{\Delta l} I_d A_i \exp\left(-\frac{B_i}{E(y)}\right) dy$$

where A_i and B_i are called the impact ionisation constants. Using the simplification made for E_m, it can be shown that I_{sub} is given [4]

$$I_{sub} = \frac{A_i}{B_i} l E_m I_d \exp\left(-\frac{B_i}{E_m}\right) = C_1 I_d \exp\left(-\frac{B_i}{E_m}\right) \qquad \textbf{(2.5)}$$

where $C_1 \approx 2$. Replacing E_m in Equation (2.5) with Equation (2.3), I_{sub} becomes

$$I_{sub} = \frac{A_i}{B_i}\left(V_{ds} - V_{dsat}\right) I_d \exp\left(-\frac{lB_i}{V_{ds} - V_{sat}}\right) \qquad \textbf{(2.6)}$$

The most common way of characterising the substrate current is to plot I_{sub} against gate voltage V_{gs} with drain voltage V_{ds} as a parameter as shown in Figure 2.4. The initial rise is due to the increase of the drain current and further increase in V_{gs} results in a decrease in I_{sub} due to an increase in V_{dsat}, thus reducing the channel electric field. The result is a bell shaped curve with a maximum occurring at a gate voltage $V_{gs} \approx (0.3...0.5)\ V_{ds}$.

In order to perform the calculated plot of Figure 2.4 it is necessary to know, in Equation (2.6), the values of the impact ionisation constants. This could be achieved indirectly by plotting $\ln[I_{sub}/I_d(V_{ds} - V_{dsat})]$ versus $1/(V_{ds} - V_{dsat})$ from experimental data on I_{sub}. Such a plot is shown in Figure 2.5 where all data points fall on a simple straight line. The slope of this line is $-lB_i$ and the intercept point with the y axis yields the value of A_i/B_i. It was shown in [5] that the line in Figure 2.5 is common for different gate voltages and different channel lengths. This means that, in order to plot the line in Figure 2.5, two I_{sub} measurements at two bias points are needed on a single device. The line, in turn, could be used to characterise I_{sub} for arbitrary channel length, width and bias conditions.

The slope of the line is $-lB_i$=23V and the intercept point with the y axis is A_i/B_i=0.05V^{-1}. It should be noted that the linear relationship in Figure 2.5 is valid only for I_{sub}/I_d ratio down to 10^{-6}, below which the assumption that $(V_{ds}-V_{dsat})/l >> E_{sat}$ is no longer valid. However, it was shown in [6] that the insertion of the E_{sat} term in Equation (2.3) will keep good linearity for I_{sub}/I_d ratio down to at least 10^{-9}. Although the exact amount of I_{sub} may not be represented by Equation (2.6), the substrate current prediction using the model developed before is in general agreement with the experimental data.

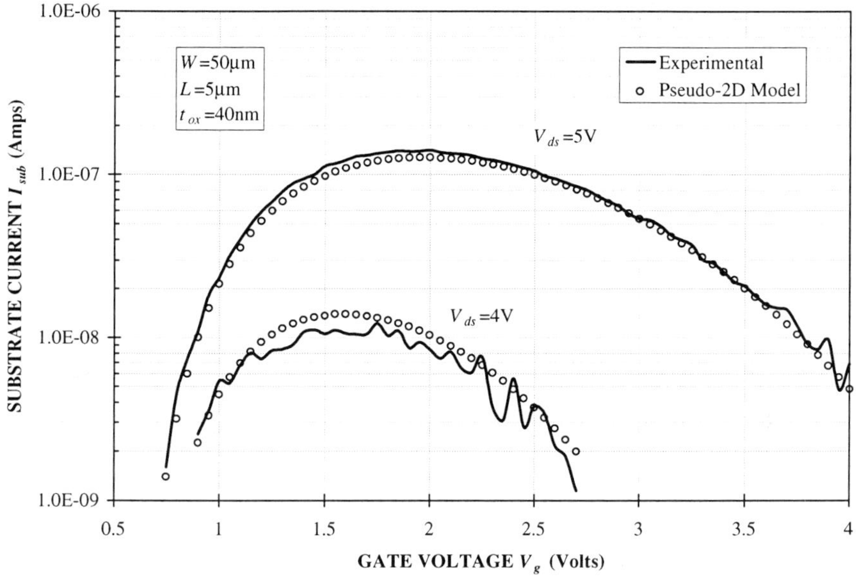

Figure 2.4: Substrate current of an n-channel device characterised as a function of V_{gs} with V_{ds} as a parameter.

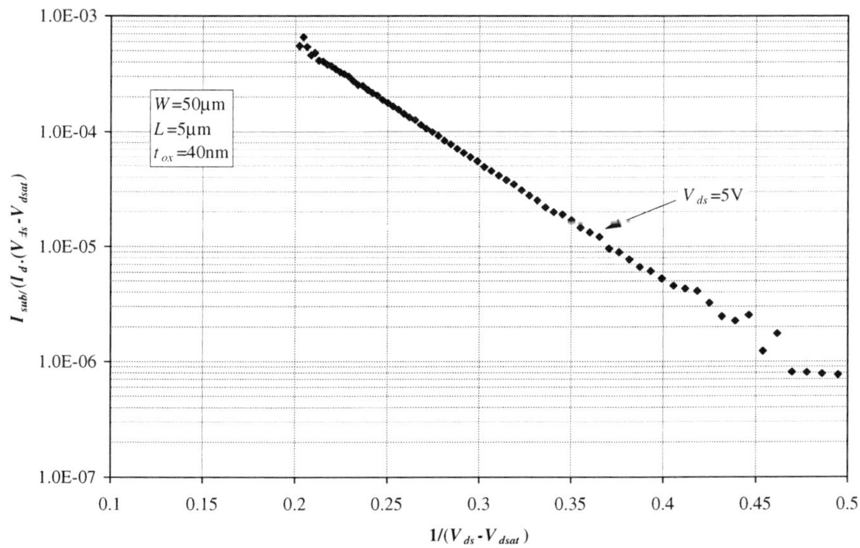

Figure 2.5: Plot of $\ln(I_{sub}/I_d(V_{ds}-V_{dsat}))$ versus $1/(V_{ds}-V_{dsat})$ for an n-channel device with L=5μm, W=50μm and t_{ox}=40nm

Finally, it was shown in [7] that Equation (2.6) applies to pMOS transistors (pMOSTs) as well. The only difference with the n-channel devices is the impact ionisation constants, as the impact ionisation is produced by hot-holes. The same measurements have been performed on a pMOST with the same size and are shown in Figure 2.6 and Figure 2.7. The resulting values are $-lB_i$=25V for the slope and $A_i/B_i = 1x10^{-5}\ V^{-1}$ for the intercept point. Applying these values to Equation (2.6) leads to the calculated curves shown in Figure 2.6.

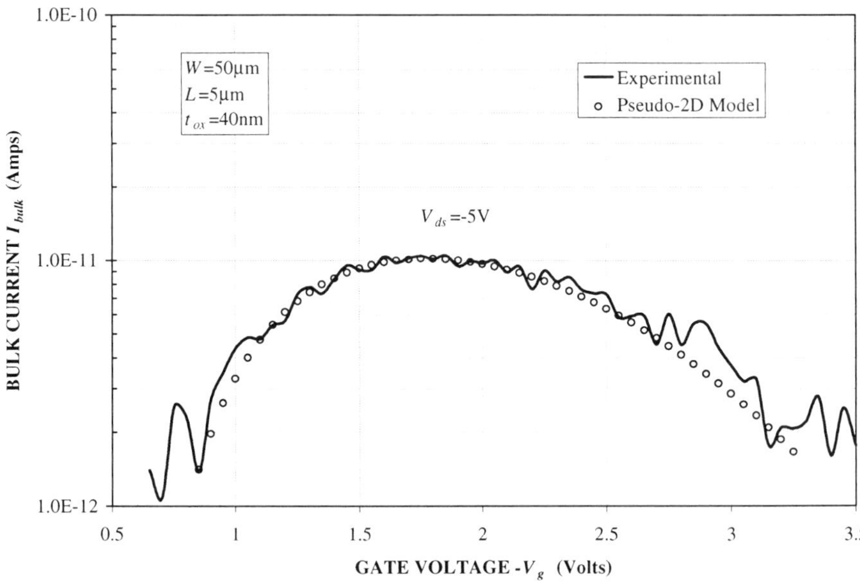

Figure 2.6: Substrate current of a p-channel device characterised as a function of V_{gs} with V_{ds} as a parameter

2.1.2 Gate current model

Two different approaches were used to model the gate current. The first approach is the effective temperature model and is based on the thermionic emission of channel heated electrons over the Si-SiO$_2$ energy barrier. The injected current density J_g is given by the Richardson's expression for thermionic emission

$$J_g = qn_s\left(\frac{kT_e}{2\pi m^*}\right)^{1/2}\exp\left(-\frac{\varphi_b}{kT_e}\right) \tag{2.7}$$

and integrated along the channel length. m^* is the effective mass of an electron, n_s the surface electron density, φ_b the energy barrier of the Si-SiO$_2$ interface and T_e the

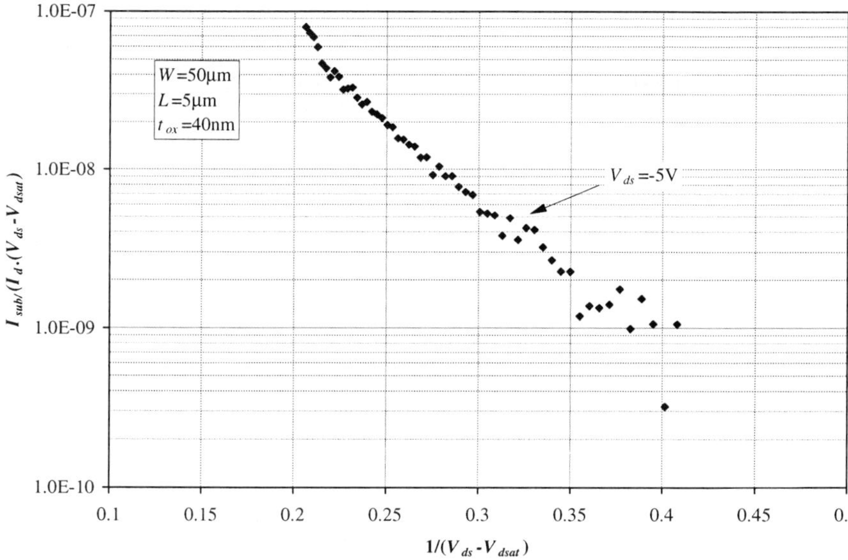

Figure 2.7: Plot of ln($I_{sub}/I_d(V_{ds}-V_{dsat})$) versus $1/(V_{ds}-V_{dsat})$ for a p-channel device with L=5µm, W=50µm and t_{ox}=40nm

effective temperature of the electron. Neglecting the charge trapping in the oxide the gate current I_g is then given by

$$I_g = W\int_0^L J_g(y)dy \tag{2.8}$$

In order to calculate the integral of Equation (2.8), $T_e(y)$ must be known and a two-dimensional device simulator must be used to evaluate the surface electron density n_s and the electric field E. Very good agreement between the measurements and this model was obtained [8].

The second approach is based on the calculation of the probability that an electron drifting in the channel reaches the gate electrode. This approach is called the lucky electron model and have been developed in [9] for the CHE injection. Three scattering events were considered, and the associated probabilities were calculated. On the first hand, the probability that an electron drifts between two points A and B in the channel and gain an energy φ_b higher than the Si-SiO_2 barrier, with a re-direction of its momentum at point B perpendicularly to the interface, is given by

$$P_{\varphi_b} = 0.25\frac{qE_y\lambda}{\varphi_b}\exp\left(-\frac{\varphi_b}{qE_y\lambda}\right) \tag{2.9}$$

where λ is the scattering mean-free-path of the electron. on the second hand, the probability P_1 that this hot electron reaches the interface without losing its energy by any inelastic collision and, the probability P_2 that it scatters in the oxide without suffering any collision, are given by complex relations that are a function of the oxide field E_{ox} [9]. Since these processes are statistically independent, the resulting probability is the product of the probability for each event. Thus, the gate current can be written as

$$I_g = I_{ds} \int_0^L P_{\varphi_b} P_1 P_2 \left(\frac{dy}{\lambda_r} \right) = I_{ds} \int_0^L P_{\varphi_b} P(E_{ox}) \left(\frac{dy}{\lambda_r} \right) \tag{2.10}$$

where λ_r is the re-direction mean-free-path. The above integral was solved by [9] and the resulting gate current can be expressed as

$$I_g = 0.5 \frac{I_d t_{ox}}{\lambda_r} \left(\frac{q\lambda E_m}{\varphi_b} \right)^2 P(E_{ox}) \exp\left(-\frac{\varphi_b}{q\lambda E_m} \right) \tag{2.11}$$

which can be approximated by

$$I_g \approx C_2 I_d \exp\left(-\frac{\varphi_b}{q\lambda E_m} \right) \tag{2.12}$$

where $C_2 \approx 2\text{x}10^{-3}$ for $V_g > V_d$ and E_m is the peak electric field at the drain end. The best fitting between the model and the experimental data was obtained for λ_r=61.6nm and λ=9.2nm. Figure 2.8 shows the measured substrate current I_{sub} and gate current I_g as a function of the gate voltage for V_{ds} values up to 11V. Note that for the considered device dimensions, the gate current remains lower than the instrument precision. Therefore, more sophisticated measurement methods, such as the charge pumping technique, are required at this level. Higher drain voltage values result in a significant increase of the substrate current leading to avalanche breakdown of the drain junction. This is not the case of the p-channel device as will be pointed out further.

The gate current characteristics as a function of V_{gs}, V_{ds} and L can be explained by using the simple models of E_m and V_{dsat} in the expression of I_g. At a fixed drain bias and for a low V_{gs}, the oxide electric field E_{oxd} at the drain end inhibits the collection of the hot electrons by the gate electrode, even if V_{dsat} is small and hence the peak channel electric field E_m is high. As V_{gs} increases towards V_{ds} the repulsion field E_{oxd} decreases and a sharp increase in the gate current is observed. For $V_{gs} > V_{ds}$, all the emitted electrons are collected by the gate electrode, but increasing V_{gs} will also decrease the peak channel electric field E_m resulting in a decrease of the gate current. Consequently, a peak of I_g is observed at $V_{gs} \approx V_{ds}$ when the repulsion field in the oxide E_{oxd} is zero. At a fixed gate bias and when $V_{ds}<V_{gs}$ the gate current increases with increasing V_{ds} because E_m is increasing and E_{oxd} is in a direction favourable to the collection of the injected electrons by the gate electrode. When $V_{ds}>V_{gs}$ the gate current remains constant with increasing V_{ds}. Although the energy of the electrons is increasing with increasing V_{ds}, the increasing repulsion field is preventing the electrons from reaching the gate.

The gate current of a pMOST as a function of V_{gs} with the drain voltage V_{ds} as a parameter is shown in Figure 2.9. It should be noted that the peak of I_g is located at low V_{gs} near the peak of the substrate current. This is due to the fact that I_g is caused by avalanche hot-electrons created by the impact ionisation of channel hot-holes.

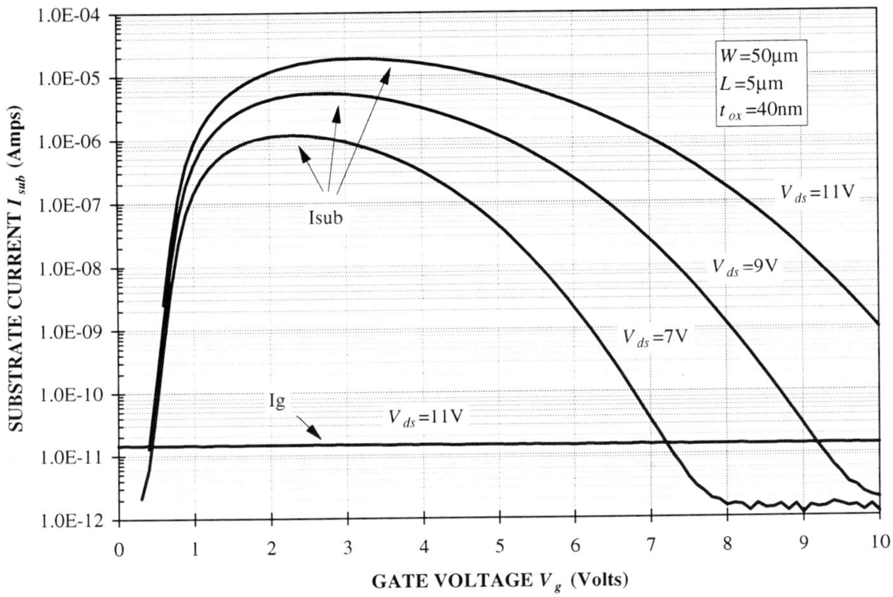

Figure 2.8: Gate and substrate currents of an n-channel device characterised as a function of V_{gs} with V_{ds} as a parameter. (L=5μm, W=50μm and t_{ox}=40nm)

The gate current of the pMOSFET in Figure 2.9 results from electron injection rather than holes. It is due to the lower Si-SiO$_2$ energy barrier, the longer mean-free-path of electrons and the favourable direction of the oxide field for electron injection. Comparing the gate currents of the nMOST and the pMOST, the electron gate current of the pMOST is higher because of the direction of the oxide field E_{ox}, although the number of electrons is several order of magnitudes smaller than in the nMOST. Since the source of electrons in a pMOST is from the impact ionisation of channel hot-holes, causing an electron substrate current I_{sub}, the gate current of a pMOSFET was derived from Equation (2.11) [7] replacing I_d by I_{sub}

$$I_g = 0.5\frac{I_{sub}t_{ox}}{\lambda_r}\left(\frac{q\lambda E_m}{\varphi_b}\right)^2 P(E_{ox})\exp\left(-\frac{\varphi_b}{q\lambda E_m}\right) \quad (2.13)$$

where the expression for $P(E_{ox})$ is the same as in the case of the nMOSFET as well as the energy barrier φ_b and the values of the mean-free-paths λ_r and λ.

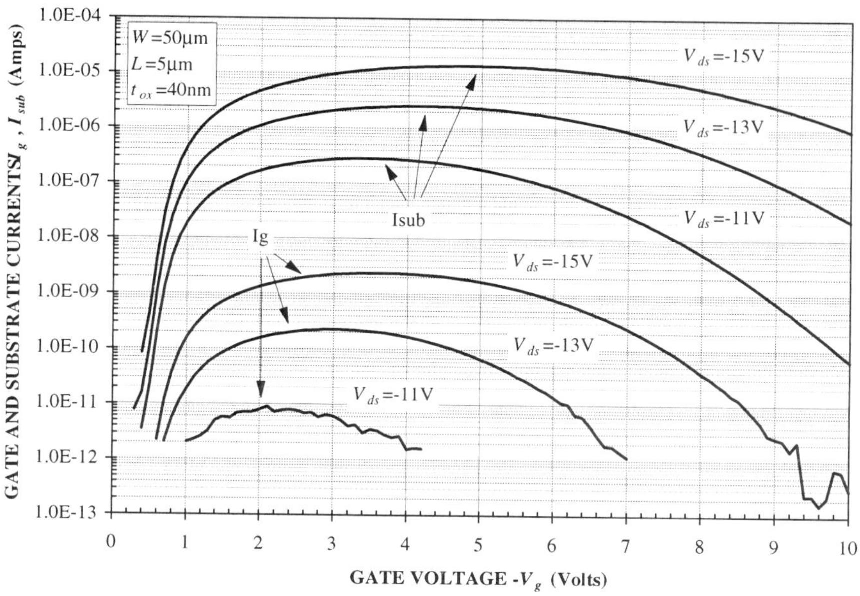

Figure 2.9: Gate and substrate currents of a p-channel device characterised as a function of V_{gs} with V_{ds} as a parameter. (L=5μm, W=50μm and t_{ox}=40nm)

2.1.3 MOSFET degradation

MOSFET degradation, by charge trapping N_{ot} (electrons/holes) or fast interface states generation N_{it}, results in a modification of the electric characteristics such as the threshold voltage V_t, the linear region transconductance g_m and the subthreshold slope ΔS. Whether charge trapping or fast interface states generation is the main cause of the device degradation, is a subject which is still being debated.

In the case of an nMOST, a net negative charge density in the oxide and the interface states is generally responsible for an increase of the threshold voltage and a decrease of g_m and ΔS. The typical I_d-V_g and I_d-V_d characteristics of an n-channel MOSFET are shown in Figure 2.10 and Figure 2.11 before and after stress. The conventional interpretation of the degradation of the device characteristics [10] is to consider, on the first hand, that any positive shift of the characteristics is due to negative trapped charges N_{ot}. This conclusion was demonstrated from uniform *substrate hot-electron* (SHE) injection experiments [11],[2]. They consist of optically generating electrons in the substrate that drift towards the Si-SiO_2 interface and gain enough energy from the high surface electric field to overcome the barrier.

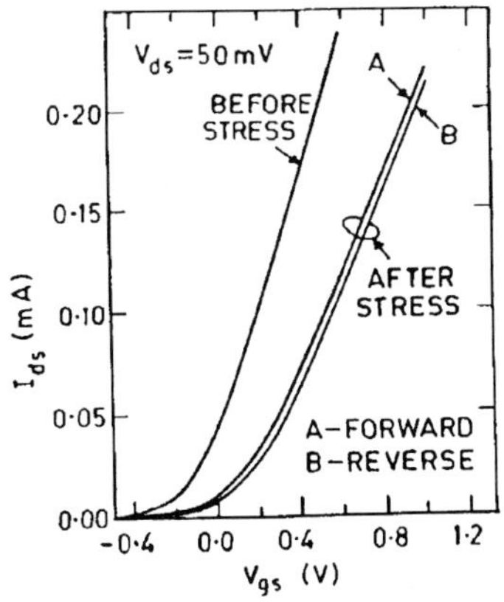

Figure 2.10: Threshold voltage shift of an nMOS transistor due to hot carrier injection (After Hu et al. [3], ©1985 IEEE)

On the second hand, any decrease in the subthreshold gradient and the linear region transconductance is mainly due to fast interface states generation N_{it} causing the channel mobility degradation.

In the conventional interpretation only CHE injection effects were considered and DC stress conditions were performed. The situation is actually more complicated than the simple interpretation presented above. It is often difficult to distinguish between the two degradation mechanisms, with any degree of accuracy, on the bases of stressed characteristics. This is because of the two dimensional nature of the electric field and the localised position of the damage in the oxide near the drain end (about 100nm length starting from the metallurgical junction of the drain).

Recent studies have focused more attention on the possible enhanced effect of AC stress conditions. It was shown in [12] that depending on the gate voltage bias, hot-holes injection is also possible and plays an important role in MOSFET's degradation. Plotting the measured transconductance degradation of a device as a function of the gate voltage V_{gs}, two peaks were observed. The first peak corresponds to a negative relative change of maximum transconductance $\Delta g_m/g_m$ already observed by [13]. It occurs at the maximum substrate current condition $V_{gs}\approx V_{ds}/2$. The second peak is located around $V_{gs}\approx V_t$ and corresponds to a positive $\Delta g_m/g_m$. This is shown in Figure 2.12a where I_{sub}, I_g^- and $\Delta g_m/g_m$ are represented as a function of V_{gs}, as well as a 2-D simulation of the different currents involved, in Figure 2.12b. The interpretation was the following: The bell-shaped positive $\Delta g_m/g_m$ is due to holes injection and trapping at low V_{gs}. The decrease followed by a sudden change of sign in $\Delta g_m/g_m$ is due to an increasing injection of electrons neutralising the trapped holes. The negative peak of $\Delta g_m/g_m$ is determined by a simultaneous injection of holes and electrons where electrons are trapped by the previously trapped holes, which not only neutralise them but also, in a still unknown way, cause interface states generation. Further decrease in negative $\Delta g_m/g_m$ is due to an injection

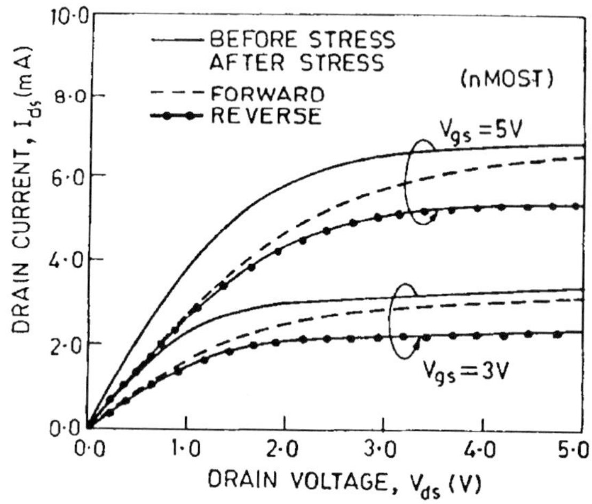

Figure 2.11: Degradation of the I_d-V_d characteristics of an nMOS transistor after stressing at V_{ds}=7.5V and V_{gs}=3V. L=1.2µm, t_{ox}=20nm, t=5minutes (After Ong et al. [7], ©1990 IEEE)

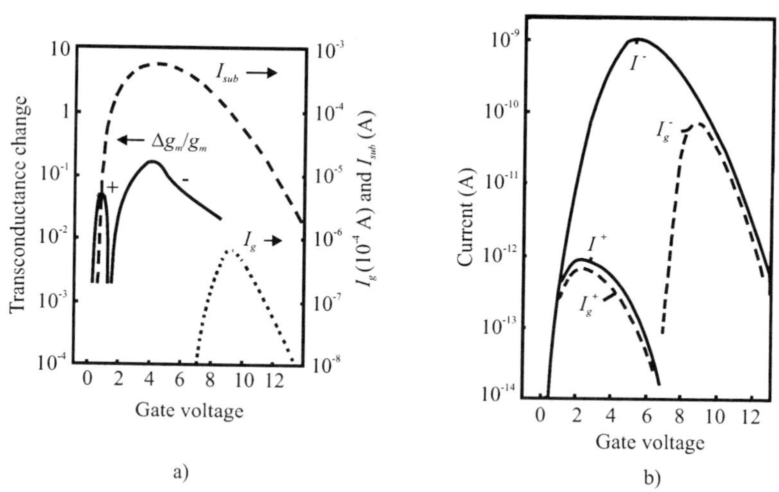

Figure 2.12: a) I_{sub}, I_g^- and $\Delta g_m/g_m$ as a function of gate voltage. L_{eff}=1.8µm, W=100µm, T_{ox}=42nm, stress conditions are V_d=8V, V_b=-2.5V, t=16h b) Total injected electron and hole currents I^-, I^+ and carriers reaching the gate electrode I_g^-, I_g^+ (2-D simulation). (After Hofmann et al. [12], ©1985 IEEE)

of electrons only, responsible for a lateral shift of the characteristics. Measurements of a holes gate current I_g^+ were not performed on the device under stress because of the extremely low values (1fA). This is why a 2-D simulation was made, and in order to validate the results of I_g^+, measurements were performed on a MOSFET with very thin gate oxide (t_{ox}=10nm). Later investigations [14],[15], using the charge pumping technique and the floating gate technique, have confirmed the existence of I_g^+ for higher gate thickness (t_{ox}=24nm) and the generation of a large amount of N_{it} due to a simultaneous injection of electrons and holes.

More recently, the previous degradation theory has been completed by other investigations using alternate stress experiments in order to point out some hidden effects. It consists in applying alternately different DC stress conditions in order to study the interaction of the different injections on the overall degradation behaviour. Therefore, at $V_{gs} \approx V_t$, the condition for hot-holes injection and trapping, neutral electron traps ($N_{ot,h}$) are created [16] and their influence on the I_d-V_g characteristics is important once electrons are injected. Moreover, at this low V_{gs} condition, interface states are also created, but their effect on the I_d-V_g characteristics is masked by trapped holes [17]. An electron injection phase is necessary to neutralise these trapped holes. Under the condition of interface states generation $V_{gs} \approx V_{ds}/2$, it has been shown by [18] that trapping of electrons $N_{ot,e}$ is also present and contribute to the g_m degradation. The effect of the trapped electrons decreases as the degradation increases. The distinction between both degradation mechanisms was performed by a complete compensation of the trapped electrons with a hot-holes injection phase. For $V_{ds}/2 < V_{gs} < V_{ds}$, interface states generation N_{it} and trapping of electrons $N_{ot,e}$ occur during the same stress condition but N_{it} is triggered at a certain finite time after $N_{ot,e}$ [19]. Finally, at $V_{gs} \approx V_{ds}$, the damage is essentially due to the trapping of electrons $N_{ot,e}$, as already stated by [2].

In the case of a pMOST, [ONG] have observed a net negative charge density as well. In contrast to the case of an nMOST, the negative charge density is responsible for a decrease of the threshold voltage and an increase of g_m and I_d. It is believed, as was discussed in section 2.1.2, that the negative charge is due to avalanche hot-electrons that are injected and trapped in the oxide. Furthermore, it is also believed that interface states are not occupied by electrons because of the small density of electrons in pMOSFETs. Therefore they remain neutral without affecting the characteristics of the device. As a consequence, the degradation in pMOSFETs is mainly due to trapped electrons in the oxide, and is maximum under stress conditions of peak gate current. The I_{ds}-V_{ds} characteristics of a pMOST before and after stressing are shown Figure 2.13. The device is 1.2μm length and t_{ox}=20nm. The stress was performed under the maximum gate current and for 5 minutes.

The degradation of device performance arising from hot-carrier effects can be suppressed by reducing the magnitude of the maximum channel electric field E_m. To reach this objective, the most common solutions in submicron technologies use a lightly doped buffer region between the heavily doped drain and the channel. The lightly doped region may be created either by adding fast diffusing phosphorus in double diffused drain (DDD) structure, or by adding a low-dose implant between the gate edge and the heavily doped drain. This is achieved by self-aligning the low-dose implant to the gate electrode, and the other to the gate electrode on which two oxide sidewall spacers have been formed. The resulting drain structure is called lightly doped drain or LDD. It reduces significantly the magnitude of the maximum electric field at the channel end of the device, increasing thus the hot-carrier reliability. The LDD structure is the most widely used solution with gate lengths smaller than 1.25μm. The main reason is that the DDD structure do not allow simultaneously an optimised lightly doped region length and shallow junctions, due to the long thermal drive in needed.

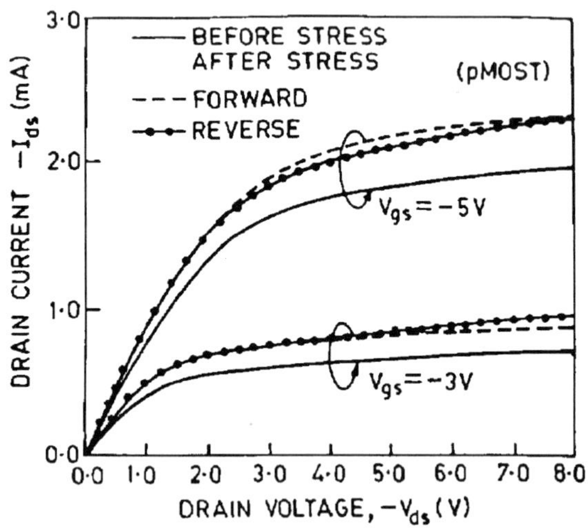

Figure 2.13: I_{ds}-V_{ds} characteristics of a pMOSFET before and after stressing. L=1.2μm, t_{ox}=20nm, V_{ds}=7.5V, V_{gs}=3V, t=5 minutes (After Ong et al. [7], ©1990 IEEE)

2.1.4 Static device lifetime model

Device lifetime τ is commonly defined as the stress time at which the electrical parameters have reached a certain failure criterion. For instance, ΔV_{th}=10mV, $\Delta I_d/I_d$=4% or $\Delta g_m/g_m$=4% are usually used as good criterions. In general, I_d is easier to measure, as V_{th} is normally determined by extrapolating the *I-V* characteristics. Actually, in some cases the threshold voltage is defined more precisely as the voltage for which the drain current I_d reaches a certain value.

In the case of an nMOSFET and under DC stress conditions, τ was found empirically [8] to follow a power-law function of the substrate current I_{sub}. This observation was justified later by *Hu et al.* [3] where device lifetime τ was expressed as

$$\tau \frac{I_d}{W} = C_1 \left(\frac{I_{sub}}{I_d} \right)^{-m} \tag{2.14}$$

where m and C_1 are technology dependent constants. The plot of $\tau I_d/W$ versus I_{sub}/I_d on a log-log scale will be a straight line where the slope is m and the intercept point with the y axis is C_1.

In the case of a pMOSFET, it was shown by *Ong et al.* [7] that device lifetime τ correlates better with the gate current I_g rather than the substrate current I_{sub}. This is due to the fact that the degradation of pMOSFET's is mainly caused by charge trapping in the gate oxide, as was discussed in section 2.1.3. Therefore, the lifetime of a pMOST can be expressed as

$$\tau \frac{I_d}{W} = C_2 \left(\frac{I_g}{I_d} \right)^{-m} \tag{2.15}$$

When the constants m and C_2 are defined for the used technology, it is then easy to determine if a MOSFET can operate under a given power supply. Considering for example a typical 1μm technology with m=3 and C_1=3, and considering a bias condition such that $I_{sub}/I_d \approx 0.01$, the device lifetime is $\tau = 3x10^6$s=0.1 year according to Equation (2.14), which is obviously not acceptable.

2.1.5 Dynamic device lifetime model

Although a good comprehension of the DC degradation mechanisms has been carried out, the static device lifetime models don't fit the transient stress conditions under which MOSFETs are normally exposed. Therefore, recent studies [20] have focused more attention on the enhanced AC degradation. They have pointed out that the discrepancy between DC and AC stress is due to the fact that DC lifetime models don't take into account all the hot-carrier degradation mechanisms. Indeed, They have shown that the enhanced AC degradation is due to the presence of three damage modes, rather than the main mode associated to the peak of the substrate current in the case of an nMOSFET. The three degradation modes are:

- Neutral electron traps $N_{ox,h}$ and interface states generation taking place at low gate voltages. The static device lifetime model associated to this degradation mechanism is given by

$$\tau_{N_{ox,h}} = B \frac{(I_{sub} / I_d)^{-n}}{I_d} \tag{2.16}$$

- Interface states generation N_{it} taking place at mid V_{gs} and the associated static device lifetime model is given by Equation (2.17) is an approximate form of Equation (2.14) where the drain current I_d was considered proportional to W only. Indeed, devices are generally stressed at various values of V_{ds} with V_{gs} adjusted for maximum I_{sub}. As can be seen in Figure 2.4, the variation of V_{gs} under such stress conditions is very week. Therefore, I_d can be considered proportional to W and constant as a function of V_{gs}.

$$\tau_{N_{it}} = A I_{sub}^{-m} \tag{2.17}$$

- At high V_{gs}, damage consists principally of electron traps $N_{ox,e}$. The lifetime due to $N_{ox,e}$ damage is modelled by

$$\tau_{N_{ox,e}} = C \frac{(I_g / I_d)^{-l}}{I_d} \tag{2.18}$$

A, B, C and m, n, l are empirical constants. In order to estimate the AC stress lifetimes, the quasi-static contributions for the three damage modes should be calculated by integrating each of the equations above over the time period T of the AC stress waveform. For example, the value of $\tau_{N_{ox,h}}$ is calculated according to

$$\frac{1}{\tau_{N_{ox,h}}} = \frac{1}{BT} \int_0^T \left(\frac{I_{sub}}{I_d} \right)^n I_d dt \tag{2.19}$$

where quasi-static values are used for all currents. The values of $\tau_{N_{it}}$ and $\tau_{N_{ox,e}}$ are similarly calculated. Finally, the lifetime model taking into account all the three damage modes is then given by

$$\frac{1}{\tau_{AC}} = \frac{1}{\tau_{N_{ox,h}}} + \frac{1}{\tau_{N_{it}}} + \frac{1}{\tau_{N_{ox,e}}} \tag{2.20}$$

2.2 Breakdown voltage

MOSFETs degradation due to channel hot-carrier effects is a long term reliability problem which arises when the drain or gate bias voltages are pushed beyond the power supply limit, or when the device is scaled down without reducing the supply voltage. However, MOSFETs degradation is not the only bad effect resulting from the increase of the bias voltages. Destructive breakdown mechanisms such as avalanche breakdown, surface breakdown, snapback breakdown and gate oxide breakdown can also be involved. These mechanisms are discussed in the following sections.

2.2.1 Avalanche breakdown

Considering a reverse biased junction, the high electric field of the depletion layer sweeps out any electron or hole entering in this region, by either space charge generation or diffusion from the neighbouring quasi-neutral regions. As stated before, if the energy of the electron is higher than 1.3eV it can generate a secondary electron-hole pair by impact ionisation. Avalanche breakdown is defined as the condition under which the impact ionisation rate becomes infinite. In order to calculate the avalanche breakdown of the junction it is first necessary to characterise the impact ionisation process by the impact ionisation coefficients, α_n for electrons and α_p for holes.

$$\alpha_n = 7x10^5 \exp\left(-\frac{1.23x10^6}{E}\right) \tag{2.21}$$

$$\alpha_p = 1.6x10^6 \exp\left(-\frac{2x10^6}{E}\right) \tag{2.22}$$

The expressions of the impact ionisation coefficients are valid for electric fields E ranging from $1.75x10^5$ to $6x10^5$V/cm, and can be approximated in many cases by one ionisation coefficient [21]

$$\alpha = 1.8x10^{-35} E^7 \tag{2.23}$$

This approximation was made in order to allow the calculation of a closed-form solution to the avalanche breakdown voltage of the theoretical abrupt and linearly graded junctions. Numerical techniques are actually used to calculate the breakdown voltage of a real diffused junction. However, such a junction type is considered as being intermediate to the abrupt and linearly graded cases.

In order to compute the avalanche breakdown voltage it is first necessary to determine when the impact ionisation process reaches an infinite rate. Consider a rectangular form N^+-P drain junction with an electron-hole pair generated in the depletion layer at a distance x from the metallurgical junction. Under the influence of the electric field E the electron will be swept toward the N^+ region and the hole will be swept toward the P region. The hole will create $\alpha_p dx$ electron-hole after travelling a distance dx and the electron will create $\alpha_n dx$ electron-hole pair. The total number of electron-hole pair $M(x)$ created by the initial electron-hole pair generated at a distance x is given by [21]

$$M(x) = \frac{\exp\left[\int_0^x \left(\alpha_n - \alpha_p\right) dx\right]}{1 - \int_0^W \alpha_p \exp\left[\int_0^x \left(\alpha_n - \alpha_p\right) dx\right] dx} \tag{2.24}$$

M(x) is also commonly known as the *multiplication coefficient*. The avalanche breakdown voltage is defined as the voltage at which *M(x)* reaches infinity. It occurs when the following condition is satisfied

$$\int_0^W \alpha_p \exp\left[\int_0^x \left(\alpha_n - \alpha_p\right) dx\right] dx = 1 \tag{2.25}$$

where the left-hand side of Equation (2.25) is known as the *ionisation integral*, which simplifies to Equation (2.26) when the approximation of Equation (2.23) is considered.

$$\int_0^W \alpha \cdot dx = 1 \tag{2.26}$$

The avalanche breakdown voltage of a device can then be calculated by evaluating the ionisation integral analytically or numerically. An analytical solution can be obtained in the case of an abrupt and linearly graded junctions as well as for the cylindrical and spherical junction termination. They are particularly useful for the comprehension of the practical diffused junctions that are usually simulated numerically.

1) Abrupt junction diode: Consider a device fabricated with shallow junctions where the substrate is lightly doped and the diffusion is heavily doped. The parallel plane of this junction can be represented by an abrupt N^+-P junction. Applying a reverse bias voltage V_a the depletion layer extends only into the lightly doped side due to the high doping level of the N^+ diffusion. Therefore, in order to calculate the parallel plane breakdown voltage BV_{pp} Poisson's equation needs to be solved only in the P side

$$\frac{d^2V}{dx^2} = -\frac{dE}{dx} = -\frac{Q(x)}{\varepsilon_s} = \frac{qN_A}{\varepsilon_s} \tag{2.27}$$

where Q is the charge of the depletion layer, q the electron charge, ε_s the dielectric constant of the semiconductor and N_A the acceptor doping density.

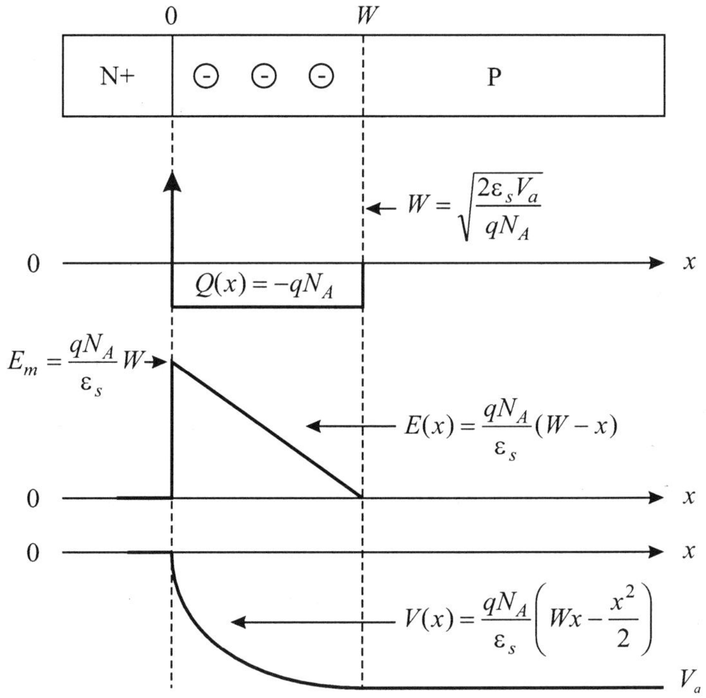

Figure 2.14: Electric Field and potential profiles for an abrupt reverse-biased P-N junction. From J. B. Baliga, "Breakdown Voltage", Chap. 3, in *Modern Power Devices*, Krieger, Copyright © 1987 John Wiley & Sons, Inc. Reprinted with permission of John Wiley & Sons, Inc.

Integrating Equation (2.27) gives the electric field distribution along the depletion layer, which in turn leads to the potential distribution once integrated. Both profiles are shown in Figure 2.14

The breakdown voltage can be calculated by using the electric field distribution in the ionisation integral of Equation (2.26). In this case, the ionisation integral of an abrupt junction diode can be expressed as

$$\int_0^W 1.8x10^{-35}\left[\frac{qN_A}{\varepsilon_s}(W - x)\right]^7 dx = 1 \tag{2.28}$$

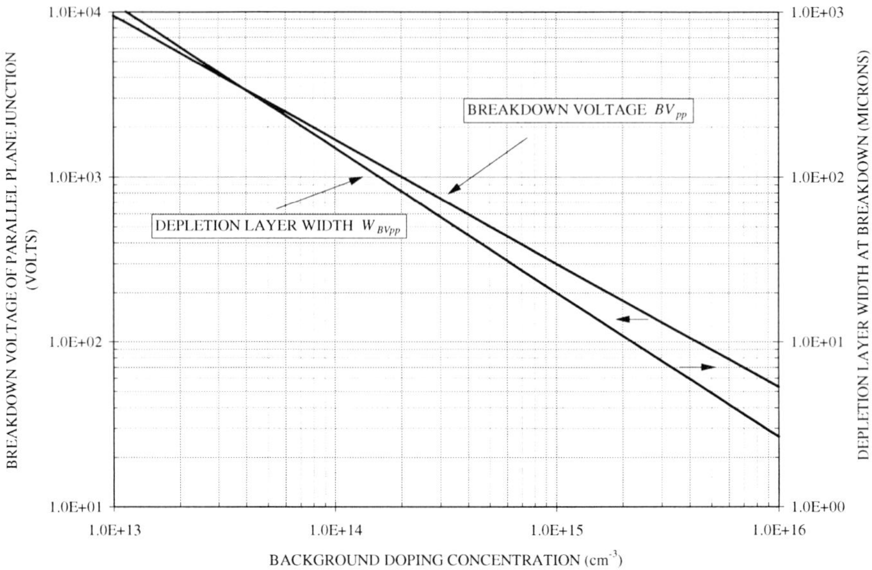

Figure 2.15: Avalanche breakdown voltage and the corresponding depletion layer width for parallel plane abrupt junction diodes. From J. B. Baliga, "Breakdown Voltage", Chap. 3, in *Modern Power Devices*, Krieger, Copyright © 1987 John Wiley & Sons, Inc. Reprinted with permission of John Wiley & Sons, Inc.

Solving Equation (2.28) gives the depletion layer width at avalanche breakdown voltage

$$W_{BV_{pp}} = 2.67x10^{10} N_A^{-7/8} \tag{2.29}$$

As a consequence, the parallel plane avalanche breakdown voltage at x=W and the maximum electric field at x=0 are then calculated.

$$BV_{pp} = 5.34x10^{13} N_A^{-3/4} \tag{2.30}$$

$$E_{BV_{pp}} = 4010 N_A^{1/8} \tag{2.31}$$

The avalanche breakdown voltage and the corresponding maximum depletion layer width for an abrupt junction diode are presented in Figure 2.15 as a function of the doping level N_A of the substrate. It should be noted that a typical substrate doping level of 10^{15} cm^{-3} results in a parallel plane breakdown voltage of 300V and depletion layer width at breakdown of 20μm.

2) Linearly graded junction diode: In the linearly graded junction the doping profile is linear from the N side to the P side of the junction. Therefore, the depletion layer extends equally into both sides as shown in Figure 2.16, where the distribution of the electric field

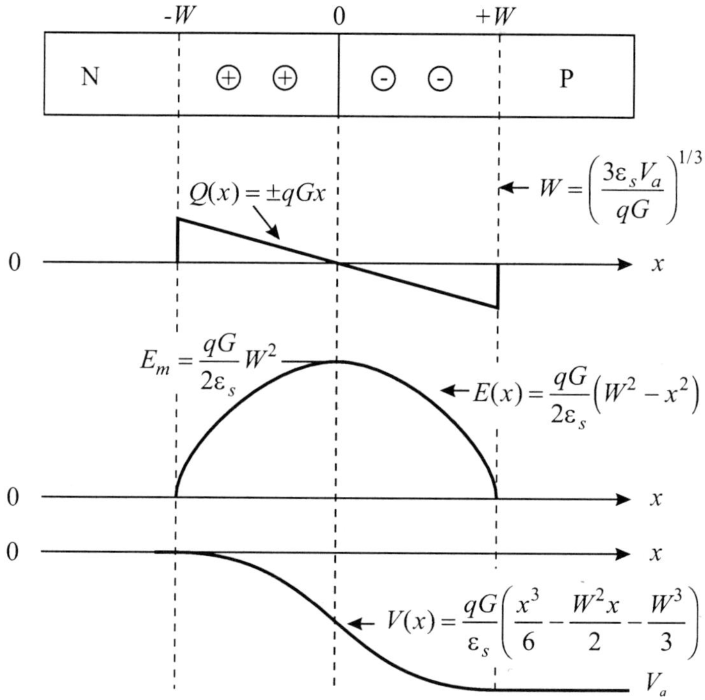

Figure 2.16: Electric Field and potential profiles for a linearly graded P-N junction. From J. B. Baliga, "Breakdown Voltage", Chap. 3, in *Modern Power Devices*, Krieger, Copyright © 1987 John Wiley & Sons, Inc. Reprinted with permission of John Wiley & Sons, Inc.

and the distribution of the potential are calculated by solving Poisson's equation along the depletion layer.

From the electric field distribution in Figure 2.16, the ionisation integral in the case of a linearly graded junction can be expressed as

$$\int_{-W}^{+W} 1.8x10^{-35}\left[\frac{qG}{2\varepsilon_s}\left(W^2 - x^2\right)\right]^7 dx = 1 \qquad (2.32)$$

Thus, the depletion layer width at breakdown on each side of the junction can be derived from Equation (2.32)

$$W_{BV_{pp}} = 9.1x10^{5} G^{-7/15} \qquad \textbf{(2.33)}$$

and the breakdown voltage at x=W as well as the maximum electric field at x=0 are calculated.

$$BV_{pp} = 9.2x10^{9} G^{-2/5} \qquad \textbf{(2.34)}$$

$$E_{BV_{pp}} \; 2.0x10^{4} G^{1/15} \qquad (2.35)$$

The parallel plane avalanche breakdown voltage and the corresponding maximum depletion layer width for a linearly graded junction are shown in Figure 2.17 as a function of the doping gradient G. It should be noted that the avalanche breakdown voltage in the case of a linearly graded junction is generally higher than the breakdown voltage of an abrupt junction. The reason is that for the same range of background doping levels the linearly graded junction can support voltage on both sides of the junction.

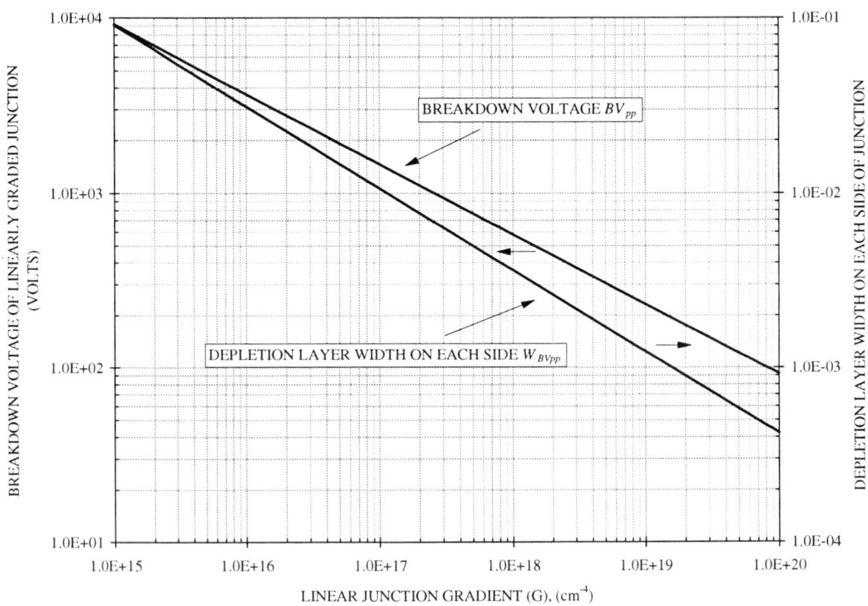

Figure 2.17: Avalanche breakdown voltage and the corresponding depletion layer width for a linearly graded junction. From J. B. Baliga, "Breakdown Voltage", Chap. 3, in *Modern Power Devices*, Krieger, Copyright © 1987 John Wiley & Sons, Inc. Reprinted with permission of John Wiley & Sons, Inc.

3) Diffused junction diode: In modern standard CMOS technologies, diffused junctions using ion implantation followed by high-temperature heat treatment are often used in device processing. Under these conditions, the doping profile of the junction takes a Gaussian shape and the concentration rises rapidly on one side of the junction. If the surface density is high ($N_s \approx 1x10^{20}$ cm^{-3}) and the diffused junction is shallow ($X_j \leq 1\mu m$), it can be considered as an abrupt one. On the second hand, if the surface density is low ($N_s \approx 1x10^{17}$ cm^{-3}) and the diffused junction is deep ($X_j \approx 10\mu m$), it can be considered as a linearly graded one. An analytical closed form solution to the avalanche breakdown voltage is not possible in the case of a diffused junction and numerical techniques were used [22] to solve the ionisation integral. The resulting parallel plane avalanche breakdown voltage and depletion layer width, for both of the latter cases, are shown in Figure 2.18 and Figure 2.19 as a function of the background doping level N_A.

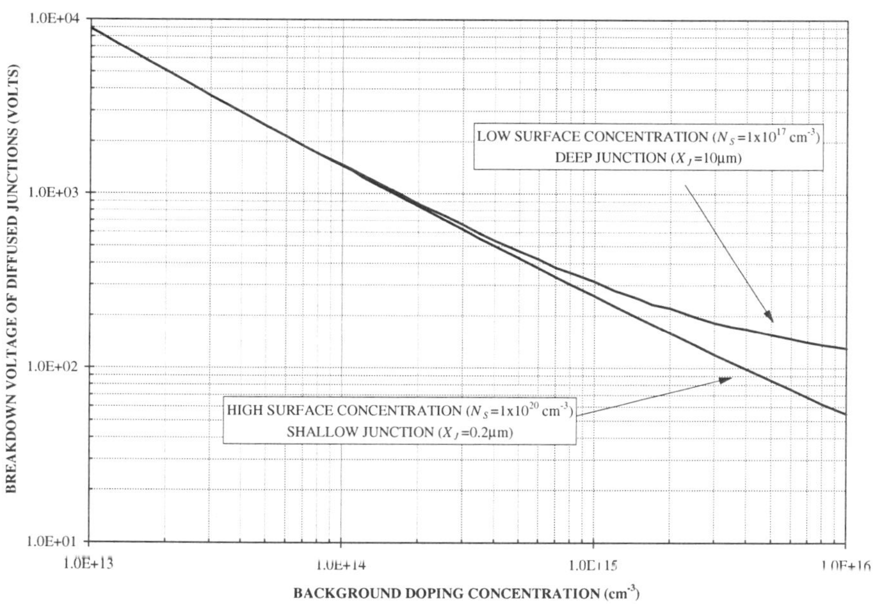

Figure 2.18: Avalanche breakdown voltage in the case of two diffused junction types that are close to a linearly graded junction and an abrupt junction. From J. B. Baliga, "Breakdown Voltage", Chap. 3, in *Modern Power Devices*, Krieger, Copyright © 1987 John Wiley & Sons, Inc. Reprinted with permission of John Wiley & Sons, Inc.

The two curves of Figure 2.18 are useful boundaries to estimate the parallel plane breakdown voltage variation as a function of the diffusion profile in most of the practical junctions. In Figure 2.19, the depletion layer width in the substrate is almost identical in both cases for substrate doping levels up to $1x10^{16}$. Considering typical substrate and diffusion doping levels of 10^{15} cm^{-3} and 10^{18} cm^{-3}, the parallel plane breakdown voltage is about 300V and the depletion layer width in the substrate is about 20μm.

From Figure 2.18, one can expect that any standard CMOS technology is able to withstand supply voltages up to 300V. Actually, in the previous sections only semi-infinite junctions were considered. In practical devices, the junction edge termination can lead to a severe reduction of the breakdown voltage. In fact, if the junction is poorly terminated, its breakdown voltage can be as low as 10%-20% of the parallel plane case. Figure 2.20 shows the shape of a practical diffused junction formed by ion implantation through a rectangular window of SiO_2 masking layer. After a heat treatment the dopant diffuses laterally at the edges of the diffusion window and extend to about 85% of the vertical depth X_j. In order to simplify the breakdown voltage analysis, this extension can be approximated as being equal to the junction depth. The resulting junction termination is cylindrical at the mask edges and spherical at the sharp corners of the diffusion window. The electric field lines along the depletion layer of the junction termination are shown in the cross section of Figure 2.20. A crowding of the lines is located at the junction curvature resulting from the charge balance on both sides of the junction. Therefore, the breakdown voltage of a practical diffused junction is expected to be smaller at the edges than in the parallel plane, as the impact ionisation is much higher.

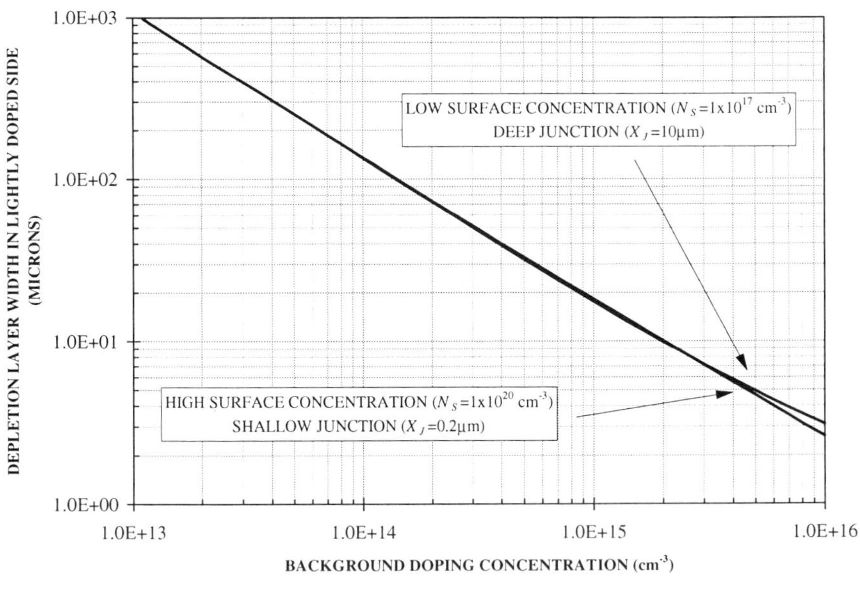

Figure 2.19: Depletion layer width in the case of two diffused junction types that are close to a linearly graded junction and an abrupt junction. From J. B. Baliga, "Breakdown Voltage", Chap. 3, in *Modern Power Devices*, Krieger, Copyright © 1987 John Wiley & Sons, Inc. Reprinted with permission of John Wiley & Sons, Inc.

4) Cylindrical junction: If the lateral diffusion is assumed to be equal to the junction depth the radius of the cylindrical junction is r_j and the radius of the depletion layer is r_d, as shown in Figure 2.20. As in the case of the parallel plane junction any electron-hole pair generated in the depletion layer at a distance r will be swept out by the electric field. The electron will create $\alpha_n dr$ electron-hole pair after travelling a distance dr and the hole will create $\alpha_p dr$ electron-hole pair. The depletion layer is assumed to extend only in the lightly doped side as well. The breakdown voltage is calculated by solving Poisson's equation in the depletion layer using cylindrical co-ordinates

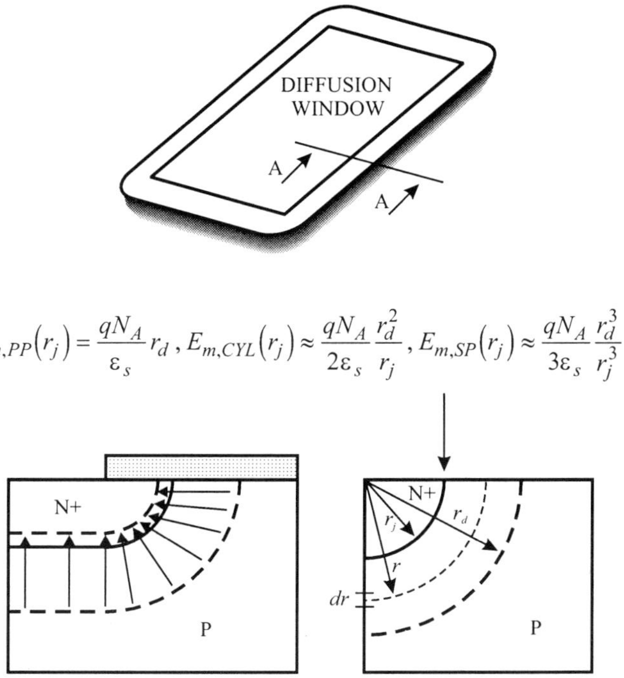

$$E_{m,PP}(r_j) = \frac{qN_A}{\varepsilon_s} r_d \, , \; E_{m,CYL}(r_j) \approx \frac{qN_A}{2\varepsilon_s} \frac{r_d^2}{r_j} \, , \; E_{m,SP}(r_j) \approx \frac{qN_A}{3\varepsilon_s} \frac{r_d^3}{r_j^3}$$

Figure 2.20: Diffused junction formed by ion implantation through a rectangular window of SiO_2 masking layer. Cross section of the junction contour and comparison of the maximum electric field E_m. From J. B. Baliga, "Breakdown Voltage", Chap. 3, in *Modern Power Devices*, Krieger, Copyright © 1987 John Wiley & Sons, Inc. Reprinted with permission of John Wiley & Sons, Inc.

$$\frac{1}{r}\frac{d}{dr}\left(r\frac{dV}{dr}\right) = -\frac{1}{r}\frac{d}{dr}(rE) = \frac{qN_A}{\varepsilon_s} \tag{2.36}$$

where the junction is assumed to be similar to an abrupt one. The potential distribution $V(r)$ and the electric field distribution $E(r)$ were calculated in [23] by integrating Equation (2.36) along the radius of the depletion layer. They are given by Equations (2.37) and (2.38).

$$E(r)=\frac{qN_A}{2\varepsilon_s}\left(\frac{r_d^2-r^2}{r}\right) \tag{2.37}$$

$$V(r)=\frac{qN_A}{2\varepsilon_s}\left[\left(\frac{r_j^2-r^2}{2}\right)+r_d^2\ln\left(\frac{r}{r_j}\right)\right] \tag{2.38}$$

In order to evaluate the breakdown voltage of the cylindrical junction an analytical closed-form solution of the ionisation integral was found by [23]. The electric field distribution was approximated by

$$E(r)=\frac{K}{r} \tag{2.39}$$

on the basis of the assumption that the impact ionisation occurs primarily at the metallurgical junction, where the actual electric field in Equation (2.37) and the approximation in Equation (2.39) are equal and are maximum. Under these conditions, the normalised critical electric field at breakdown and the normalised breakdown voltage of the cylindrical junction are given by

$$\frac{E_{c,CYL}}{E_{c,PP}}=\left(\frac{3W_{c,PP}}{4r_j}\right)^{1/7} \tag{2.40}$$

$$\frac{BV_{CYL}}{BV_{PP}}=\left\{\frac{1}{2}\left[\left(\frac{r_j}{W_c}\right)^2+2\left(\frac{r_j}{W_c}\right)^{6/7}\right]\ln\left[1+2\left(\frac{W_c}{r_j}\right)^{8/7}\right]-\left(\frac{r_j}{W_c}\right)^{6/7}\right\} \tag{2.41}$$

where the normalisation is performed to the parallel plane case. The normalised breakdown voltage of the cylindrical junction is shown in Figure 2.21 as a function of the normalised radius of curvature $r_j/W_{c,PP}$.

The breakdown voltage of a cylindrical junction is always smaller than the semi-infinite parallel plane case. Moreover, for a given substrate doping level N_A, the breakdown voltage depends on the radius of curvature of the junction in comparison to the depletion layer width at breakdown for the parallel plane case. Therefore, if we consider two diffused junctions one being shallow and the second deep, with the same doping profiles and the same background doping level. The depletion layer widths for both cases are identical in the lightly doped side and for a given applied reverse bias voltage. The breakdown voltage of the shallow junction will be lower than the deep junction one, because the electric field lines in the latter case will spread out along a wide junction contour rather than being concentrated at a point close to the surface. As a consequence, shallow junctions are generally used to fabricate low-voltage devices whereas high-voltage devices require deep junctions. In addition, if the deep junction is lightly doped the breakdown voltage will be maximised, as the resulting junction is closer to the linearly graded one.

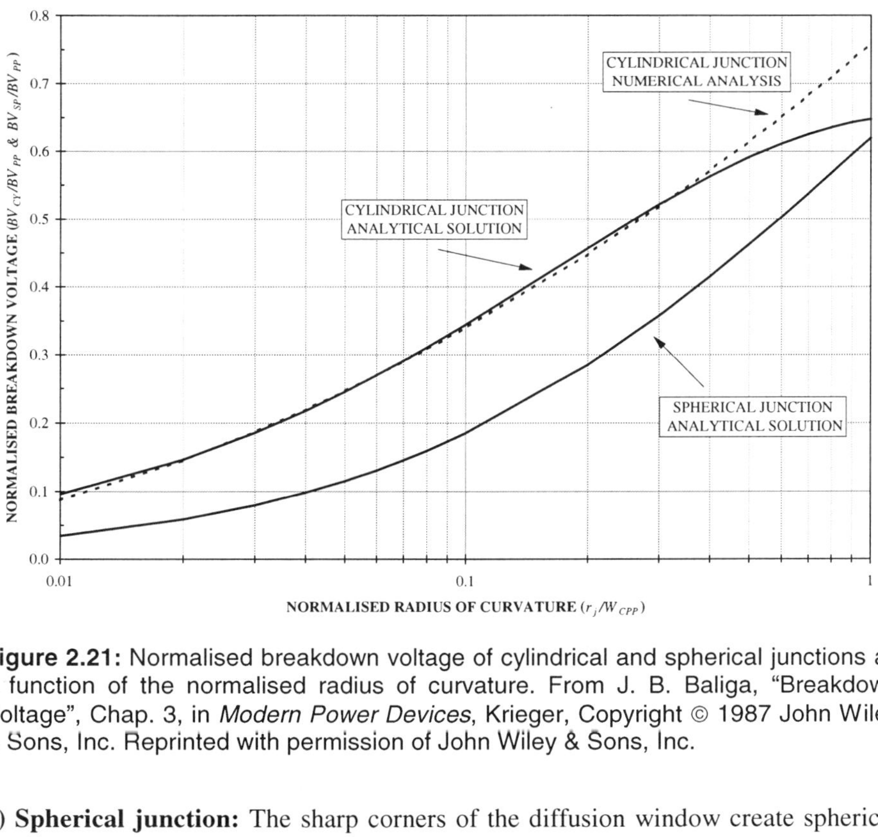

Figure 2.21: Normalised breakdown voltage of cylindrical and spherical junctions as a function of the normalised radius of curvature. From J. B. Baliga, "Breakdown Voltage", Chap. 3, in *Modern Power Devices*, Krieger, Copyright © 1987 John Wiley & Sons, Inc. Reprinted with permission of John Wiley & Sons, Inc.

5) Spherical junction: The sharp corners of the diffusion window create spherical junction termination at the corners of the diffused junction. As in the case of the cylindrical junction the depletion layer is assumed to extend only in the lightly doped side. Hence, the breakdown voltage is calculated by solving Poisson's equation in the depletion layer using the following spherical co-ordinates.

$$\frac{1}{r^2}\frac{d}{dr}\left(r^2\frac{dV}{dr}\right)=-\frac{1}{r}\frac{d}{dr}\left(r^2E\right)=\frac{qN_A}{\varepsilon_s} \tag{2.42}$$

where the junction is assumed to be similar to an abrupt one. The potential distribution $V(r)$ and the electric field distribution $E(r)$ were calculated in [23] as well, and are given by

$$E(r) = \frac{qN_A}{3\varepsilon_s}\left(\frac{r_d^3 - r^3}{r^2}\right) \tag{2.43}$$

$$V(r) = \frac{qN_A}{3\varepsilon_s}\left[\left(\frac{r_j^2 - r^2}{2}\right) + r_d^3\left(\frac{1}{r_j} - \frac{1}{r}\right)\right] \tag{2.44}$$

Using the approximation of the electric field distribution given in Equation (2.45) the avalanche breakdown voltage can be calculated by solving the ionisation integral. As in the case of the cylindrical junction, the approximation was based on the assumption that the impact ionisation occurs primarily at the metallurgical junction where the actual electric field and the approximation are equal and are maximum.

$$E(r) = \frac{K}{r^2} \tag{2.45}$$

Using this approximation, a closed-form analytical solution of the ionisation integral was derived by *Baliga et al.* [23] and the normalised critical electric field at breakdown and the normalised breakdown voltage of the spherical junction are given by

$$\frac{E_{c,SP}}{E_{c,PP}} = \left(\frac{13}{8}\frac{W_c}{r_j}\right)^{1/7} \tag{2.46}$$

$$\frac{BV_{SP}}{BV_{PP}} = \left(\frac{r_j}{W_c}\right)^2 + 2.14\left(\frac{r_j}{W_c}\right)^{6/7} - \left[\left(\frac{r_j}{W_c}\right)^3 + 3\left(\frac{r_j}{W_c}\right)^{13/7}\right]^{2/3} \tag{2.47}$$

The normalised spherical breakdown voltage is shown in the same plot as for the case of the cylindrical junction for comparison. As can be noted, the spherical avalanche breakdown voltage is generally 2 times lower than the cylindrical junction one. Therefore, spherical junctions in device terminations must be avoided. This is possible by rounding the sharp corners of the rectangular diffusion window, where the radius of curvature R of the rounded corner must be several times larger than the depletion layer width at breakdown. In this case, the loss in the active junction area is given by

$$A = (4 - \pi)R^2 \tag{2.48}$$

2.2.2 Surface breakdown

Avalanche breakdown in a diffused junction diode occurs first in the cylindrical part of the junction because of its radius of curvature. In fact, it

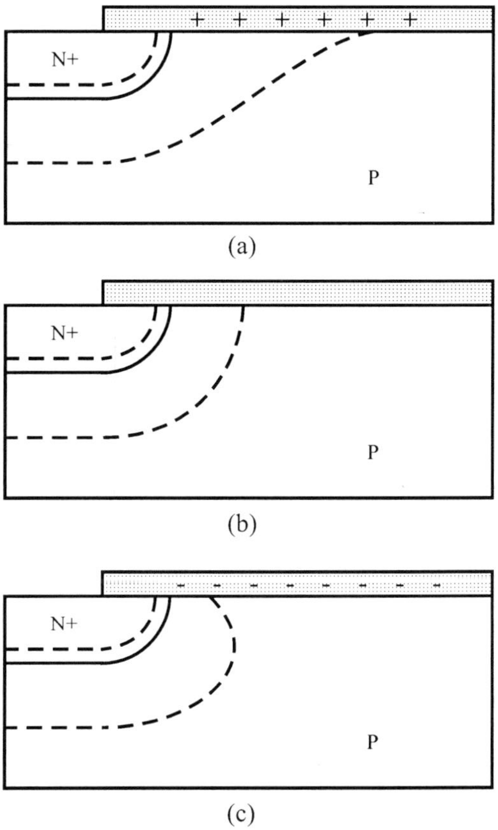

Figure 2.22: Influence of surface charge on depletion layer spreading at the edge of a diffused junction diode: a) positive charge b) zero charge c) negative charge From J. B. Baliga, "Breakdown Voltage", Chap. 3, in *Modern Power Devices*, Krieger, Copyright © 1987 John Wiley & Sons, Inc. Reprinted with permission of John Wiley & Sons, Inc.

was assumed that there is no charge at the surface of the semiconductor over the depletion layer edge. The presence of surface charge has a strong influence on the depletion layer spreading at the surface as shown in Figure 2.22. It results in an increase or decrease of the breakdown voltage depending on the polarity of the charge.

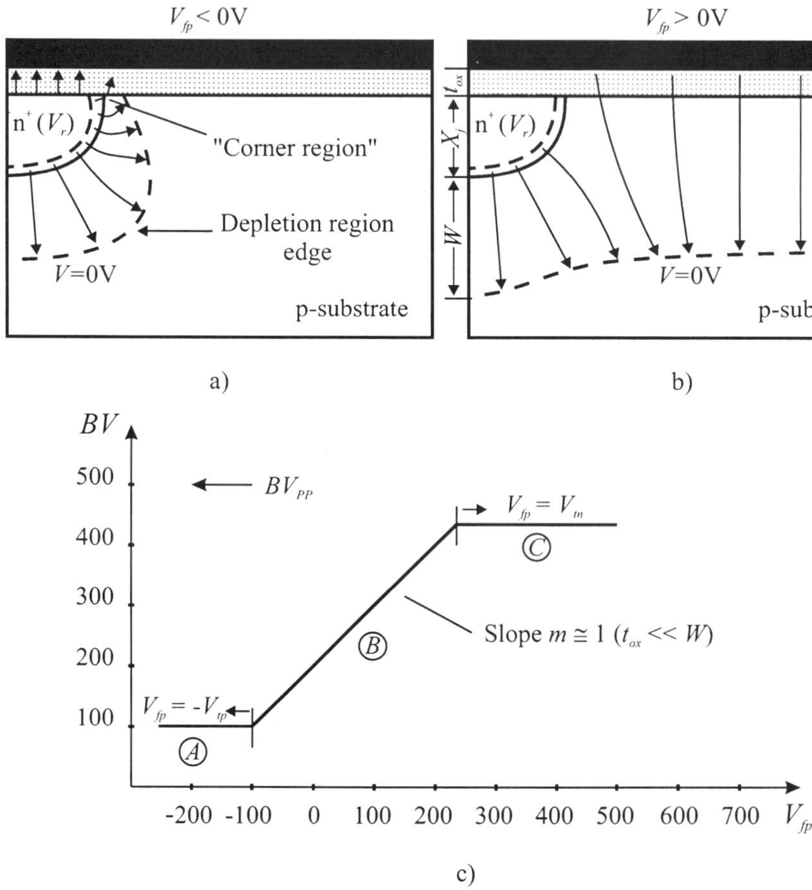

Figure 2.23: Planar diffused junction with field-plate at the edge a) effect of negative field-plate voltage b) effect of positive field-plate voltage c) asymptotes of the breakdown voltage characteristic vs. field-plate voltage (After Grove et al. [24], ©1967 IEEE)

The presence of positive surface charge will cause the depletion layer at the surface of the lightly doped side to extend further as it compensates the charge of the depletion layer. This leads to a decrease in the crowding of the electric field lines and results in an increase of the breakdown voltage. The presence of negative charge at the surface will cause the opposite effect. In, practice, the surface charge density of a thermally grown SiO_2 ranges between 10^{10} to $10^{12}/cm^2$. The use of field-plates is a reliable way to overcome such limitation. In fact, placing a metal field-plate at the edge of the planar junction will influence the depletion layer curvature by altering the surface potential. This is shown schematically in Figure 2.23, where an n^+ diffusion is implanted in a p-type substrate. When the junction is reversed biased and a negative bias is applied to the field-plate with respect to the substrate, an accumulation layer is formed at the surface of the substrate forcing the depletion layer to shrink as illustrated in Figure 2.23a. The result is an increase of the electric field concentration in the "corner region" decreasing the breakdown voltage of the junction. When a positive bias is applied to the field-plate, a depletion layer is formed at the surface of the substrate forcing the depletion layer of the junction to expand as illustrated in Figure 2.23b. The result in this case is the spreading of the electric field concentration in the "corner region" increasing the breakdown voltage. Under particular geometrical and bias conditions the breakdown voltage of the junction can be made to approach the parallel plane value. These conditions are detailed in the following paragraphs. The previous observations were made experimentally by *Grove et. al* in [24]. Numerical solutions of the two dimensional potential distribution in the "corner region" gave results which are in general agreement with the experimental observations. It was found that the two dimensional electric field distribution in the "corner region" could be modelled with a reasonable fitting of the simulation data by the following expression

$$E_{corner} = \frac{V_r}{W} + \frac{V_r - V_{fp}}{\frac{\varepsilon_s}{\varepsilon_{ox}} t_{ox}} \tag{2.49}$$

where W is the depletion region width far from the surface, V_r the reverse bias voltage of the junction and V_{fp} the field-plate voltage. For a given field-plate voltage, the breakdown condition is reached when $E_{corner}=E_c$, where E_c is the critical electric field at breakdown. Thus, solving Equation (2.49) for V_r at breakdown, yields an approximate relationship between breakdown voltage and field-plate voltage

$$BV = m \cdot V_{fp} + m \cdot \frac{\varepsilon_s}{\varepsilon_{ox}} \cdot E_c \cdot t_{ox} \tag{2.50}$$

where BV is the breakdown voltage and m is given by

$$m = \frac{1}{1 + \frac{\varepsilon_s t_{ox}}{\varepsilon_{ox} W}} \tag{2.51}$$

where $m \leq 1$. Thus, the enhancement of the breakdown voltage of the junction is linear as a function of the field-plate voltage as illustrated in region B of Figure

2.23c. The efficiency of this enhancement increases with decreasing oxide thickness and substrate doping concentration level, since m is closer to unity. In other words, the field-plate is more efficient under such conditions. The validity domain of Equation (2.50) is given by the threshold voltages of the n^+ diffusion V_{tp} and the p-type substrate V_{tn}. In fact, if the field-plate voltage is sufficiently negative the underlying surface of the diffused region can be inverted. Thus, the depletion layer shape in the "corner region" becomes independent of the field-plate voltage, leading to the saturation of the breakdown voltage as illustrated in region A of Figure 2.23c. Similarly, if the field-plate voltage is increased up to the substrate threshold value V_{tn}, an inversion layer is formed in the substrate side of the junction. Beyond this point, the width of the depletion layer in the substrate reaches its maximum value X_{dmax}, leading to the saturation of the breakdown voltage as illustrated in region C of Figure 2.23c. Thus, the upper limit of the breakdown voltage of a planar junction with a field-plate, approaches the parallel plane one for sufficient bias on the field-plate.

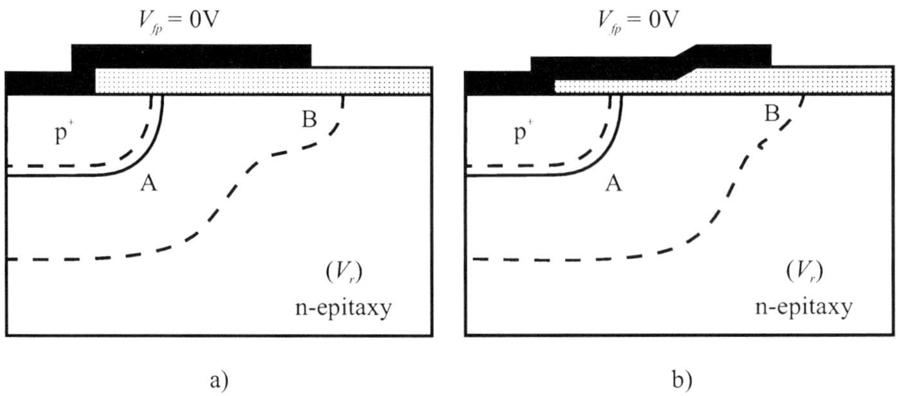

Figure 2.24: Planar diffused junction with field-plate formed by the metallisation extension over the oxide

In actual devices, it is impractical to provide a separate electrode to bias the field-plate. Therefore, it is preferable to create the field-plate by extending the junction metallisation over the oxide, as shown in Figure 2.24a. The junction type in this figure is opposite to the previous case as we anticipate the typical example of a lateral high-voltage nMOS device, where the lightly doped drain consists of an n-type epitaxial layer. Thus, assuming that the p^+ diffusion is grounded, i.e. V_{fp}=0V, the depletion layer of the junction termination extends beyond the field-plate edge with increasing reverse bias voltage. The result is a shift of the electric field concentration from the junction curvature at point A to the field-plate edge at point B, depending on the oxide thickness t_{ox} and the depletion layer width X_d under the field-plate. This was observed by *Conti et al.* in [25] for a planar junction with a field-plate tied to its highly doped side. Numerical analysis has shown that a high concentration of the electric field is located at the field-plate edge for small oxide thickness. Precisely, avalanche breakdown occurs in the silicon at the edge of the field-plate for X_d/t_{ox} higher than 12. Therefore, this ratio must be kept lower than 12 by increasing the oxide thickness, assuming a fixed doping concentration level. However, as the oxide thickness is increased the field-plate becomes less efficient resulting in an increase of the electric field concentration at the junction curvature, where avalanche breakdown can now occur. Therefore, the optimum oxide thickness balances the electric field concentration between points *A* and *B*.

A better solution, that meets the opposite requirements of a thin oxide near the junction and a thick oxide at the field-plate edge, consists of using a stepped field-plate as illustrated in Figure 2.24b. Particular care must be taken to achieve a smooth transition of the oxide in order to avoid a new breakdown region.

2.2.3 Snapback breakdown

Consider a long n-channel MOSFET device with diffused drain and source junctions. If the drain bias voltage is large enough to create impact ionisation in the depletion

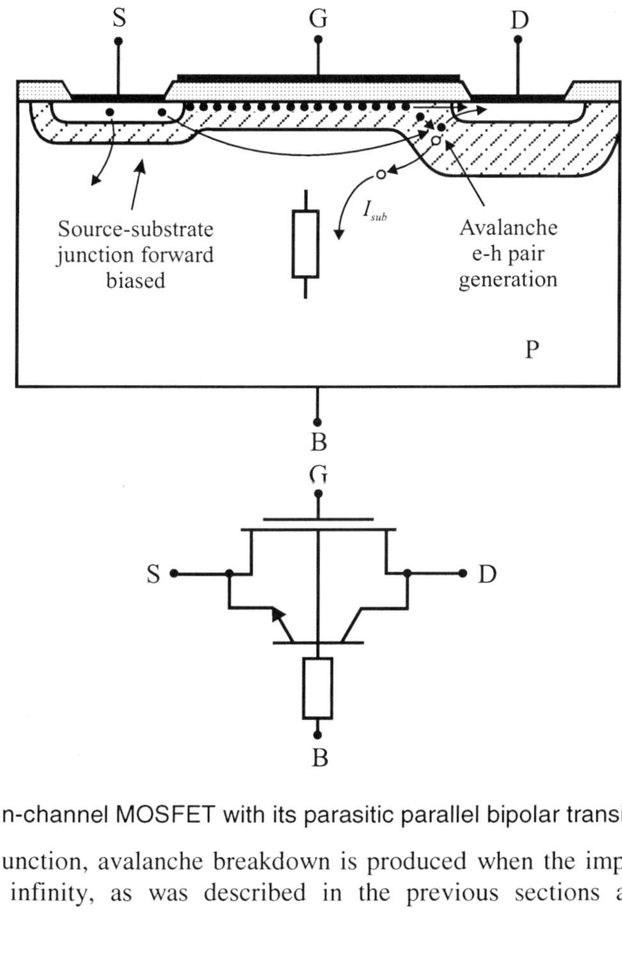

Figure 2.25: n-channel MOSFET with its parasitic parallel bipolar transistor

layer of the junction, avalanche breakdown is produced when the impact ionisation rate reaches infinity, as was described in the previous sections and is shown schematically in Figure 2.26. If the device is now scaled down, an additional destructive mechanism occurs. The substrate current I_{sub}, resulting from avalanche-generated holes, can develop a voltage drop across the substrate resistance R_{sub}, as shown in Figure 2.25. This voltage drop can raise the substrate potential near the source and forward bias the source-substrate junction. The source of the MOSFET acts then as an emitter, and electrons are injected into the substrate. If these electrons are collected by the drain depletion layer they can create further electron-hole pairs. As a consequence, the substrate current is increased and the source-substrate junction becomes further biased, leading to further electron injection. Hence, a positive feedback loop is triggered, and the current rises exponentially due to the resulting npn bipolar transistor in parallel with the MOSFET. When this transistor switches on, a negative-resistance or snapback behaviour appears on the *I-V* characteristics of the transistor, as shown in Figure 2.26. This behaviour is typically observed in open-base bipolar transistors, which is obviously the case of the present parasitic transistor as the substrate potential near the source (emitter) is determined by the avalanche generated substrate current and not an externally applied voltage.

In short-channel devices, snapback breakdown is generally triggered before avalanche breakdown due to the condition under which the current reaches infinity in the parasitic bipolar transistor. First, consider a simple bipolar transistor with a floating base. In such device, the base current is zero and the collector current is given by

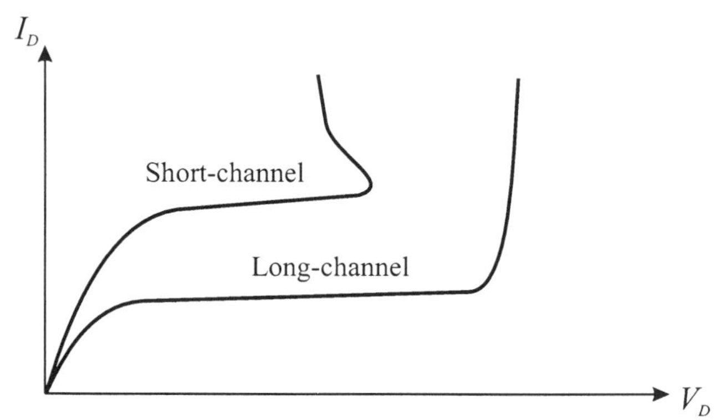

Figure 2.26: Schematic diagram of the *I-V* characteristics for long-channel and short-channel MOSFET's

$$I_C = \alpha I_E + I_{CB0} = \alpha I_C + I_{CB0} \tag{2.52}$$

where α is the common base current gain and I_{CB0} is the leakage current of the open-base collector junction. When the breakdown voltage is reached, the collector current becomes

$$I_C = M(\alpha I_C + I_{CB0}) \tag{2.53}$$

Solving for I_C gives

$$I_C = \frac{MI_{CB0}}{1-\alpha M} \tag{2.54}$$

Breakdown is defined as the condition under which I_C reaches infinity. It is the case when

$$M = \frac{1}{\alpha} \tag{2.55}$$

This condition is lower than $M=\infty$ for a simple *pn* junction where the current in avalanche breakdown is given by $I=MI_{leak}$. The snapback behaviour can be explained assuming that the multiplication factor is related to the collector-emitter voltage V_{CE} by the following approximation [21]

$$M = \frac{1}{1-\left(\frac{V_{CE}}{V_{BV}}\right)^4} \tag{2.56}$$

When V_{CE} reaches the breakdown condition for which $M=1/\alpha$, the collector current rises to infinity causing an increase of common base current gain α. In order to maintain the breakdown condition M must decrease by forcing the collector emitter voltage V_{CE} to decrease. Thus, any increase of the collector current is followed by a decrease of the collector-emitter voltage leading to a negative resistance or snapback behaviour in the *I-V* characteristics.

In the case of a short-channel MOSFET, the parasitic bipolar transistor behaves very much like an open base-transistor. An analytical expression for the source current (emitter) was found in [26], as well as a condition for the multiplication factor at breakdown. Assuming an injection efficiency γ of the source junction close to 1, the source current is given by

$$I_s = \frac{(M-1)I_D(R_{sub}+R_{ext})-V_{SB}-0.65}{[1-\alpha-(M-1)k\alpha](R_{sub}+R_{ext})} \tag{2.57}$$

where R_{sub} is the substrate spreading resistance, R_{ext} is any external resistance connected between the substrate and the supply voltage, V_{SB} the external source-to-substrate bias, k the fraction of electrons collected by the drain depletion layer and causes additional impact ionisation to the total number of electrons collected by the drain and finally, 0.65 is switch on voltage of the base-emitter junction. The source current I_s becomes infinite when the following condition is reached

$$M - 1 = \frac{1-\alpha}{k\alpha} = \frac{1}{k\beta} \tag{2.58}$$

where β is the common emitter current gain of the parasitic lateral bipolar transistor when the MOSFET is in the saturation region. The total drain current, when the parasitic bipolar transistor is switched on, is given by the sum of the substrate current (base), the source current (emitter) and the MOSFET drain current in the saturation region

$$I_{DT} = M\left(I_D + k\alpha I_s\right) + \alpha(1-k)I_s \tag{2.59}$$

The I-V characteristics of the MOSFET can be plotted using Equation (2.59) and assuming that the multiplication factor M is known. This is achieved in [26] by solving the ionisation integral in the space charge region near the drain end. The spreading substrate resistance R_{sub} is also necessary. It was estimated in [26] as a function of the width and length of the transistor and the resistivity of the substrate. For a given substrate doping level, the MOSFET must be designed so that this type of breakdown does not occur.

2.2.4 Gate-oxide breakdown

In the previous sections, breakdown mechanisms in the silicon part of the MOSFET structure have been discussed. Thin gate-oxide breakdown is also possible if the gate bias is pushed beyond the supply voltage limit. Such a bias condition would first lead to channel hot-carrier effects, already discussed in section 2.1, because of the high electric field in the oxide. As was mentioned, the degradation of the device characteristics is the main issue of hot-carrier effects, but the device may still work. If the stress condition is further increased, the oxide integrity degradation may cause catastrophic failure. In this paragraph the phenomenon of oxide breakdown is discussed as a function of the electric field strength.

In modern standard CMOS technologies, thin oxide films can withstand electric fields as high as 12MV/cm before breaking down instantaneously. Such dielectric strength is only possible in SiO_2 films that are defect free, which is unfortunately not always the case. There are three groups of oxide failure modes as a function of the electric field strength.

The first group is called the C-mode group and can withstand the highest electric fields (8-12 MV/cm). The high-field failure mechanism exhibited by this group is referred to as intrinsic failure and the oxide is assumed to be defect free. Considering a MOSFET with a typical 15nm gate oxide thickness, and assuming that it lies in the

C-mode group, the breakdown voltage of such gate oxide would be 15V if the typical highest-electric field strength is considered to be 10 MV/cm.

The second group is called B-mode group and suffers failure at electric fields of medium strength (3-7 MV/cm). The oxides of this group are assumed to contain weak spots that do not produce instant breakdown but may fail during normal operating conditions. Various types of defects have been proposed as being responsible for the B-mode failures, including *Sodium contamination in the oxide film*, *Substrate metal contamination*, *Surface roughness*, *Localised regions where the oxide is thinner* and *Crystalline defects in the substrate*. B-mode failure is responsible for the early breakdown of oxides and represents a serious problem for MOSFETs with thin gate oxide. For instance, 5V supply voltage applied across a 12.5 nm gate oxide thickness, would lead to an oxide field that lies in the B-mode failure group (~4 MV/cm). These parts are generally screened out and excluded from use by using the burn-in technique. It consists in making the devices operate for some period of time, during which most of those defective parts actually do fail. It is obviously preferable that B-mode group parts fail during burn-in rather than failing after they are installed in a given system. Accelerated burn-in conditions such as high temperature and electric field strength, are generally selected in order to accelerate the B-mode failure mechanisms.

Finally, the A-mode group fail instantly after applying a relatively small electric field (<1 MV/cm). Such oxide films are believed to have big defects so that they are already shorted before applying the low strength field.

2.3 Drain Induced Barrier-Lowering (DIBL)

Hot-carrier effects (HCE) such as channel hot-electron emission, avalanche breakdown and snapback breakdown can be controlled by choosing the appropriate technological parameters. These are, bulk doping level, junctions depth X_j, gate oxide thickness t_{ox} and effective channel length L_{eff}. If MOSFET devices are designed by optimising these parameters for minimum hot-carrier effects, care must be taken to avoid drain induced barrier-lowering (DIBL). Table 2.1 illustrates the dependencies of HCE and DIBL upon MOSFET design parameters.

Hot-carrier effects have been discussed and characterised in the previous sections of this chapter. In this section the barrier-lowering effects are analysed as a function of the gate and drain bias conditions, as well as the set of technological parameters mentioned in Table 2.1. But first, the nature of DIBL is discussed.

The concept of DIBL was introduced by *Troutman* in [27] and consist in determining the change in the potential energy barrier between the source and channel at the Si-SiO_2 interface. This change is caused by the junction depletion region charges, which arise from both the built-in *pn* junction electric field and the reverse-bias drain voltage, and the gate charge.

In order to illustrate this concept, the surface potential $\varphi_s(y)$ along the channel for different MOSFET lengths is shown in the graph of Figure 2.27. Channel current is controlled by carrier emission over a potential barrier at the source, and for long channel devices, the potential is constant over a large fraction of the channel length (L_{eff}=5μm). As a consequence, the minimum value of the potential can be calculated by solving the one-dimensional Poisson's equation in a direction normal to the interface. When the channel length is reduced and V_{ds} is increased, assuming that all the other parameters are kept constant, the minimum value of $\varphi_s(y)$ is increased, thus lowering the potential energy barrier for electron flow between source and drain.

Tech. parameter	characteristic	Main limiting effect
Bulk doping level	High	Hot-carrier effects (HCE)
	Low	Drain induced barrier-lowering (DIBL)
Junctions depth X_j	Shallow	HCE
	Deep	DIBL
Effective channel length L_{eff}	Short	HCE/DIBL
	Long	--------
Oxide thickness t_{ox}	Thin	HCE
	Thick	DIBL

Table 2.1: hot carrier-effects and drain induced barrier lowering upon MOSFET design parameters (After Troutman [27], ©1979 IEEE)

The one-dimensional solution to Poisson's equation is obviously no longer adequate. Numerical analysis must be used to find solutions of the two dimensional Poisson's equation for calculating the electric field and potential in the channel region. But this approach is not realistic for circuit simulations as it is time consuming. Therefore, an analytical solution to Poisson's equation is preferable using a quasi 2-D approach based on the DIBL model, although it is less accurate. It consist in approximating the two dimensional Poisson equation in the channel with the following expression [28]

$$\frac{\varepsilon_s X_{d\max}}{\eta}\frac{d^2\varphi_s(y)}{dy^2}+\varepsilon_{ox}\frac{\left[V_t-V_{FB}-\varphi_{surf}(y)\right]}{t_{ox}}=qN_A X_{d\max} \tag{2.60}$$

where η is a fitting parameter, V_{FB} the flat band voltage and X_{dmax} the maximum width of the depletion region in the channel of a long-channel device, which is given by

$$X_{d\max} = \sqrt{\frac{2\varepsilon_s}{qN_A}\left(\varphi_{s,lc} + V_{sb}\right)} \tag{2.61}$$

where V_{sb} is the source-to-bulk voltage and $\varphi_{s,lc}$ is the surface potential of a long-channel MOSFET in strong inversion

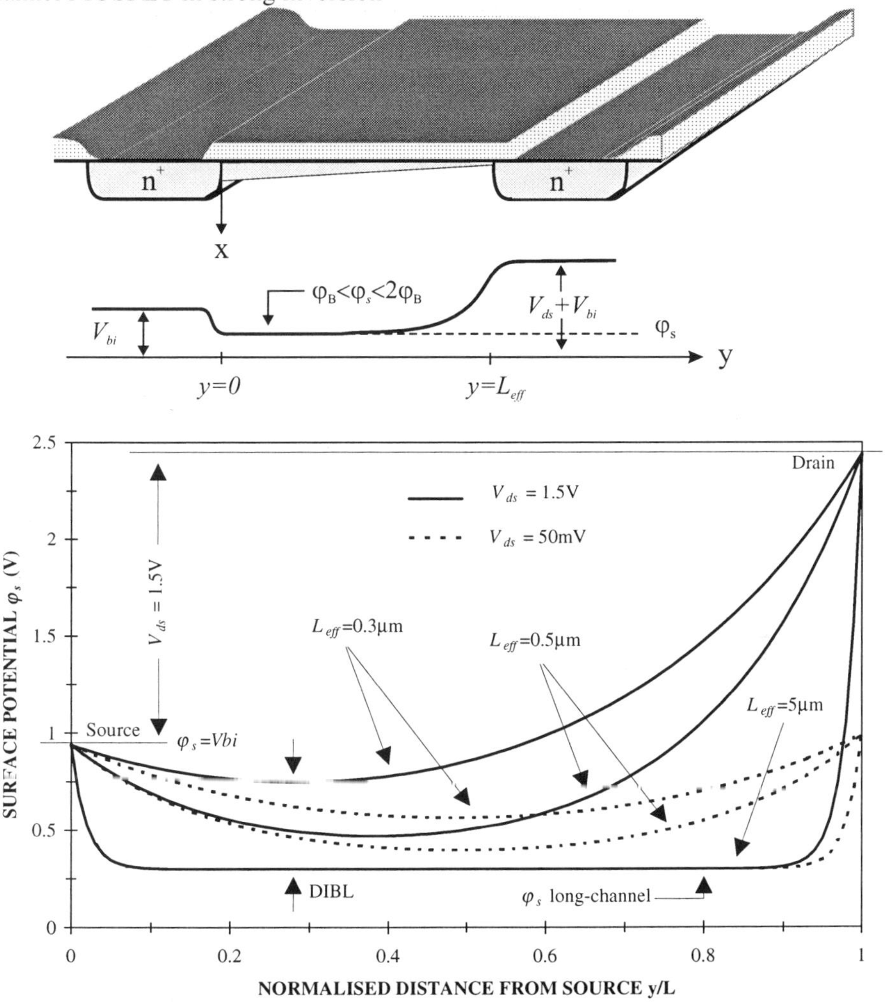

Figure 2.27: Surface potential distribution for different channel lengths and drain voltage. The device parameters are t_{ox}=10nm, N_A=10^{16}cm^{-3}, n_j=10^{20}cm^{-3}

$$\varphi_{s,lc} = 2\varphi_B = 2U_T \ln\left(\frac{N_A}{n_i}\right) \tag{2.62}$$

The solution to equation (2.60), under the boundary conditions $\varphi_s(0)=V_{bi}$ and $\varphi_s(L)=V_{bi}+V_{ds}$ is given by

$$\varphi_s(y) = V_{ON} + \left(V_{bi} + V_{ds} - V_{ON}\right)\frac{\sinh\left(\frac{y}{l}\right)}{\sinh\left(\frac{L}{l}\right)} + \left(V_{bi} - V_{ON}\right)\frac{\sinh\left(\frac{L-y}{l}\right)}{\sinh\left(\frac{L}{l}\right)} \tag{2.63}$$

where $V_{ON}=V_{gs}-V_t$, V_{bi} the built-in potential of the source-substrate and drain-substrate junctions and l is the characteristic length defined as

$$l = \sqrt{\frac{\varepsilon_s t_{ox} X_{d\max}}{\varepsilon_{ox}\eta}} \tag{2.64}$$

The DIBL results in a decrease of the threshold voltage leading to a higher surface drain current I_{ds} than the one expected with long-channel devices. Such I_{ds} behaviour is called surface-DIBL effects and is usually modelled by subtracting a term ΔV_t from the threshold voltage expression. ΔV_t is obtained by subtracting the minimum value of the long-channel surface potential $\varphi_{s,lc}$ from the minimum value calculated using Equation (2.63). Surface-DIBL in MOSFET structures with a fixed L_{eff} can be empirically characterised by plotting log I_{ds} as a function of V_{gs} with V_{ds} as a parameter. A quantitative estimate can be extracted from the shift in gate voltage ΔV_{gs} at a fixed drain current (typically 0.1μA/μm) as V_{ds} is increased. Hence, the shift in threshold voltage ΔV_t as a function of the drain bias is estimated by ΔV_{gs}, which is a good first order estimation. Normalising ΔV_{gs} to ΔV_{ds}, an acceptable value of $\Delta V_{gs}/\Delta V_{ds}$ is approximately 25mV/V, depending on the circuit application. A set of log I_{ds} versus V_{gs} curves, characterising surface-DIBL, is shown in Figure 2.28a where the drain voltage is increased in 1V increments. Note that surface-DIBL does not affect the subthreshold slope ΔS. In Figure 2.28b, ΔS increases as V_{ds} is increased beyond ~4V, indicating that another phenomenon than surface-DIBL is setting in. In fact, the altering of the subthreshold slope is a typical manifestation of subsurface-DIBL or punchthrough. It is associated with the merging of the source and drain depletion regions, and generally takes place below the surface. This is due to the V_t adjust implant which is used to raise the doping of the surface channel region above the doping of the substrate.

Once punchthrough occurs, any increase in V_{ds} will lower the potential energy barrier for carriers in the source. Thus, considering the case of an n-channel MOSFET, a subsurface electron flow between source and drain will take place. This

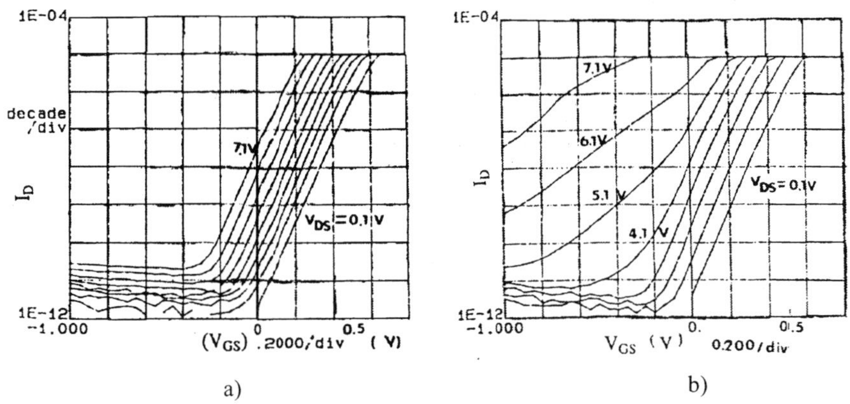

Figure 2.28: Log I_{ds} versus V_{gs} curves with V_{ds} varying in 1V increments. a) V_{gs} shift indicating surface-DIBL. b) Punchthrough taking place at V_{ds}~4V as ΔS is increased beyond this value. From S. Wolf, "The Submicron MOSFET", Chap. 5, in *Silicon Processing for the VLSI ERA*, Volume 3, Copyright ©1995 by Lattice Press. Reprinted with permission.

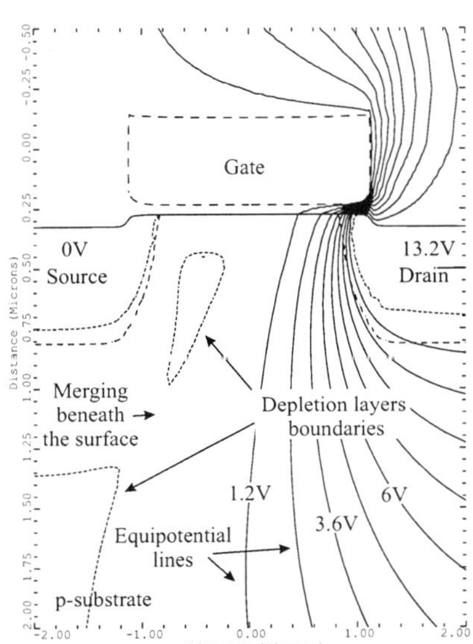

Figure 2.29: cross section of an n-channel MOSFET under punchthrough condition with V_{ds}=13V showing the merging of the depletion layers of the drain and source junctions

component of I_{ds} is called punchthrough current and must be insignificant under normal bias conditions. Therefore, punchthrough is characterised empirically by defining a punchthrough voltage V_{PT} as being the value of V_{ds} at which I_{ds} reaches some specific magnitude (1nA/μm) when V_{gs}=0. Note that V_{PT} can be roughly estimated by determining the value of V_{ds} at which the sum of the depletion layers of the source and drain junctions equals the channel length, using the following expression

$$V_{PT} = \frac{qN_A}{2\varepsilon_s}\left(L_{eff} - r_j\right)^2 \tag{2.65}$$

Nevertheless, no accurate analytical model exists for determining V_{PT} as a function of the implant dose and depth, and the 2-D Poisson's Equation must be solved numerically to predict V_{PT}. Figure 2.29 shows the cross section of an n-channel MOSFET under punchthrough condition with V_{ds}=13V, where the merging of the depletion layers of the drain and source junctions occurs beneath the surface.

2.4 References

[1] Y. A. El-Mancy, D. M. Caughey, "Modelling weak avalanche multiplication currents in IGFETS and SOS transistors for CAD", *IEEE IEDM*-75, *Dig. Tech. Papers,* pp.31-34

[2] P. E. Cottrell, R. R. Troutman, "Hot-electron Emission in N-channel IGFET's", *IEEE Trans. Elec. Dev.*, Vol. ED-26, NO. 4, April 1979, pp.520-533

[3] C. Hu, S. C. Tam, P. K. Ko, T. Y. Chan, K. W. Terrill, "Hot-electron-Induced MOSFET Degradation—Model, Monitor, and improvement", *IEEE Journal of Solid-State circuits.*, Vol. SC-20, NO. 1, February 1985, pp. 295-305

[4] C. Hu, "Hot-electron Effects in MOSFET's", *IEEE IEDM*-83, pp.176-181

[5] C. Hu, "Hot Carrier Effects", in *Advanced MOS Device Physics* (N. G. Einspruch and G. Gildenblat, eds.), VLSI Electronics Vol. 18, pp.119-157, Academic Press Inc., New York, 1989

[6] T. Y. Chan, P. K. Ko, C. Hu, "A Simple Method to characterize substrate current in MOSFET's", *IEEE Trans. Elec. Dev. Lett.*, EDL-5(12), pp.505-507 (1984)

[7] T. C. Ong, P. K. Ko, C. Hu, "Hot-Carrier Current Modelling and Device Degradation in Surface-Channel p-MOSFET's", *IEEE Trans. Elec. Dev.*, Vol. ED-37, NO. 7, July 1990, pp.1658-1666

[8] E. Takeda, H. Kume, T. Toyabe, S. Asai "Submicrometer MOSFET Structure minimizing hot-carrier generation", *IEEE Trans. Elec. Dev.*, Vol. ED-29, 611(1982)

[9] S. Tam, P. K. Ko, C. Hu "Lucky-Electron Model of Channel Hot-Electron Injection in MOSFET's", *IEEE Trans. Elec. Dev.*, Vol. ED-31, NO. 9, September 1984, pp.1116-1125

[10] F. C. Hsu, S. Tam, "Relationship Between MOSFET Degradation and Hot-Electron-Induced Interface-State Generation", *IEEE Trans. Elec. Dev. Lett.*, EDL-5, NO. 2, February 1984, pp.50-52

[11] T. H. Ning, H. N. Yu, "Optically Induced Injection Of Hot Electrons into SiO_2", *J. Appl. Phys.*, Vol. 45, pp.5373, 1974

[12] K. R. Hofmann, C. Werner, W. Weber, G. Dorda, "Hot-Electron and Hole-Emission Effects in Short n-Channel MOSFET's", *IEEE Trans. Elec. Dev.*, Vol. ED-32, NO. 3, March 1985, pp.691-699

[13] E. Takeda, N. Susuki, "An Empirical Model for Device Degradation Due to Hot-Carrier Injection", *IEEE Trans. Elec. Dev. Lett.*, EDL-4, NO. 4, April 1983, pp.111-113

[14] P. Heremans, H. E. Maes, N. Saks, "Evaluation of Hot Carrier Degradation of N-Channel MOSFET's with the Charge Pumping Technique", *IEEE Trans. Elec. Dev. Lett.*, EDL-7, NO. 7, July 1986, pp.428-430

[15] N. S. Saks, P. L. Heremans, L. V. D. Hove, H. E. Maes, R. F. De Keersmaecker, G. J. Declerck, "Observation of Hot-Hole Injection in NMOS Transistors Using a Modified Floating-Gate Technique", *IEEE Trans. Elec. Dev.*, Vol. ED-33, NO. 10, October 1986, pp.1529-1533

[16] B. S. Doyle, M. Bourcerie, C. Bergonzoni, R. Benecchi, A. Bravis, K. R. Mistry, A. Boudou "The Generation and Characterisation of Electron and Hole Traps Created by Hole Injection During Low Gate Voltage Hot-Carrier Stressing of n-MOS Transistors", *IEEE Trans. Elec. Dev.*, Vol. ED-37, NO. 8, August 1990, pp.1869-1876

[17] R. Bellens, P. Heremans, G. Groesneken, H. E. Maes, "Hot-carrier Effects in n-Channel MOS Transistors Under Alternating Stress Condidtions", *IEEE Trans. Elec. Dev. Lett.*, EDL-9, NO. 5, May 1988, pp.232-234

[18] T. Tsuchiya "Trapped-Electron and Generated Interface-Trap Effects in Hot-Electron-Induced MOSFET Degradation" *IEEE Trans. Elec. Dev.*, Vol. ED-34, NO. 11, November 1987, pp.2291-2296

[19] B. S. Doyle, M. Bourcerie, J. C. Marchetaux, A. Boudou "Interface State Creation and Charge Trapping in the Medium-to-High Gate Voltage Range ($V_d/2 \geq V_g \geq V_d$) During Hot-Carrier Stressing of n-MOS Transistors", *IEEE Trans. Elec. Dev.*, Vol. ED-37, NO. 3, March 1990, pp.744-754

[20] K. Mistry, B. Doyle, "A Model for AC Hot-Carrier Degradation in n-Channel MOSFET's", *IEEE Trans. Elec. Dev. Lett.*, EDL-12, NO. 9, September 1991, pp.492-494

[21] B. J. Baliga, "Modern Power Devices", Krieger Publishing Company, Florida, 1987

[22] M. S. Adler, V. A. K. Temple "Semiconductor Avalanche Breakdown Design Manual", GE Technology Marketing Operation, Schenectady, New York 1979

[23] B. J. Baliga, S. K. Ghandhi, "Analytical Solutions for the Breakdown Voltage of Abrupt Cylindrical and Spherical Junctions", *Solid State Electronics*, Vol. 19, pp. 739-744(1976)

[24] A. S. Grove, O. Leistiko, W. W. Hooper, "Effect of Surface Fields on the Breakdown Voltage of Planar Silicon p-n Junctions" *IEEE Trans. Elec. Dev.*, Vol. ED-14, NO. 3, March 1967, pp.157-162

[25]F. Conti, M. Conti, "Surface Breakdown in Silicon Planar Diodes Equipped with field-plate", *Solid State Electronics*, Vol. 15, pp. 93-105(1972)

[26] F. C. Hsu, P. K. Ko, S. Tam, C. Hu, R. S Muller, "An Analytical Breakdown Model for Short-Channel MOSFET's" *IEEE Trans. Elec. Dev.*, Vol. ED-29, NO. 11, November 1982, pp.1735-1740

[27] R. R. Troutman, "VLSI Limitations from Drain-Induced Barrier Lowering" *IEEE Trans. Elec. Dev.*, Vol. ED-26, NO. 4, April 1979, pp.461-469

[28] S. Wolf, "The Submicron MOSFET", in *Silicon Processing for the VLSI Era*, Lattice Press, California, 1995

3 MOSFET HIGH-VOLTAGE TECHNOLOGIES

Abstract:
The existing solutions avoiding the different undesirable effects presented in the previous chapter are discussed here. These solutions are distinguished first at a device level where the isolated vertical and lateral DMOSFET structures are presented. Then, the corresponding high-voltage technologies are discussed according to the isolation technique used between devices. A rough presentation of the high-voltage technologies based on the p-n junction isolation and the dielectric isolation techniques is performed. The former are more commonly known as BCD (Bipolar CMOS and DMOS) technologies and start with a bipolar basis. The latter require tricky mechanical processing of the starting material. Interest is focused on the high-voltage technologies using a CMOS basis, where the required supplementary masks and processing steps are added to implement the high-voltage devices.

During the last three decades, there has been an increasing interest in power MOSFET technologies for use in the various high-voltage (HV) applications presented in the first chapter which require power device technologies. The motivation for the development of these devices arose from the large base drive current required by power bipolar transistors and their limited switching speed capability. Two MOSFET structures were developed in parallel as a function of the current ratings and breakdown voltage requirements.

The first structure is the so-called vertical MOSFET structure used for the DMOS and VMOS devices shown in Figure 3.1. They are generally known to display a high switching speed and a very high input impedance. Unfortunately, these merits are offset by a higher on-resistance per unit area and a larger processing cost when compared to the bipolar transistor. As a consequence, power DMOS and VMOS FETs are particularly suited for high-frequency circuits such as switched-mode regulators, and for low voltage applications (<100V) where the on-resistance reaches acceptable values. Thus, voltage and current ratings of such devices are breakdown voltages up to 1000V at maximum current levels of about 1A and current handling capability of over 20A at breakdown voltages below 100V. However, because of their non isolated structure they are generally used as discrete devices and are not to be compared with lateral HV MOSFETs but with bipolar power device technologies [1].

The second structure is an isolated MOSFET structure used for the lateral DMOS and vertical buried layer DMOS devices that are presented in section 3.1. They are more adequate for *High-Voltage Integrated Circuits* (HVICs) and *Power Integrated Circuits* (PICs). PICs and HVICs are both defined as monolithic chips containing low-voltage (LV) and HV devices. The distinction between them is made at the voltage and current ratings level [2]. On the one hand, HVICs are generally designed to deliver low output currents (up to 100mA) at supply voltages up to 1000V. They typically find applications in electrostatic and ink-jet printers and plasma displays. On the second hand, PICs are designed to handle currents between 0.5A and 30A and operate between 50V and 500V. They find applications in linear and switching regulators, ac motor controls, automotive control circuits and electronic ballasts for fluorescent lights.

In order to combine HV devices with LV analog and digital circuitry, a common solution used in PICs and HVICs technologies consists in starting with a HV device technology and to try to include LV circuit elements into it. This approach requires a large number of processing steps resulting in an expensive power device technology. Moreover, it usually performs complementary LV devices but not HV ones. It generally includes low-voltage CMOS devices for digital circuits, a LV bipolar transistor for analog circuits and a power DMOS or bipolar transistor for HV circuits. These technologies are briefly reviewed in section 3.2 in order to point out their complexity.

Another common solution consists in modifying an existing LV CMOS technology to implement HV devices [3]. This approach usually requires the addition of at least two supplementary masks and two implantation steps, resulting in an expensive LV process. One cost effective solution is the use of the same standard

LV CMOS technology and to extend the channel near the drain with a lightly doped buffer region, without any modification of the process steps [4]. This approach represents the technological basis of this book and will be called *Smart Voltage eXtension* (SVX). It has the advantage of keeping a standard threshold voltage V_t and

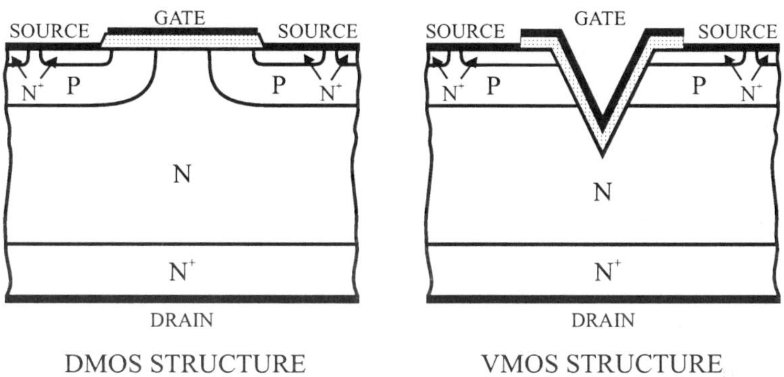

Figure 3.1: Vertical power MOSFET structures mostly used for discrete devices

high current gain in comparison to PICs and HVICs technologies. As a consequence, the required surface to achieve equal current driving capabilities is optimised. Moreover, it offers an increased design flexibility as the extension of the channel made for the HV devices is completely transparent for the LV parts of the circuit. Hence, existing standard LV libraries and computer aided design tools are still usable as far as special layout rules and models are added for the HV parts of the circuit. The SVX approach has the merit of covering part of the PICs and HVICs applications with the cost of a standard CMOS technology. In fact, HV devices designed with the SVX method are able to deliver output currents up to 500mA with supply voltage ranging between 5V and 170V. They typically find applications in flat panel display drivers, ink-jet printers, power amplifiers, power supplies and automotive. The limits of these applications are a trade off between surface and circuit performance in comparison to PICs and HVICs technologies.

The example of an actual implementation of lateral HV transistors in a standard low-voltage CMOS technology is discussed in section 3.3, while the SVX method is discussed more in detail in the next chapter.

3.1 Isolated highvoltage MOSFETs

The vertical buried layer DMOS structure is very similar to the vertical DMOSFET shown in Figure 3.1, except that the drain, source and gate electrodes are on the same side. A buried layer is used to collect the current flowing vertically from the source to the lightly doped drain. In the lateral DMOS structure, the lightly doped drain extends along the surface, and the current flows laterally. The characteristics of these devices are discussed in the following paragraphs.

3.1.1 Vertical buried layer DMOSFET

Vertical buried layer DMOSFETs are generally fabricated in an n-type epitaxial layer, as shown in Figure 3.2. The channel is located in the p-type diffusion layer under the gate electrode. When the device is on and the drain voltage V_{ds} is zero, an inversion layer is formed but no current is flowing. As V_{ds} is increased the electrons in the inversion layer drift with a velocity proportional to the electric field along the channel $v=\mu_n E_y$, as in the case of a low voltage device. When V_{ds} is such that the voltage drop across the channel V_{Ch} reaches the value V_{gs}-V_t, the channel is pinched-off. It should be noted at this point that, in vertical DMOSFETs, the total drain-to-source voltage drop V_{ds} is not only along the channel but can be separated into four components

$$V_{ds} = V_R + V_{Ch} + V_{depl} \tag{3.1}$$

where V_R is the voltage drop across the lightly doped epitaxial layer and any parasitic resistance (buried layer, interconnections etc...), V_{Ch} the voltage drop across the channel and V_{depl} the voltage drop across the depletion layer that may form around the junction between the channel and the drain drift region (epitaxial layer). Further increases in V_{ds} do not change V_{Ch} or I_d, but serve to expand the depletion layer at the p-n junction. The resulting I_d-V_{ds} characteristics are similar to that of low voltage devices. Detailed analysis of the drain current control in DMOSFETs can be found in the specialised literature [5],[6].

The principal function of the epitaxial drain drift region is to block and support the full forward voltage held off by the transistor. Indeed, the mechanism of avalanche breakdown of the drain junction, discussed in chapter 2 section 2.2.1, can be improved significantly by the use of a lightly doped drain. By the choice of an appropriate relative doping level between the channel region and the drain drift region, the spreading of the depletion layer can be controlled to occur mainly along the depth of the epitaxial layer. In Figure 3.2, the depletion layer edge between the p-channel region and the epitaxial layer is shown schematically, as well as a typical example of doping profiles along the surface and depth of the device. Since the doping level of the epitaxial layer is higher at the surface (region A) than the body region (region B) the depletion layer width is grater in the latter. Hence, avalanche breakdown voltage of the drain junction depends on the thickness of the epitaxial layer and its doping concentration. The breakdown voltage increases with lighter doping and higher thickness of the epitaxial layer. However, since the full forward current in the ON state of the device is flowing in the lightly doped drain, it is most important to minimise the parasitic resistance of the epitaxial layer in order to minimise power dissipation and maximise current handling capabilities. Therefore, the epilayer should not be thicker than it has to be to support the blocking voltage. Voltage range for a vertical DMOSFET such that shown in Figure 3.1 range from about 5μm for 50V device to about 50μm for 1000V device. Nevertheless, in vertical buried layer DMOSFETs epitaxial layer thicker than 20μm are costly and impractical to produce, as deep n^+ diffusions take many hours to be achieved. A first order theory based on a parallel plane $p^+n^-n^+$ diode structure is developed in [5], where the optimum doping profile and thickness of the epitaxial layer are specified

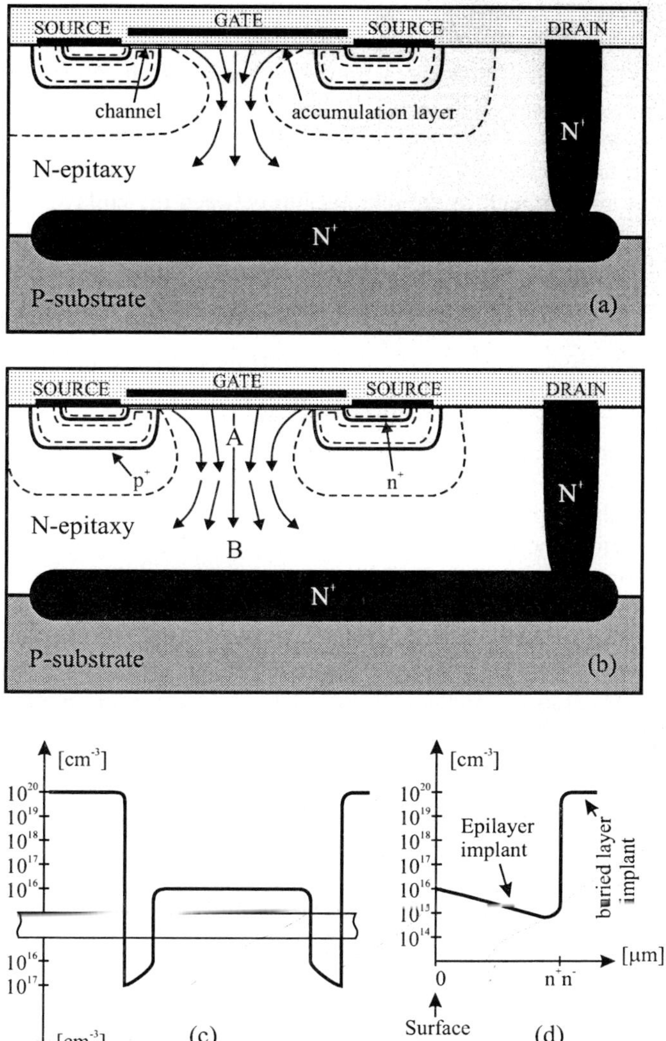

Figure 3.2: Vertical buried layer DMOSFET a) uniform doping of the epitaxial layer. b) non uniform doping of the epitaxial layer along the depth of the device. c) typical doping concentration along the surface. d) typical doping profile from the surface towards the buried layer

for minimum resistance. An optimised relationship between ON resistance and breakdown voltage of the junction is also given by

$$R_{EPI} \approx \frac{6x10^{-13}}{A} BV_{pp}^{2.5} \tag{3.2}$$

where BV_{pp} is the parallel plane breakdown voltage of the junction in volts, A the area of the diode in m^2 and R_{EPI} is in Ω. The relationship in Equation (3.2) is based on the assumption of a uniform distribution of the current in the drain epitaxial layer. Actually, lateral non uniformity of the current distribution, due to source and channel diffusions, is more dominant in the relationship between the epilayer resistance and the drain breakdown voltage. It can be shown that most benefit is obtained when the lateral doping concentration is as uniform as the one shown in Figure 3.2c.

In comparison to uniformly doped epitaxial layers, a non uniform doping profile, in a direction perpendicular to the surface, is desirable in vertical DMOS structures. On the one hand, it reduces the effect of the parasitic JFET located between the two channel diffusions, assuming that the doping concentration level at the surface is higher than that of the uniformly doped case, as shown in Figure 3.2a and Figure 3.2b. On the other hand, it minimises the contribution of the epitaxial layer resistance caused by a higher current density in this region. Therefore, starting from the surface of the epilayer, the doping profile decreases towards the buried layer. It reaches a minimum just before the metallurgical n^+n^- junction where the doping concentration rises rapidly from typically less than 10^{15} cm^{-3} to 10^{19} cm^{-3}. It is found that such a profile does not significantly affect the breakdown voltage of the drain junction.

Since the scope of this section was to present briefly the high-voltage capabilities of isolated vertical DMOSFETs, parasitic components as well as reliability and temperature considerations are not discussed here. The reader interested in more detailed analysis can refer to the specialised literature [5],[6].

3.1.2 Lateral DMOSFETs

In lateral-double diffused MOS transistors, the thickness and doping profile of the lightly doped drain are not the only controlling parameters for high-voltage capabilities. Since the lightly doped drain extends laterally, the amplitude of surface electric fields becomes as important as determining the high voltage capabilities of the device. The different techniques used to overcome this limitation are the scope of this paragraph.

The originality of the lateral DMOS (LDMOS) structure is its ability to achieve short channel (L=0.4..2μm) and HV operation at the same time. It should be noted that at the time where DMOSFETs were introduced [7],[8],[9], photomasking and diffusion tolerances were such that typical channel length of MOSFETs was not smaller than 8 microns. Hence, LDMOS transistors were used as HV switches and most particularly as microwave devices. They were the rival discrete devices of bipolar microwave transistors. A cross section of the two basic structures used in DMOSFETs is shown in Figure 3.3.

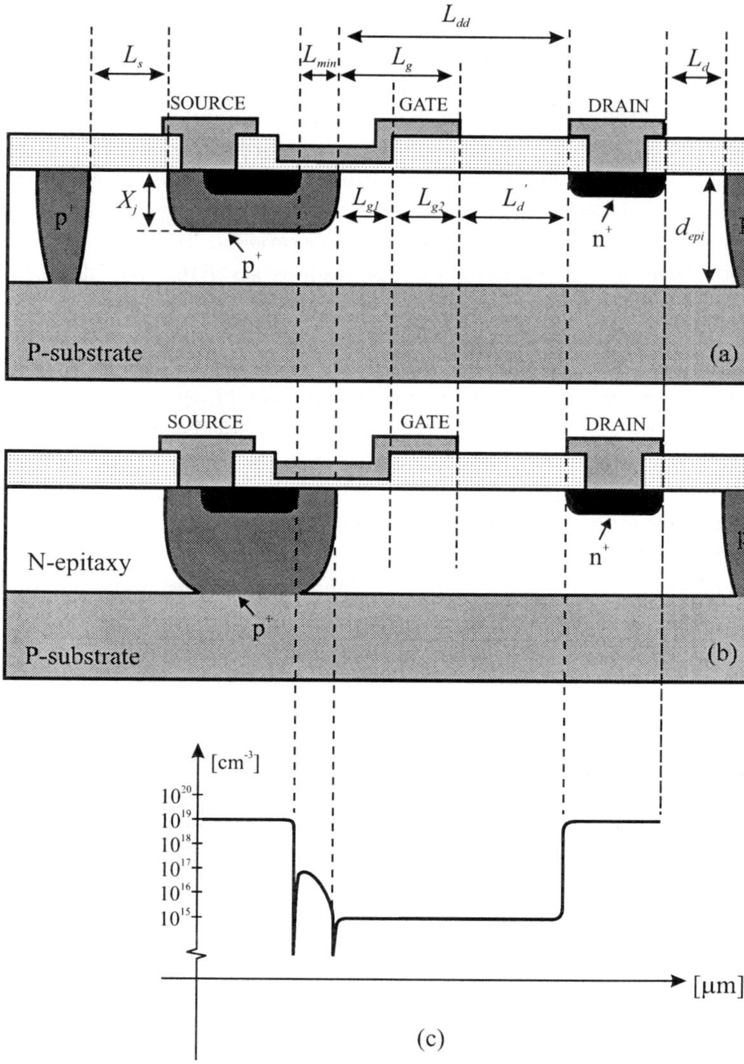

Figure 3.3: Cross section of the two basic lateral DMOSFET structures

The difference between the LDMOS structures of Figure 3.3 resides in the channel biasing. In Figure 3.3a the channel and the source are connected to the same voltage but are isolated from the substrate. This option is very interesting for applications requiring a high-voltage source follower configuration. In Figure 3.3b the channel is automatically grounded via the p^- substrate. Although the difference in the channel bias voltage, the fabrication method is identical in both cases. The starting material is a p-type substrate. An n^- epitaxial layer is then deposited followed by an oxide growth. The latter is opened, and the first p-channel region is diffused through this window to a thermally controlled depth. In case (a) the diffusion is typically from 0.5

to 4μm depth, while in case (b) it is as deep as the epitaxial layer. Particular care must be taken in order to carry out the p diffusion step, because the exact channel length is determined by the difference in lateral diffusions between the p and n^+ regions. Therefore, the oxide opening cut must be very clean and the p diffusion must be achieved in an non oxidising atmosphere in order to retain the oxide edge. The step following the p diffusion is the drain contact opening while the initial channel opening is protected. Then, the n^+ drain and source regions are diffused in an oxidising atmosphere as in conventional MOS process. The superimposition of the different n-type and p-type doping concentrations gives the final doping profile of the device along the surface, as shown in Figure 3.3c. The following steps are identical to a conventional metal gate MOS process. That is, an opening in the oxide above the channel region where the gate oxide is grown, followed by a contact opening at the drain and source locations where the metal layer is deposited. During the metal deposition step, the gate is formed over the thin oxide layer and steps on the thick oxide over the lightly doped drain, to form the filed plate for HV devices. In modern LDMOSFETs a polysilicon gate is used to perform self-alignment of the gate with the source diffusion as well as threshold adjustment, as in conventional polysilicon gate CMOS process. The basic steps described above are the basic steps required to fabricate lateral double-diffused MOSFETs. These steps are shown in Figure 3.4 in the case of a metal gate process.

As already mentioned, the high-voltage capabilities of the lateral DMOS transistor are due to the presence of an n^- drift region between the channel end and the n^+ drain diffusion. The critical lateral and vertical dimensions to the HV capability of the device are shown in Figure 3.3 and are detailed in the following paragraphs for both cases (a) and (b).

1) The lateral spacings L_d and L_s: The drain to isolation and source to isolation spacings are necessary to prevent punchthrough breakdown between junctions. Punchthrough of the $p^+n^-p^+$ source to isolation structure occurs with the merging of the p^+n^- depletion layers. Assuming a uniform doping of the epitaxial layer, the worst case punchthrough condition is when the source is connected to the substrate, leading to equal depletion layer width for each junction. Thus, the source to isolation spacing L_s at punchthrough is twice the depletion layer width $W_{PT(s)}$ in the epitaxial layer

$$L_s = 2W_{PT(s)} \tag{3.3}$$

where $W_{PT(s)}$ is given by the relationship between the reverse bias voltage and the depletion layer width of an abrupt junction diode (chapter 2 section 2.2.1)

$$W_{PT(s)} = \sqrt{\frac{2\varepsilon_s}{qN_{epi}} V_{PT}} \tag{3.4}$$

and where V_{PT} is the punchthrough voltage. For instance, considering a 200V device, the minimum spacing L_s required to avoid punchthrough breakdown is about 32μm for a typical 10^{15} cm^{-3} doping concentration of the epitaxial

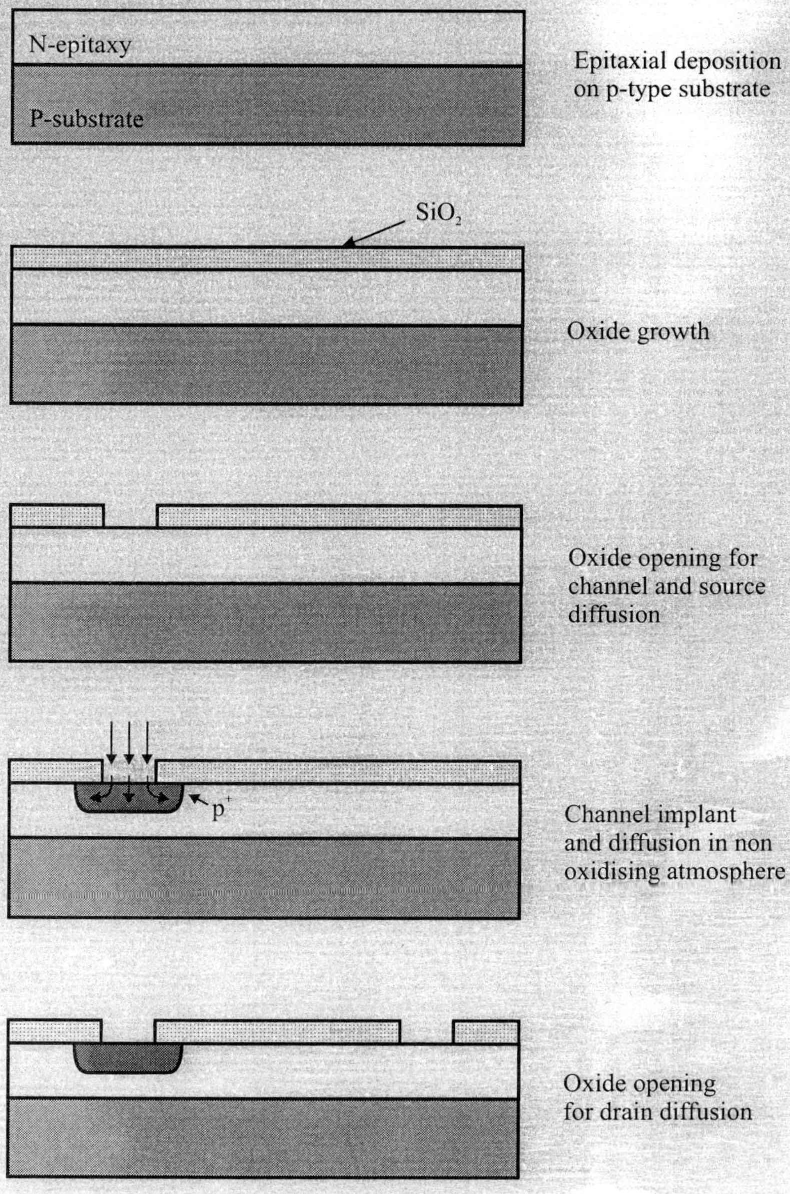

Figure 3.4: Basic steps required for the fabrication of HV lateral DMOSFETs in metal gate process

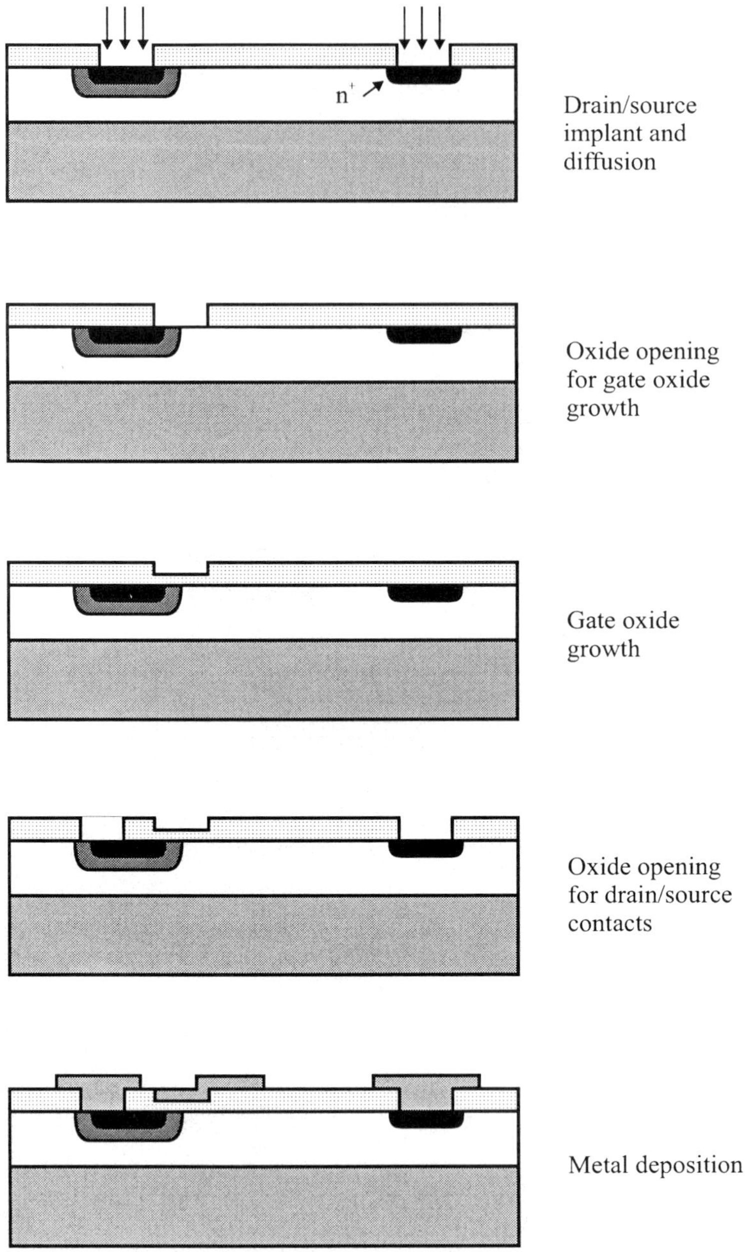

Figure 3.5: Basic steps required for the fabrication of HV lateral DMOSFETs in metal gate process (*continued*)

layer. In the case of the drain to isolation spacing L_d, punchthrough occurs when the depletion layer of the p^+n^- junction reaches the drain diffusion contact. Since the depletion layer extends from the p^+n^- side only, the spacing L_d is equal to the expression of the depletion layer given in Equation (3.4), resulting in

$$L_d = W_{PT(s)} \tag{3.5}$$

Thus, considering the same 200V device, the minimum spacing required to avoid punchthrough between isolation and drain diffusion contact is about 16μm for the same doping level of the epitaxial layer.

2) The epitaxial layer thickness d_{epi}: The epitaxial layer must be chosen such that punchthrough between source (back gate) and substrate does not occur before avalanche breakdown of the device, as in the case of the lateral spacings L_d and L_s. It should be noted that this constraint is only present in the lateral DMOSFET of case (a), as the channel of the device in case (b) is automatically grounded via the substrate. This holds for the lateral spacing L_s as well. In order to estimate the minimum thickness of the epitaxial layer, the punchthrough condition must be defined. The structure to be considered is very similar to the one in the lateral spacing case L_s, except that the isolation diffusion is replaced by the substrate material. Thus, punchthrough breakdown occurs when the depletion layer of the p^+n^- junction merges with the p^-n^- junction one. The depletion layer width of the former, at punchthrough voltage, is already given by Equation (3.4). In the case of the p^-n^- junction, it can be easily calculated by considering the electric field and potential distributions along the depletion layer on both sides of the junction. The resulting depletion layer width in the n^- side of the junction is given by

$$W_{PT(sub)} = \sqrt{\frac{2\varepsilon_s N_A}{qN_{epi}(N_{epi} + N_A)} V_{PT}} \tag{3.6}$$

where N_A is the doping concentration of the epitaxial layer. The epitaxial layer thickness is then given by the sum of both depletion layer widths at punchthrough and the depth of the p-channel diffusion

$$d_{epi} = W_{PT(s)} + W_{PT(sub)} + X_j \tag{3.7}$$

Considering the 200V device example with a substrate doping level N_A=10^{14} cm^{-3} and a channel diffusion depth X_j=4μm, the minimum required epitaxial layer thickness is about 25μm.

3) The minimum channel length L_{min}: A minimum channel length is necessary to prevent punchthrough between drain and source. Since the doping profile of the channel varies in the vertical and lateral directions as in Figure 3.3, the following simplifying assumptions must be made in order to perform a rough calculation of the required channel length.

- The source n^+ diffusion is shorted to the channel diffusion so that the depletion layer in the channel extends only from the drain side.
- Relatively deep junctions are considered so that the depth of the depletion layer under the gate is small compared to the junction depth, otherwise, punchthrough voltage would depend on the cylindrical radius of curvature r_j of the junction, as the case is in LV technologies (chapter 2 section 2.3).
- The drain-channel p^+n^- junction is linearly graded in the channel side of the metallurgical junction and abrupt in the drain side.

Considering the channel diffusion as a linearly grade one, the depletion layer width in the channel is given by (Figure 2.16 of chapter 2)

$$W_{PT(ch)} = \left(\frac{3\varepsilon_s}{qG} V_{PT} \right)^{\frac{1}{3}} \tag{3.8}$$

where G is the slope of the graded junction and V_{PT} is the reverse voltage at punchthrough. In the case of the doping profile shown in Figure 3.3, the slope G is defined as the ratio of the difference between the doping concentrations at the boundaries of the channel to the channel length at punchthrough

$$G = \frac{N_{ch(s)} - N_{ch(d)}}{L_{\min}} \tag{3.9}$$

where $N_{ch(s)}$ is the doping concentration at the source end and $N_{ch(d)}$ is the one at the drain end. Assuming that the punchthrough condition is reached when the channel becomes completely depleted by the drain junction, the minimum channel length required to avoid punchthrough is given by

$$L_{\min} = W_{PT(ch)} = \left(\frac{3\varepsilon_s}{qG} V_{PT} \right)^{\frac{1}{3}} \tag{3.10}$$

Replacing G in Equation (3.10) by the expression given in Equation (3.9), the resulting minimum channel length can be expressed as

$$L_{\min} = \sqrt{\frac{3\varepsilon_s}{q\left[N_{ch(s)} - N_{ch(d)}\right]} V_{PT}} \tag{3.11}$$

Thus, considering a typical doping concentration at the source end of about 10^{17} cm^{-3} and about 10^{15} cm^{-3} at the drain end (epitaxial layer doping), the minimum channel length required to avoid punch through is about 2µm for a 200V device. It should be noted at this point that since the doping concentration at the metallurgical junction does not change instantaneously, the peak is not located at the source edge but is slightly shifted into the channel. Therefore, the minimum channel length calculated above must be increased by the same amount as the distance required to reach the peak (0.3 to 0.5µm). Moreover, it should be noted that the doping concentration profile in the channel controls the threshold voltage of the device as well. Since a minimum amount of boron doping must be present to avoid punchthrough between drain and source, it sets a minimum value of the threshold voltage. The controllability of the latter as a function of the doping levels can be found in [10].

4) The lightly doped drain extension length Ldd and the field-plate lengths L_{g1} and L_{g2}: As was mentioned in the previous paragraphs, the presence of a lightly doped drain extension allows lateral devices to withstand the voltage drop between the high-voltage drain contact and the channel. Nevertheless, the simple extension of the drain length L_{dd} is not sufficient to overcome the high-voltage constraints. In fact, because of surface effects (chapter 2 section 2.2.2), avalanche breakdown would occur at the surface of the drain-to-channel junction without the presence of a gate field-plate, whose utility was already discussed in the same section. Therefore, the gate must extend over the thin oxide layer by L_{g1} and then it must step over the thick oxide layer by L_{g2}. Hence, the depletion region of the drain-to-channel junction terminates under thick oxide. It is important that the gate overlap stops short of the drain n^+ contact, otherwise avalanching of the surface near the n^+ contact degrades the voltage capability of the device [11]. Because of the two dimensional distribution of the electric field and potential in the depletion region located under the field-plate, the determination of the lightly doped drain length L_{dd} and field-plate lengths L_{g1} and L_{g2} is a complex task. Numerical methods are usually used to solve Poisson's equation in this region. However, first order considerations can lead to a rough estimation of these parameters. Indeed, the drain-to-channel junction with a simple field-plate of minimum length L_g, is shown in Figure 3.6. The optimal field-plate length is considered when, at maximum reverse drain voltage, the cylindrical portion of the depletion layer created by the field-plate, intersects the one of the junction curvature. In this way, the electric field lines at the surface are spread along the resulting depletion layer, as shown by the hatched area in Figure 3.6. Note that in the previous consideration a cylindrical approximation was made on the depletion layer at the edge of the field-plate. Hence, the maximum efficiency of the field-plate effect is obtained when the field-plate length L_g satisfies the following Pythagoras law

$$L = r_j + L_g = \sqrt{r^2 - X_d^2} \tag{3.12}$$

Replacing r by r_j+W, the minimum field-plate length can be expressed as

$$L_g = \sqrt{\left(r_j + W\right)^2 - X_d^2} - r_j \quad \textbf{(3.13)}$$

where W is the depletion layer width of the reverse biased drain-to-channel junction. The depletion layer X_d can be calculated considering the MOS structure formed by the field-plate, the oxide layer and the n-type epitaxial layer. Assuming that the surface under the field-plate is strongly depleted for large values of V_d, but is never inverted for $V_g \geq 0$, and a one dimensional field in a direction perpendicular to the surface, the integration of Poisson's equation across the space charge region gives the expression of the surface electric field and potential

$$E_s = \frac{qN_D X_d}{\varepsilon_s} \quad (3.14)$$

$$\varphi_s = \frac{qN_D X_d^2}{2\varepsilon_s} \quad (3.15)$$

where N_D is the doping of the epitaxial layer. In the absence of any charges located at the Si-SiO_2 interface, considering the Gauss' law for the continuity of the electric field at this interface

$$\varepsilon_{ox} E_{ox} = \varepsilon_s E_s \quad (3.16)$$

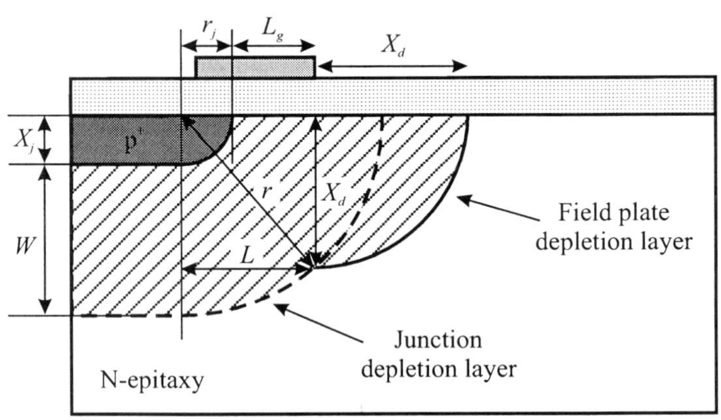

Figure 3.6: Geometrical consideration of the optimal field-plate length

the voltage drop across the oxide in the absence of charges, is given by

$$V_{ox} = E_{ox} t_{ox} = \frac{qN_D X_d t_{ox}}{\varepsilon_{ox}} \tag{3.17}$$

Considering the substrate potential at drain voltage V_d, the applied voltage V_g divides between oxide and space charge region according to

$$V_g - V_{FB} = V_g' = V_d - V_{ox} - \varphi_s \tag{3.18}$$

where V_{FB} is the flat band voltage. Replacing V_{ox} and φ_s by Equations (3.15) and (3.17), Equation (3.18) can be written as

$$V_d - V_g' = \frac{qN_D X_d t_{ox}}{\varepsilon_{ox}} + \frac{qN_D X_d^2}{2\varepsilon_s} \tag{3.19}$$

Solving Equation (3.19) for X_d, the depletion layer depth as a function of the drain bias voltage is given by

$$X_d = \sqrt{\left(\frac{\varepsilon_s}{\varepsilon_{ox}} t_{ox}\right)^2 + \frac{2\varepsilon_s}{qN_D}\left(V_d - V_g'\right)} - \frac{\varepsilon_s}{\varepsilon_{ox}} t_{ox} \tag{3.20}$$

Replacing X_d and W in Equation (3.13) by their respective expressions the resulting minimum field-plate length can be expressed as

$$L_g = \sqrt{\left(r_j + \sqrt{\frac{2\varepsilon_s}{qN_D} V_d}\right)^2 - \left(\sqrt{\left(\frac{\varepsilon_s}{\varepsilon_{ox}} t_{ox}\right)^2 + \frac{2\varepsilon_s}{qN_D}\left(V_d - V_g'\right)} - \frac{\varepsilon_s}{\varepsilon_{ox}} t_{ox}\right)^2} - r_j \tag{3.21}$$

Note that for $V_g'=V_d$, $X_d=0$ and the field-plate length is as long as the junction depletion layer width W. However, the minimum field-plate length given by Equation (3.21) is represented in Figure 3.7 as a function of the reverse drain voltage, with four reasonable gate voltage values as a parameter. The same 200V device is considered here with a gate oxide thickness of 0.4µm, a uniform epitaxial doping level of 10^{15} cm^{-3} and a channel diffusion radius $r_j=X_j=4$µm. Since the depletion layer depth X_d decreases with increasing gate voltage the minimum field-plate length must increase. Typically, a worst case configuration occurs at breakdown, and when V_g' is maximum at 30V leading to a minimum field-plate length L_g of about 10.7µm.

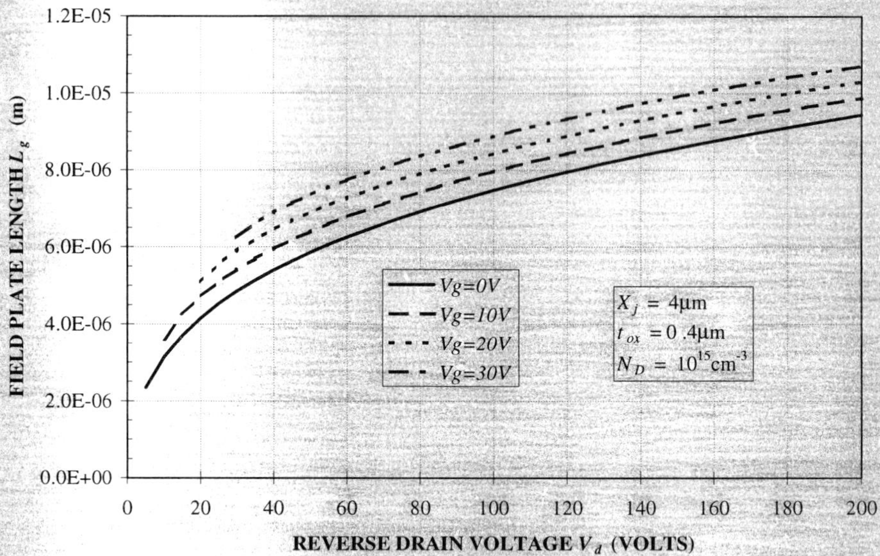

Figure 3.7: Minimum field-plate length L_g for different gate voltage values V_g

According to the field-plate theory of section 2.2.2 in chapter 2, the location of avalanche breakdown under the field-plate is a function of the ratio of the depletion layer depth X_d to the oxide thickness t_{ox}. If the ratio is higher than 15 avalanche breakdown occurs at the edge of the field-plate, otherwise it occurs at the junction curvature. In Figure 3.8, X_d/t_{ox} is shown for different oxide thickness. Note that for the high-voltage device discussed, with a thin gate oxide of 0.4µm, the field-plate is more efficient in spreading the electric field lines away from the surface, but avalanche breakdown would mainly occur at the edge of the field-plate. As the oxide thickness increases, the field-plate loses its efficiency and avalanche breakdown moves from the edge of the field-plate to the cylindrical curvature of the drain-to-channel junction. Therefore, the field-plate must start over the 0.4µm oxide layer in order to optimise its efficiency, and steps over a thicker oxide, at least a 1.5µm thickness for 200V breakdown, to minimise avalanche breakdown at the edge of the field-plate. The amount of field-plate length over the thin gate oxide L_{g1} and over the thick one L_{g2}, must be carried out by computer simulations. An optimisation of the electric field distribution must be performed as a function of the field-plate efficiency, the sharpness of the step between thin and thick oxide and the ratio between oxide thickness. In Figure 3.9, an example of the loss in field-plate efficiency due to a change in oxide thickness is shown for $V_g'=V_{gmax}'=30V$. As the oxide thickness increases, the field-plate length L_g must be increased to keep an optimum shaping of the underlying depletion region.

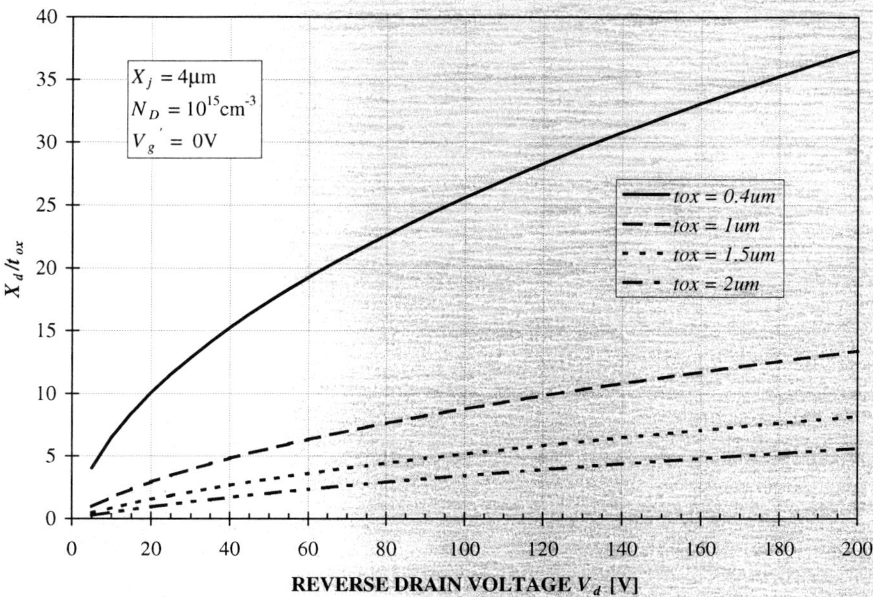

Figure 3.8: Ratio of the depletion region depth X_d to the oxide thickness t_{ox}

The lightly doped drain extension length L_{dd} is given by the sum of the total field-plate length L_g and the distance L_d' from the edge of the field-plate to the n^+ drain contact. The minimum distance L_d' is forced by the avalanche breakdown mechanism described by Declercq *et al.* in [11]. Any electron-hole pair generated by avalanche in the depletion region created by the field-plate, are swept towards the source and drain because of the induced high electric field. These injected carriers give rise to the avalanche multiplication phenomenon in this region, lowering the breakdown voltage of the device. The closer the field-plate edge is from the drain contact, the higher is the electric field concentration, leading to a higher avalanche multiplication rate. As a first order approximation, the n^+ drain contact can be placed at the edge of the depletion layer created by the field-plate. Thus, the electric field concentration is reduced significantly in this region, since the potential distribution is almost identical at the edge of the depletion layer to that of the drain contact. Assuming the cylindrical approximation and using Equation (3.20), the depletion layer width X_d is 15µm at V_g'=0V and V_d=200V. From the previous approximation it follows that L_d' can be fixed at 15µm as well. As a consequence, considering the minimum field-plate length of 10.7µm calculated above, the lightly doped drain extension L_{dd} must be about 25µm length.

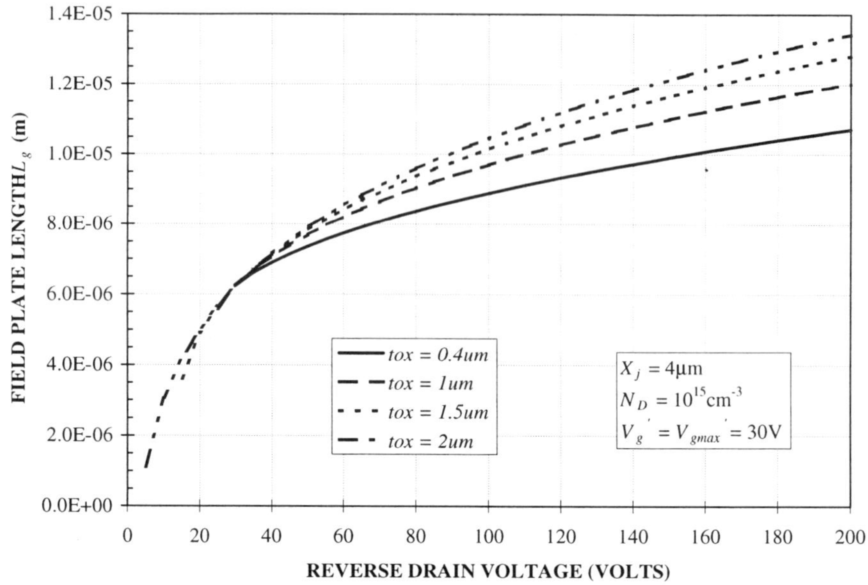

Figure 3.9: field-plate length L_g for different oxide thickness showing the loss in field-plate efficiency by an increase of L_g with increasing t_{ox}

3.2 PICs and HVICs in HV device technologies

The high-voltage device technologies used in PICs and HVICs are based on isolated HV structures such that discussed in section 3.1. The most commonly used isolation technique in these technologies is the p-n junction isolation. It involves creating in the same substrate separate epitaxial islands in which HV and LV devices will be implemented. Another interesting approach is the dielectric isolation technique which involves creating dielectrically isolated islands. This method is generally costly to produce as it requires a tricky mechanical processing of the wafers.

3.2.1 P-N junction isolation

The p-n junction isolation technique consists in starting with a lightly doped p-type substrate and implanting at specific locations n^+ type impurities, which are going to form the buried layer of vertical devices. This step is followed by a lightly doped n-type epitaxial layer growth of an appropriate thickness. Then a p^+ type implant is performed through the epitaxial layer in order to isolate the n-type islands from each other and to contain the transistors. Finally, an additional n^+ implant is performed through the epitaxial layer as well, in order to contact the buried n^+ regions. The resulting structure is shown in Figure 3.10 where different p-n junction isolation technologies are presented.

The first one, shown in Figure 3.10a, is produced by SGS Semiconductor and includes LV bipolar and CMOS transistors with vertical double diffused MOS and bipolar power transistors. They can handle output currents up to 10A and operating voltages up to 60V. The second technology is called Smart Power technology and was introduced by Motorola. It is also based on the p-n junction isolation approach but allows only one vertical pnp power device having a high current electrode at the back of the die. The LV part is formed by CMOS and npn bipolar transistors. The breakdown voltage of the power device depends on the thickness of the p-type epitaxial layer. It offers much higher current capabilities since the bottom series-resistance is quite small. The power device can be a vertical DMOSFET as well. This technology can handle currents up to 20A and operating voltages up to 110V.

Since the HV capabilities in the previous technologies depend on the epitaxial layer thickness, and since deep diffusions are costly and impractical to produce, power devices based on the vertical buried layer structure are voltage limited. This limitation was overcome by the introduction of the reduced field concept (RESURF DEVICES) [12]. It consists of a two dimensional control of the electric field at the surface of the device by selecting the appropriate doping concentration and thickness of the epitaxial layer. This concept was extended further to the lateral DMOS transistors as shown in Figure 3.10d. The device was fabricated at Philips laboratories using an epitaxial layer thickness of only 5 to 10μm with a breakdown voltage of about 500V. In Figure 3.10c and Figure 3.10e two other PICs technologies based on the RESURF technique were developed by Siliconix and General Electric respectively. In the first case, lateral double-diffused MOS devices capable of switching up to 20A and handling few hundred volts. The technology includes LV CMOS devices as well as vertical npn transistor. Finally, in the second case the technology combines lateral power DMOS and bipolar transistors capable of sustaining voltages grater than 500V but at a relatively low current level (about 100mA). It also includes a full complementary LV MOS and bipolar process.

3.2.2 Dielectric isolation

Despite its higher production cost, the dielectric isolation technique offers interesting advantages, the first being a latchup free technology since devices are dielectrically isolated from each other, as shown in Figure 3.11. It also offers reduced crosstalk between elements and very good protection from leakage current, especially with increasing temperature. The starting material of this technique is a single crystal wafer was n^+ or p^+ layers are diffused locally. Then an initial oxidising is performed at high temperature. The grooves are patterned, anisotropically etched and oxidised. A thick polysilicon layer is then deposited followed by a wafer lapping and polishing from the single crystal side. This results in isolated islands in which devices are fabricated as in Figure 3.11. Some experimental devices were developed using this technique, where breakdown voltages ranging between 500V and 1000V have been measured at current levels up to 500mA. Considerable efforts are now undertaken to lower the cost of the starting material by developing alternate methods to produce dielectrically isolated tubs. One of the most promising method is the *Bond and Etch Back Silicon On Insulator* (BESOI), which is not discussed here.

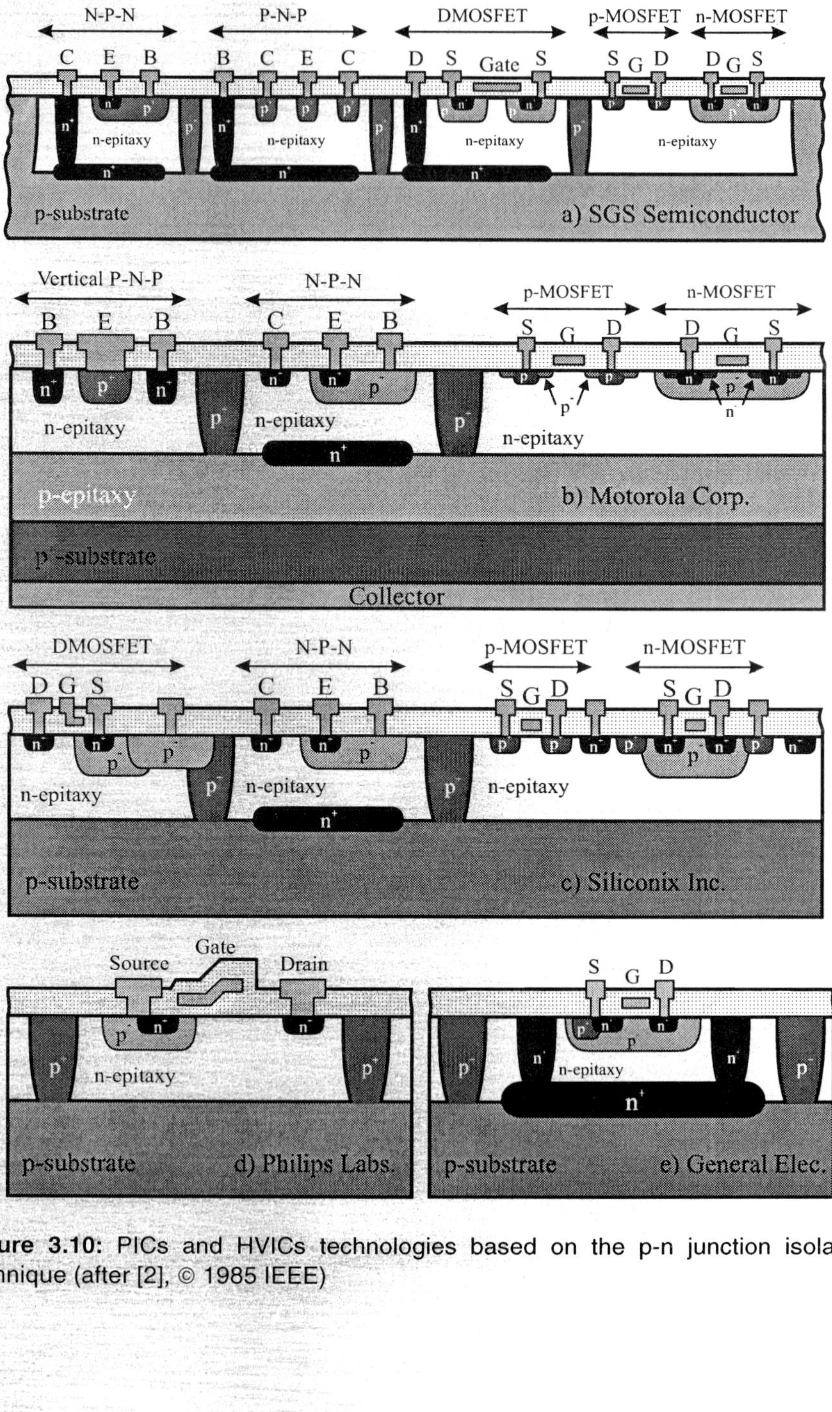

Figure 3.10: PICs and HVICs technologies based on the p-n junction isolation technique (after [2], © 1985 IEEE)

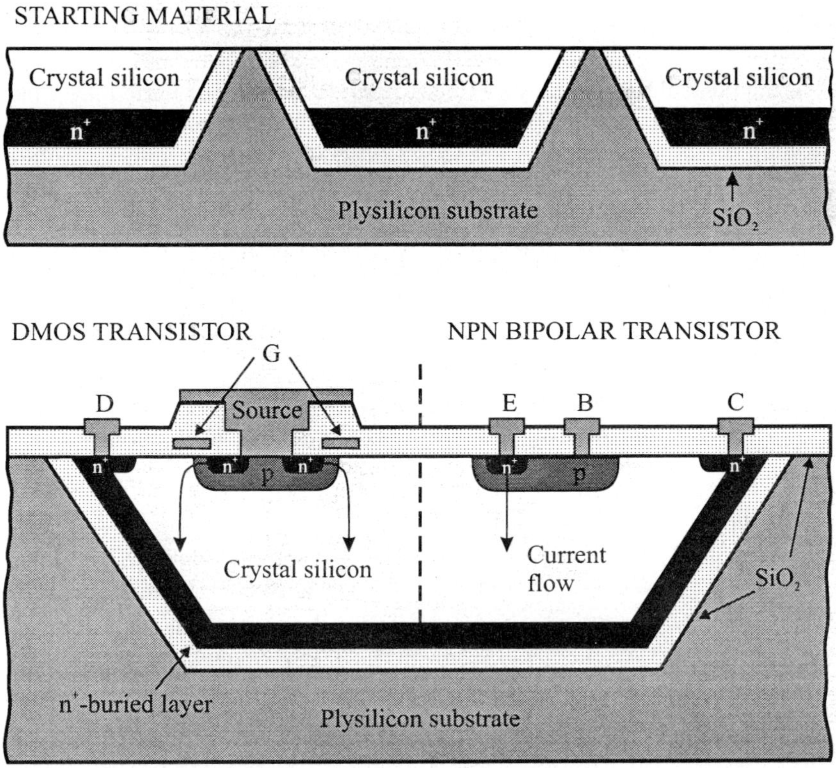

Figure 3.11: Vertical DMOS and bipolar transistors fabricated with dielectric isolation (after [2], © 1985 IEEE)

3.3 PICs and HVICs in LV CMOS technologies

As was mentioned in the introduction of this chapter, another common solution used for PICs and HVICs consists of starting with a standard low-voltage CMOS process where the supplementary masks and processing steps necessary to implement the HV features are added. In this section an actual example of an advanced HV technology based on a standard low-voltage 0.7μm CMOS process is presented. The technology is produced by Alcatel Mietec [3] and is called I2T, short for *Intelligent Interface Technology*. It uses 5 additional masks and extra processing steps to implement, LV and HV bipolar transistors, complementary HV lateral DMOSFETs and an isolated HV lateral DMOS transistor. The starting material of such a high-voltage technology is a p-substrate over which a p-type epitaxial layer is grown, as shown in Figure 3.12 a). In the case of the low-voltage process, n-WELL and p-WELL diffusions are implanted to form the substrate of the low-voltage pMOS and nMOS transistors, as shown in Figure 3.12 b). In the case of the high-voltage process, an n^+ buried layer is implanted before the growth of the epitaxial layer, at the locations of the high-voltage devices. It is mainly used to lower the resistance of these devices and to increase the punchthrough voltage of the floating LDMOS. The buried layer implant is one of the extra masks and processing steps used with respect to the low-voltage

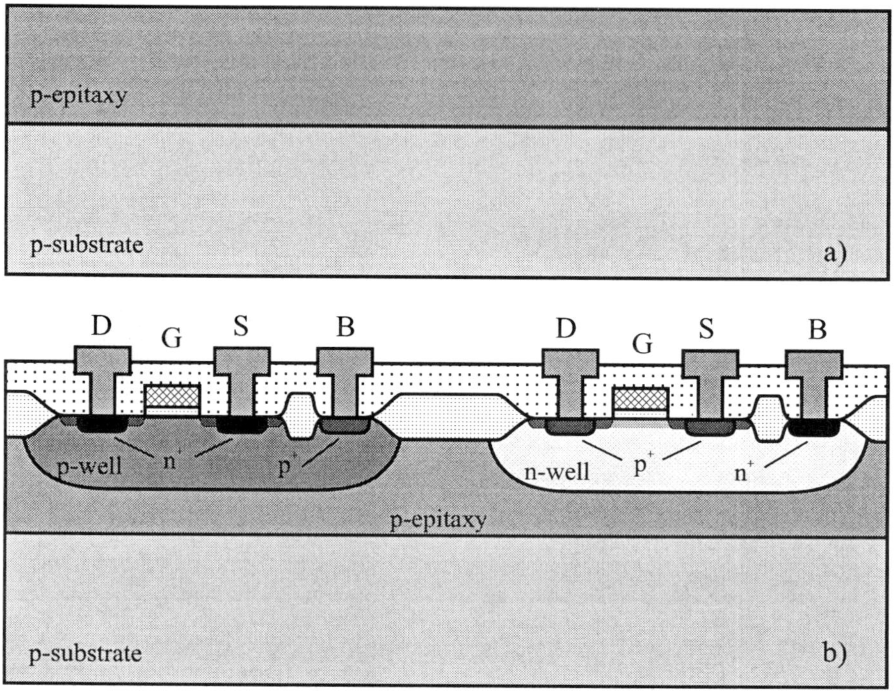

Figure 3.12: Cross section of the a) starting material of the I2T technology and the corresponding LV CMOS process b) LV CMOS process

CMOS process. The cross section of the bipolar features available in the I2T technology is shown in Figure 3.13, while the DMOSFET features are shown in Figure 3.14. The additional masks and processing steps required to implement these features can be summarised as follows

1. Buried layer: used to increase the punchthrough voltage between the P-body of the floating LDMOS devices and the substrate. The buried layer is an extra mask step and extra processing steps.

2. N-tub: used as a lowly doped n-type well for the high breakdown voltage of the HVNMOS devices. This is an extra mask step and extra processing steps.

3. P-WELL: used to isolate the N-tub/n-WELL and the p-WELL. This is an extra mask step but no extra processing steps.

4. P-body: used to implement the body diffusion (back gate source) of the floating LDMOS devices. Note that the source of these devices is self-aligned. This is an extra mask and extra processing steps.

5. Poly0: used for all the high-voltage DMOS devices in order to enable a thicker gate oxide. it is required for the higher gate voltage and the longer drive in of the P-body diffusion. This is an extra mask step and extra processing steps.

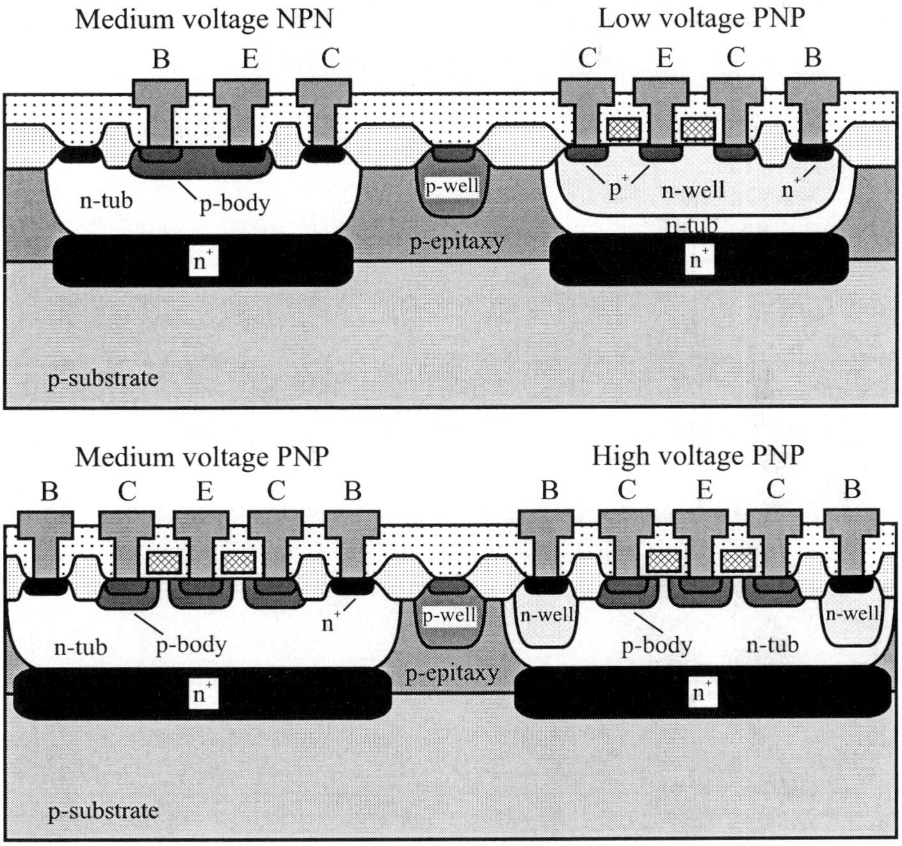

Figure 3.13: Cross section of the bipolar features available in the high-voltage I2T technology.

This increases up to 20 the total number of masks used by the I2T technology. The medium-voltage isolated NPN transistor is specified as a general purpose minimum area bipolar transistor. The maximum collector-to-emitter voltage V_{CE} may not exceed 25V when the transistor is on, and 60V when the transistor is off. Note that such a transistor can be implemented thanks to the additional P-body diffusion. While the NPN transistor is a vertical device, all the PNP transistors are lateral devices. Since the base region is on the surface of the silicon, it is protected by a poly shield. Moreover, the large base width and the high transit times, result in a low transition frequency f_t. Consequently, these devices can only be used for low speed applications. The low-voltage transistor is specified as a general purpose PNP with a

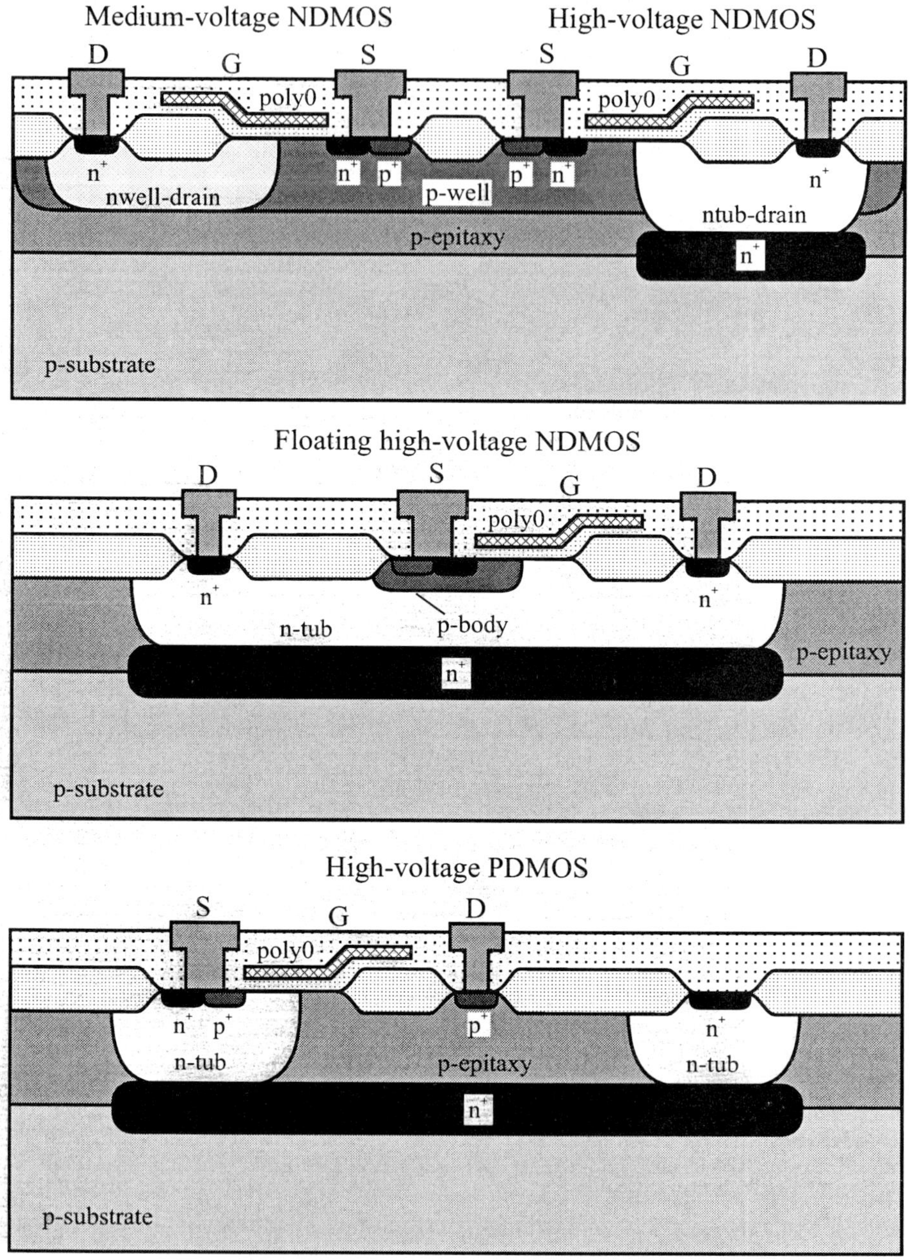

Figure 3.14: Cross section of the DMOSFETs features available in the high-voltage I2T technology.

maximum current drive capability of 100µA. The maximum V_{CE} voltage may not exceed 5.5V when the transistor is on and off. The medium and high-voltage options have the same current ratings, but their maximum V_{CE} voltage is 40V and 100V in the off state. When they are turned on, V_{CE} may not exceed 25V in both cases.

Different DMOSFET options are implemented as well, in order to cover a wide range of applications. The high and medium-voltage NDMOS transistors, are non floating devices. They exhibit a maximum drain-to-source voltage V_{dsmax} of 100V and 30V. While the former is optimised for maximum high-voltage capabilities, the latter is optimised for a minimum on state resistance (1.3Ωmm^2 against 2.3Ωmm^2). The corresponding channel lengths are fixed to 5µm and 4µm, and may not be changed. The threshold voltage is typically 0.96V and the maximum allowed continuous gate-to-source voltage V_{gsmax} is 12V. Note that the devices cannot operate simultaneously at V_{ds}=100V and V_{gs}=12V, since hot-carrier effects can result from the thin gate oxide value (about 400Å). Therefore, the maximum gate-to-source voltage can be applied in the linear region of the transistors, and must be limited to 5V in the saturation mode. An isolated high-voltage NDMOS option is available for applications where floating nodes must be switched. The maximum allowed V_{ds} value is 100V, and the typical threshold voltage value is 2.5V, given the high doping concentration level of the P-body diffusion. Note that V_{gsmax} is unchanged since the device uses the same gate oxide thickness. The channel length is fixed to 1.2µm and may not be changed as well. The resulting total on state resistance is 3.7Ωmm^2. Finally, the high-voltage PDMOS transistor exhibits a maximum drain-to-source voltage V_{dsmax} of 100V and a typical threshold voltage of -1V. The channel length of the transistor is fixed to 3µm leading to a total on state resistance of 4.9Ωmm^2.

3.4 References

[1] M. S. Adler, K. W. Owyang, B. J. Baliga, R. A. Kokosa, "The Evolution of Power Device Technology", *IEEE Trans. Electron Devices,* vol. ED-31, pp. 1570-1591, November 1984.

[2] V. Rumennik, "Power Devices Are in the Chips", *IEEE Spectrum,* vol. 22, pp. 42-48, July 1985.

[3] Alcatel Mietec I2T Design and layout Manual 1997

[4] M. Declercq, F. Clement, M. Schubert, "Design and Optimization of High-voltage CMOS Devices Compatible with Standard 5V CMOS Technology", *CICC proceedings,* paper 24.6, May 1993.

[5] D. A. Grant, J. Gowar, "Power MOSFETs theory and applications, John Wiley, New York.

[6] B. J. Baliga, "Modern Power Devices" Krieger, Malabar Florida.

[7] Y. Tarui, Y. Hayashi, T. Sekigawa, "Diffusion Self-aligned MOST: A New Approach for High Speed Device", *Proceedings of the 1st Conference On Solid State Devices*, pp. 105-110, Tokyo, 1969.

[8] H. J. Sigg, G. D. Vendelin, T. Cauge, J. Kocsis, "D-MOS Transistor for Microwave applications", *IEEE Trans. Electron Devices,* vol. ED-29, pp. 45-53, January 1972.

[9] T. P. Cauge, J. Kocsis, H. J. Sigg, G. D. Vendelin, "Double-Diffused MOS Transistor achieves microwave gain", *Electronics*, pp. 99-104, February 15, 1971.

[10] M. D. Pocha, A. G. Gonzalez, R. W. Dutton, "Threshold Voltage Controllability in Double-diffused-MOS Transistors", *IEEE Trans. Electron Devices,* vol. ED-21, pp. 778-784, December 1974.

[11] M. J. Declercq, J. D. Plummer, "Avalanche Breakdown in High-Voltage D-MOS Devices", *IEEE Trans. Electron Devices,* vol. ED-23, pp. 1-4, January 1976.

[12] J. A. Appels, H. M. Vaes, "High-Voltage Thin Layer Devices (RESURF Devices)", *IEEE Int. Trans. Electron Devices Meeting,* pp. 238-241, 1979.

4 DESIGN OF HIGH-VOLTAGE DEVICES USING THE SVX TECHNIQUE

Abstract:
A new high-voltage approach using a CMOS basis to extend the operating voltage without requiring any modification of the process steps is detailed in this chapter. It is the technological basis of this book, and will be called Smart Voltage eXtension (SVX). In order to carry out the design of the complementary high-voltage devices, the possibilities of implementing such devices by the use of existing technological layers are examined, and the main processing steps are presented. Two dimensional computer simulations are also performed to optimise the scaling of the high-voltage parameters, and to consider the reliability aspects.

The power device technologies discussed in the previous chapter start with a special high-voltage process, usually a bipolar one to implement the double diffused MOS transistors, or with a standard CMOS process in which the supplementary required masks are added to achieve lateral HV devices. In this chapter, the possibilities of implementing high-voltage transistors in a standard CMOS process are examined, keeping in mind that the steps of the process must remain unchanged. The present book focuses on this approach which will be called *Smart Voltage eXtension* (SVX) and which offers the advantages previously discussed. Therefore, the different steps required for a standard CMOS process are recalled first in section 4.1, then the design of high-voltage devices using the SVX technique is discussed in section 4.2. The scaling of the basic geometrical parameters allowing the transistor to withstand the high-voltage power supply as a function of the drain voltage is discussed in section 4.3. Finally, In order to optimise the scaling of the transistors, two dimensional computer simulations are performed for both devices in section 4.4.

4.1 Main steps of a standard n-WELL CMOS process

The starting material of a standard n-WELL CMOS process, is a high resistivity p-type substrate. The total processing steps of an entire silicon wafer are numerous, but actually, these can be summarised to the modules discussed in this section. The technology considered here is a 2μm n-WELL CMOS process with a single polysilicon layer and two metal levels. The basic process requires 12 masks to implement the complementary devices. Extra options are available to implement EEPROMs (Electrically Erasable Programmable Read Only Memories) and other analog features. These extra options are not discussed here.

4.1.1 The n-WELL implant and drive-in (Mask M26)

The n-WELL implant phase starts with an oxide growth of 4800Å followed by a mask definition delimiting the n-well location. The oxide is etched at this location where the photoresist has been previously removed. After the etching of the oxide, a thin buffer oxide of 400Å is grown to protect the surface of the silicon, and a phosphorus implant dose of 2.25×10^{12} at/cm^2 is performed at 160KeV energy, as shown in Figure 4.1a. Then, a first diffusion phase of the phosphorus is performed at 1180°C in non oxidising atmosphere such as N_2 during 735 minutes, followed by a diffusion phase in dry O_2 atmosphere during 60 minutes at the same temperature. The resulting n-well depth is about 5μm (Figure 4.1b).

4.1.2 Active areas definition (Mask M10)

After the diffusion phase of the n-well, the initial oxide is completely etched and a new thin oxide layer of 400Å is grown in a dry O_2 atmosphere at 1000°C during 35 minutes. An initial nitride layer of 1600Å is then deposited, and the active areas are delimited by the corresponding mask. Hence, the nitride is stripped outside these areas and, in a final step, the photoresist is completely removed. This main step is illustrated in Figure 4.1c.

4.1.3 Isolation between transistors (Mask M11)

The isolation between the complementary transistors uses a field doping mask and two different p-type implants. Thus, the photoresist is first etched outside the mask region, then a "channel stop" implant is performed using a boron dose of $2x10^{13}$ at/cm^2 at 80KeV energy. This step is followed by a second boron implant dose of $6.65x10^{11}$ at/cm^2 at 160KeV energy, called the "p-TUB" implant. Note that the nitride, where the photoresist has been removed, is not totally opaque to the second high energy implant, as shown in Figure 4.1d. In this region, the boron concentration reaches a peak of about $2.5x10^{16}$ cm^{-3} at 0.4μm depth.

4.1.4 Field oxide growth (LOCOS)

Once the isolation is achieved, the remaining photoresist is removed, but the nitride is kept during the growth of a 10000Å field-oxide. The growth is performed in two steps, the first being in an N_2 atmosphere at 1100°C during 130 minutes, and the second in a wet H_2O atmosphere at 950°C during 485 minutes. The resulting field-oxide shape is shown in Figure 4.2e. After the field-oxide growth, the initial nitride layer is stripped and the buffer oxide of 400Å is etched (Figure 4.2f). Note that during the thermal oxidation, the n-well depth increases of about 1μm and the peak of the boron concentration of the "p-TUB" and "channel stop" implants is reduced to 10^{16} cm^{-3} over 1-2μm depth.

4.1.5 Polysilicon gate deposition and gate oxide growth (Mask M40)

After the field-oxide growth, the next steps consist of growing and depositing the gate oxide and polysilicon gate respectively. The thin gate oxide of 400Å is grown in a dry O_2 atmosphere at 1050°C during 18 minutes. When the gate oxide growth is achieved, a boron implant dose of $4x10^{11}$ at/cm^2 is performed at 40KeV energy to adjust the threshold voltage of the devices. It is obvious that this low energy implant is stopped by the field-oxide, localising the V_t adjust under the gate oxide regions only, as shown in Figure 4.2g. This step is followed by a 3800Å polysilicon layer deposition and doping. The gates are defined by the corresponding mask, which allows the etching of the polysilicon layer. Finally, the photoresist is removed and the gate oxide is etched outside the polysilicon gate areas (Figure 4.2h).

4.1.6 Source and drain implant (Masks M25 and M24)

The next steps are associated with the implantation of the source and drain diffusions. First, a spacer oxide growth of 250Å is performed at 900°C in dry atmosphere during 15 minutes, then in a wet atmosphere during 5 minutes and finally in a non oxidising one during 15 minutes. After the definition of the implant mask of the p^+ diffusion a boron dose of 8.4×10^{14} at/cm^2 is implanted at 80KeV energy. The photoresist is then removed and the mask of the n^+ diffusion implant is defined. A phosphorus dose of 8.4×10^{14} at/cm^2 is then implanted at 80KeV. These steps are shown in Figure 4.3i and Figure 4.3j.

4.1.7 CVD SiO_2 and metal1 deposition (Masks M50 and M60)

After the drain and source implant steps, the photoresist is completely removed and a first 9000Å CVD SiO_2 is deposited and densified (Figure 4.3k). The mask of the source, drain and substrate contacts is then defined, and the etching of the oxide is performed. This step is shown in Figure 4.3l. A 6000Å aluminium layer deposition follows the SiO_2 opening, and the interconnection mask is defined for the etching of the metal1 layer (Figure 4.4m). When this is achieved, the photoresist is removed for the following steps which involves the deposition of the metal2 layer.

4.1.8 Metal2 deposition and passivation (Masks M51 and M61)

A second 9000Å CVD SiO_2 is deposited after the first metal layer deposition followed by a chemical planarisation process, as shown in Figure 4.4n. The via mask is then defined and the SiO_2 is etched at these locations. The photoresist is removed and a 9500Å aluminium layer is deposited. The etching of the metal2 layer is performed using the corresponding mask, followed by a passivation layer deposition (Figure 4.4o).

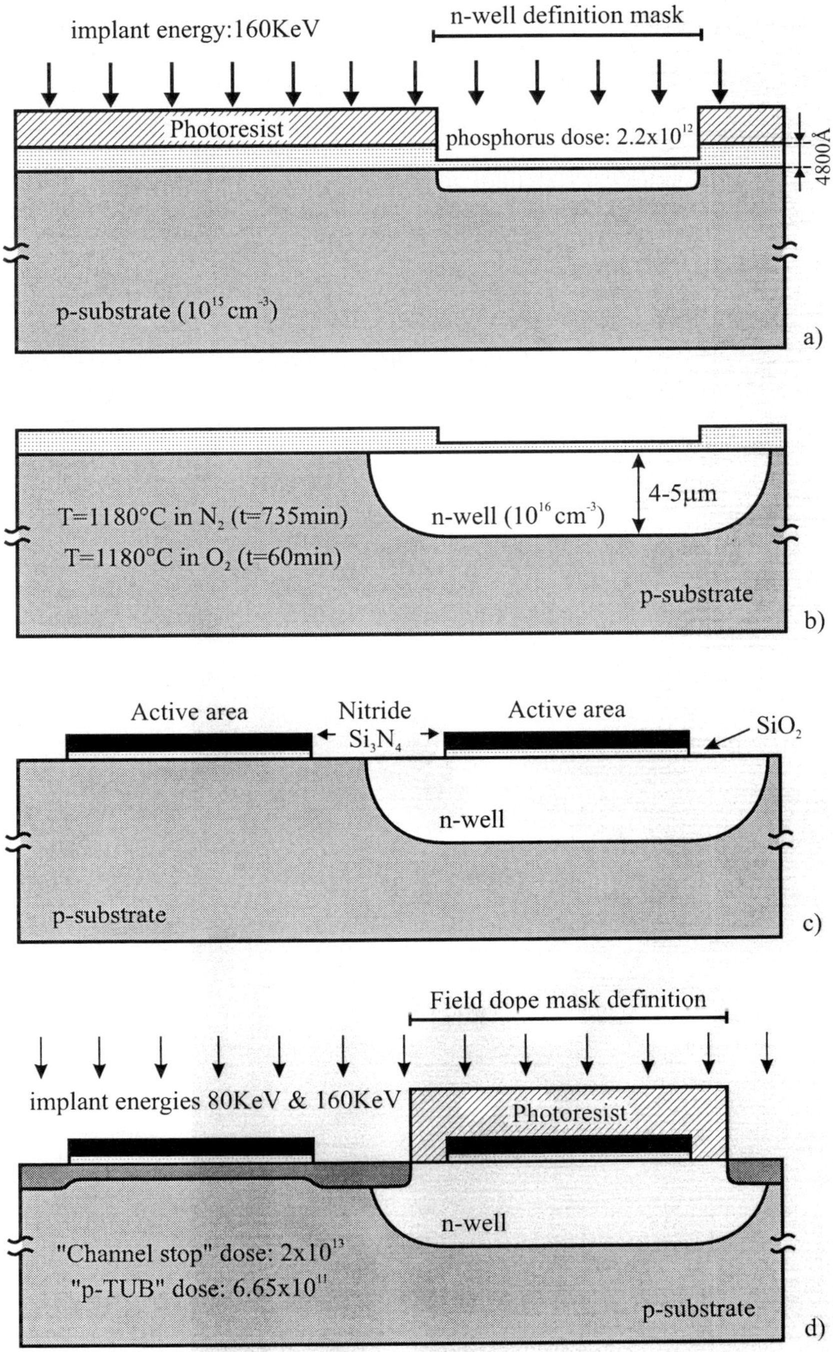

Figure 4.1: Processing steps of a standard low-voltage CMOS process

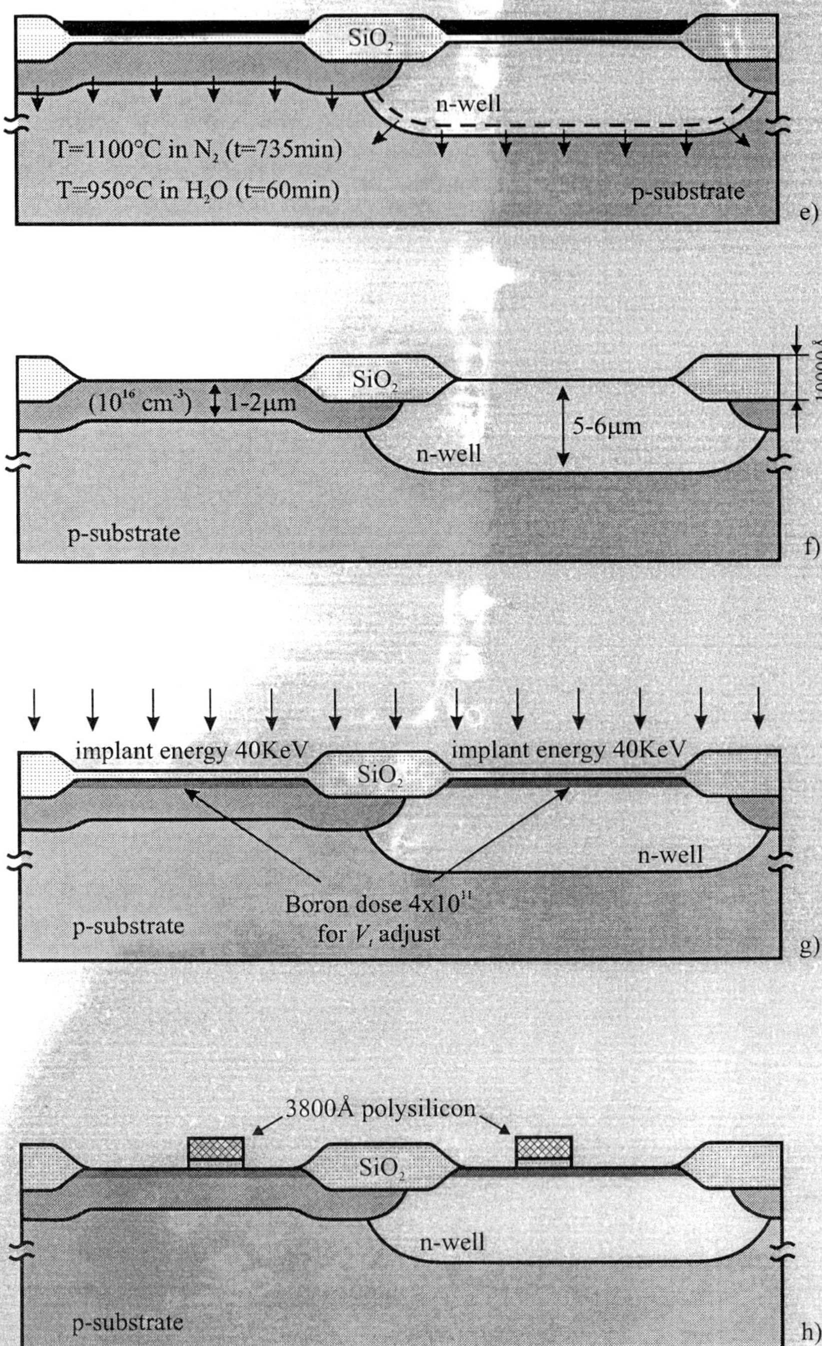

Figure 4.2: Processing steps of a standard low-voltage CMOS process (*cont.*)

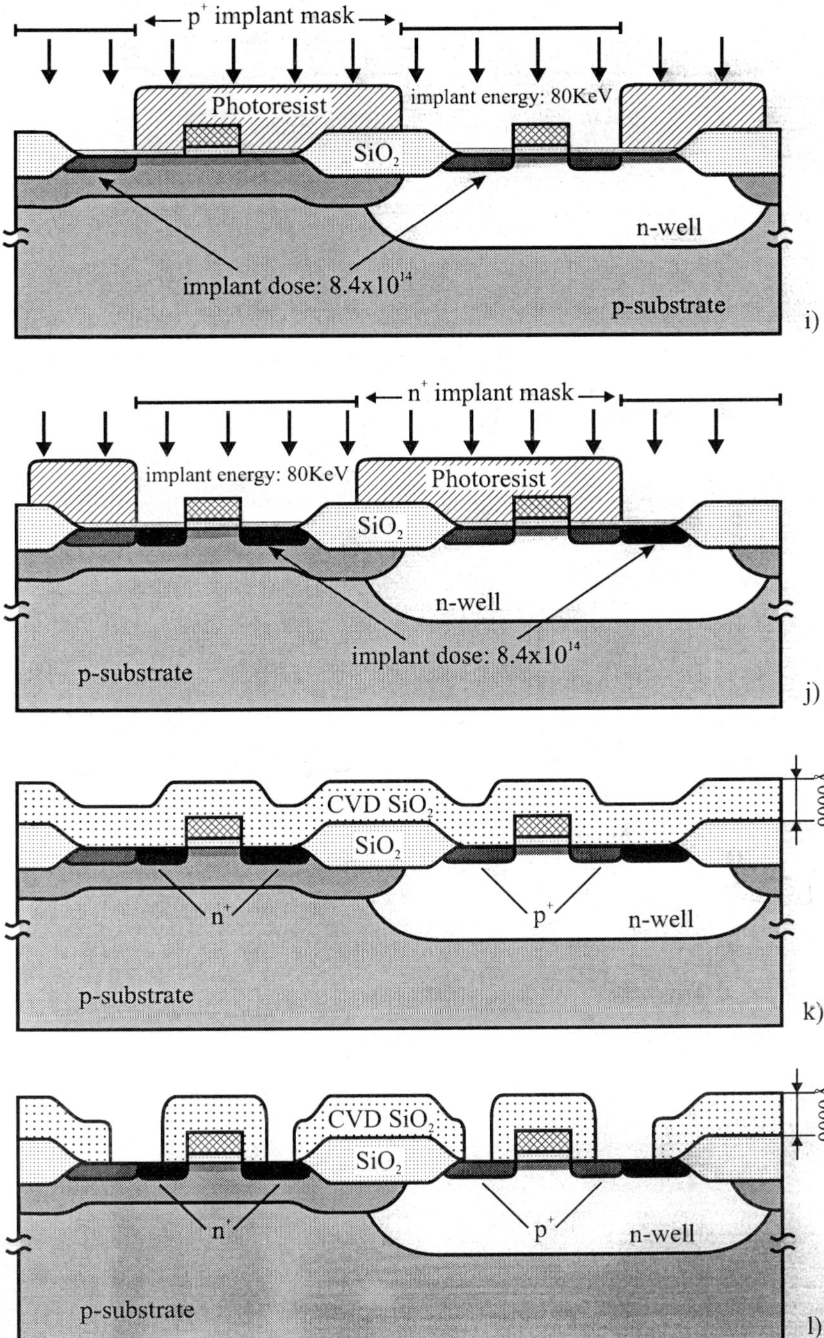

Figure 4.3: Processing steps of a standard low-voltage CMOS process (*cont.*)

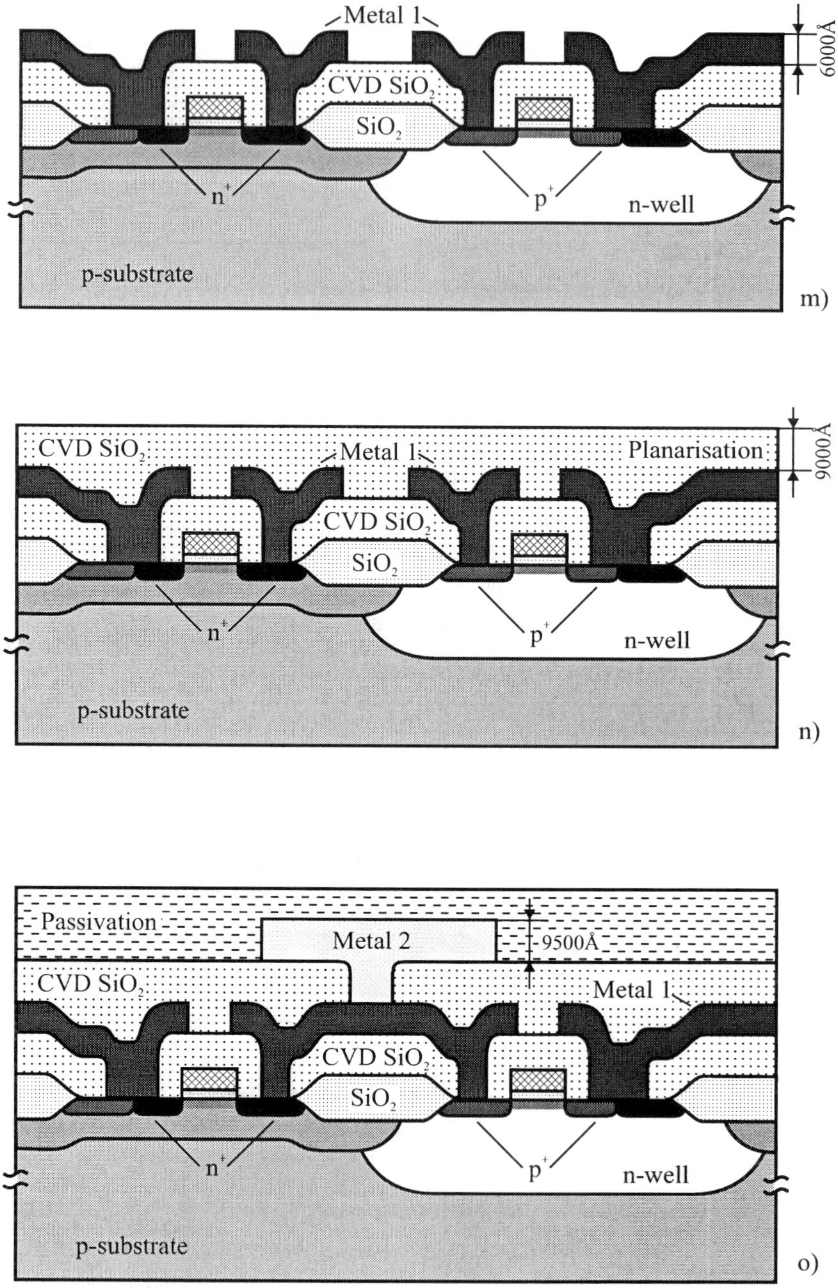

Figure 4.4: Processing steps of a standard low-voltage CMOS process (*cont.*)

4.2 Design of the SVX devices

The required high-voltage structure is shown in Figure 4.5, where the cross section of the lateral high-voltage nMOS and pMOS transistors includes ideal lightly doped drains and field-plates. Actually, these layers are not explicitly present in standard low-voltage CMOS technologies, especially when the main scope is to keep the process steps unchanged (in contrast to the CMOS technologies discussed in chapter 3 section 3.3). Therefore, existing layers should be examined and combined in order to carry out equivalent lightly doped drains. It is obvious that the issue of this operation depends strongly on the chosen technology to the extent that the feasibility of complementary high-voltage devices is sometimes impossible. The fabrication of high-voltage nMOS and pMOS transistors in the standard n-well CMOS process presented in section 4.1 is discussed in the following paragraphs.

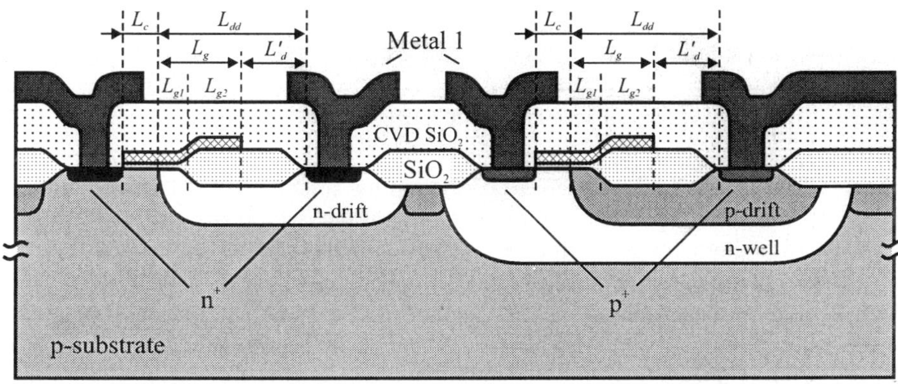

Figure 4.5: Cross section of the required complementary high-voltage devices

4.2.1 High-Voltage nMOS transistor (HVNMOS)

Considering the CMOS process flow discussed in section 4.1, the availability of an n-type lightly doped layer is straightforward. Indeed, the n-well implant is very convenient to form such layers, since the depth and doping concentration level are close to that used for the epitaxial layer of lateral DMOS devices. Hence, the lightly doped drain of the HVNMOS transistor is simply defined by the designer when drawing the masks of the circuit. It is obvious that the layout rules in this case are not respected anymore. The main processing steps of such a device are shown in Figure 4.6, Figure 4.7 and Figure 4.8 and are discussed below.

1) The lightly doped drain (Figure 4.6a): The drawn lightly doped drain is implanted through the n-WELL window, defined by the mask M26, followed by a diffusion phase at high temperature as in the case of the standard process flow. The lateral diffusion of the n-WELL (about 3-4μm) must be taken into account when designing the drain of the HVNMOS transistor, otherwise the effective channel length of the device is severely reduced.

2) Active areas definition (Figure 4.6b): A nitride (Si_3N_4) layer is deposited over a thin oxide layer and the active areas are delimited by the corresponding mask defining the source channel and drain locations. The nitride is stripped outside these locations and the next steps consist in defining the field doped mask and implanting the isolation layers. Note that the nitride layer covers part of the lightly doped drain (3-4µm) because of the n-WELL lateral diffusion.

3) "P-TUB" and "channel stop" implants (Figure 4.6c): The field doped mask M11 is called a clear mask because it removes the photoresist outside the mask definition. Since the field doped mask is derived from the n-WELL one plus an over-sizing constant of 1.5µm, the photoresist is etched outside the location of the n-WELL implant window plus 1.5µm at each side. A simplified form of the resulting equation of mask M11 is thus given by

$$M11 = M26@1.5\mu m = CNW@1.5\mu m \tag{4.1}$$

where *CNW* is the logic variable corresponding to the n-WELL layer, and @ represents the sizing operator. After the etching of the photoresist, the "p-TUB" and "channel-stop" implants are performed. As in the case of the low-voltage devices, the nitride layer being partially permeable to the second high energy implant of 160KeV, the n-WELL layer beneath is "etched" by the "P-TUB" implant. The resulting shape at the extremities of the n-WELL layer is shown in Figure 4.6c. Actually, the n-WELL layer is not etched, but a metallurgical junction is formed at this location between the "p-TUB/channel stop" implants and the n-WELL layer. Hence, the resulting junction will be referred to as being the "etching" of the n-WELL layer for convenience.

4) field-oxide growth (LOCOS) (Figure 4.7d): Since the field-oxide growth is performed at high temperature, the "p-TUB" and "channel-stop" layers diffuse of about 1µm, leading to an "etching" of the n-WELL layer at a depth of 1-2µm. Consequently, the resulting lightly doped drain junction at the channel end exhibits a small radius of curvature and extends laterally under the channel. Therefore, a high concentration of the electric field lines should be expected in this location, and a lower limit of the minimum achievable channel length would be fixed. The imperfection of the lightly doped drain at the channel end is the first consequence of an unchanged process steps condition.

5) Polysilicon gate deposition and gate oxide growth (Figure 4.7e): After the removal of the nitride layer, the thin gate oxide of 400Å is grown and the V_t adjust implant is performed. Since the energy level of this implant is very low (40KeV), it penetrates only through the thin oxide layer. As a consequence, an additional "etching" of the n-WELL layer is performed under the "bird's beak" of the field-oxide, as shown in Figure 4.7e. The etching is performed at a depth of about 0.4μm. Following the V_t adjust implant, a 3800Å polysilicon layer is deposited and doped. The gate is defined by the mask M40 which allows the etching of the polysilicon layer. A smooth stepping of the gate field-plate is allowed by the "bird's beak" of the field-oxide. Note that the etching of the lightly doped drain under the "birds beak" associated with the smooth stepping, forces the field-plate parameter L_{gl} to zero (L_{gl}=0). This condition represents the second consequence of not changing the process flow steps.

6) Source and drain contact implants (Figure 4.7f and Figure 4.8g): After the etching of the polysilicon layer the gate oxide is also etched outside the gate area. Then a spacer oxide of 250Å is grown and the n^+ contacts of the drain and source are implanted. As shown in Figure 4.8g, the drain contact diffusion is implanted in the n-WELL forming the lightly doped drain. After the n^+ implant, a p^+ implant is performed to contact the substrate of the resulting high-voltage nMOS device.

7) CVD SiO2 and metal1 deposition (Figure 4.8h and Figure 4.8i): The remaining steps, such as CVD SiO_2 deposition, metal1 and metal2 deposition and passivation of the wafer are achieved in the same manner as for the low-voltage devices.

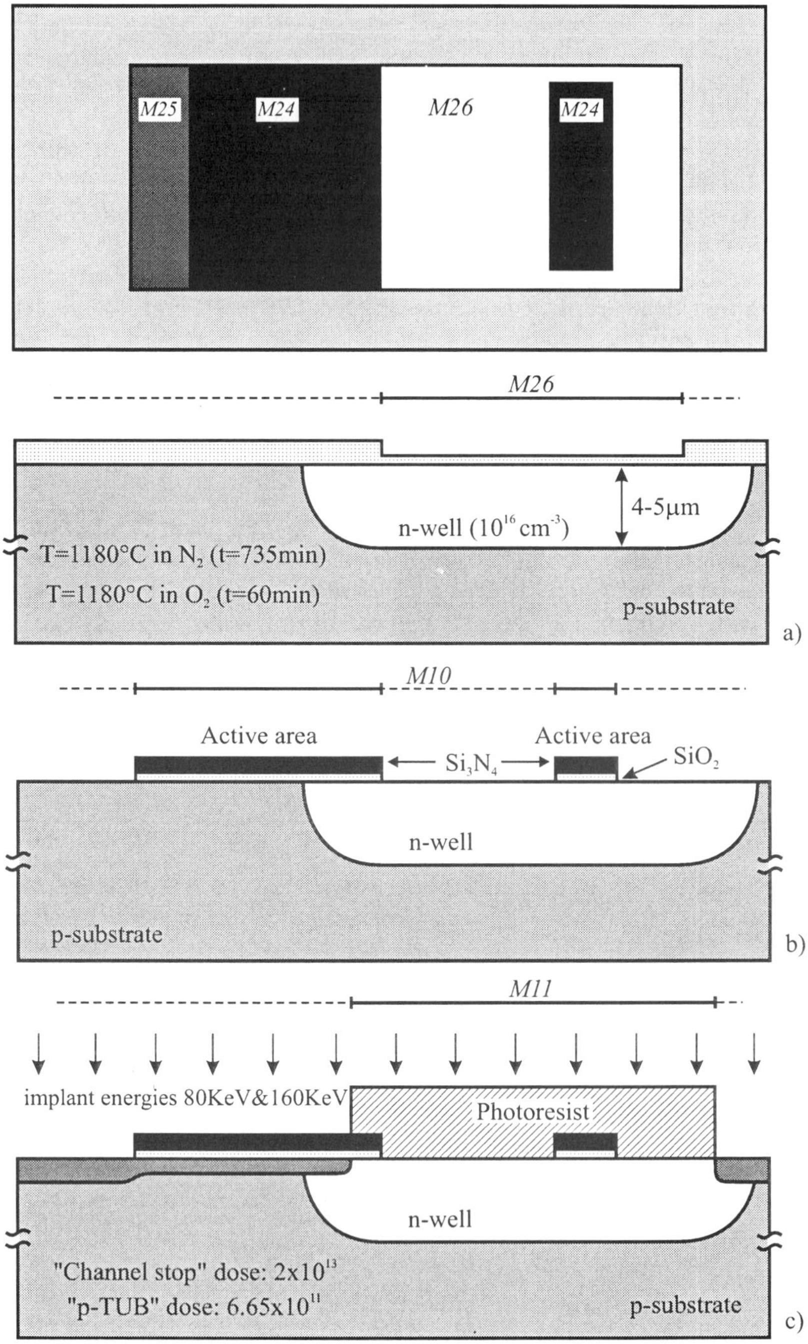

Figure 4.6: Processing steps of the high-voltage nMOS transistor

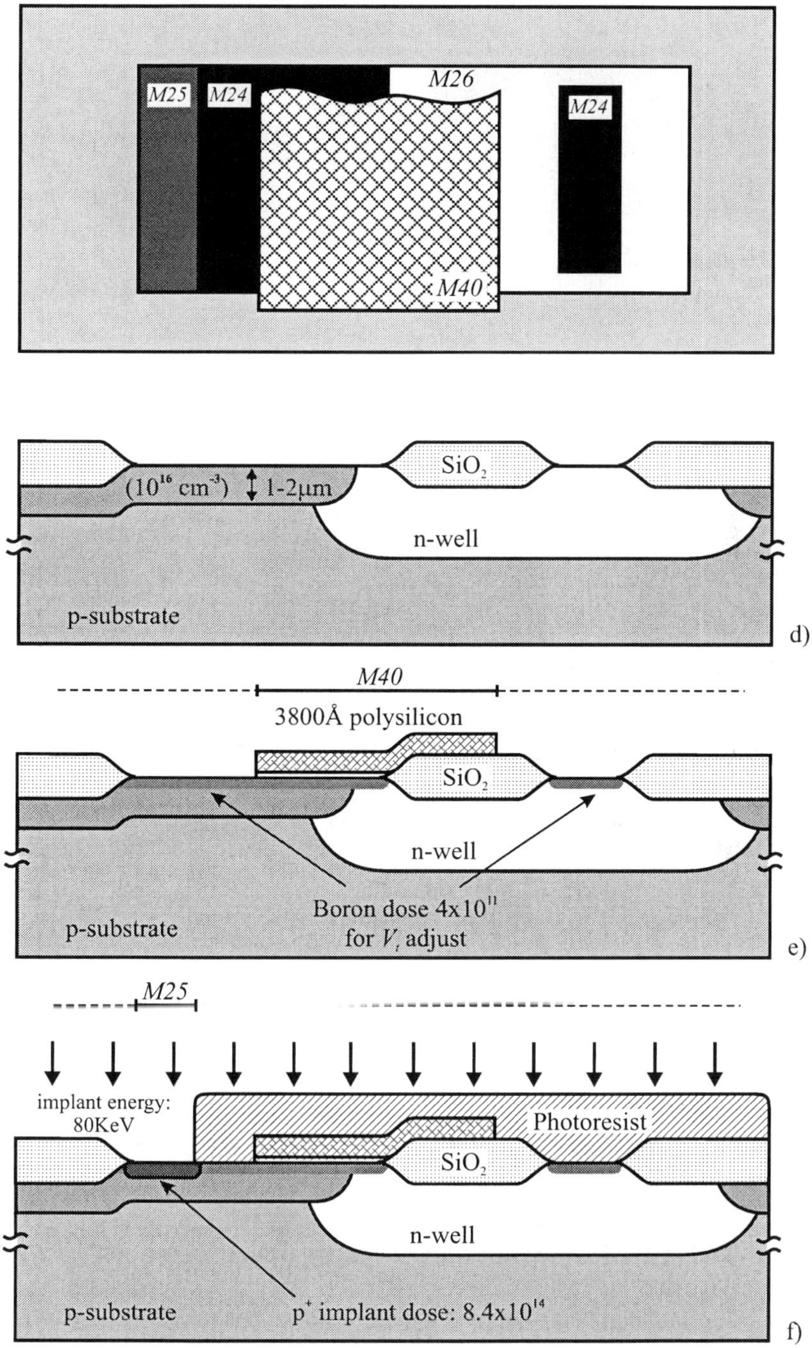

Figure 4.7: Processing steps of the high-voltage nMOS transistor (*continued*)

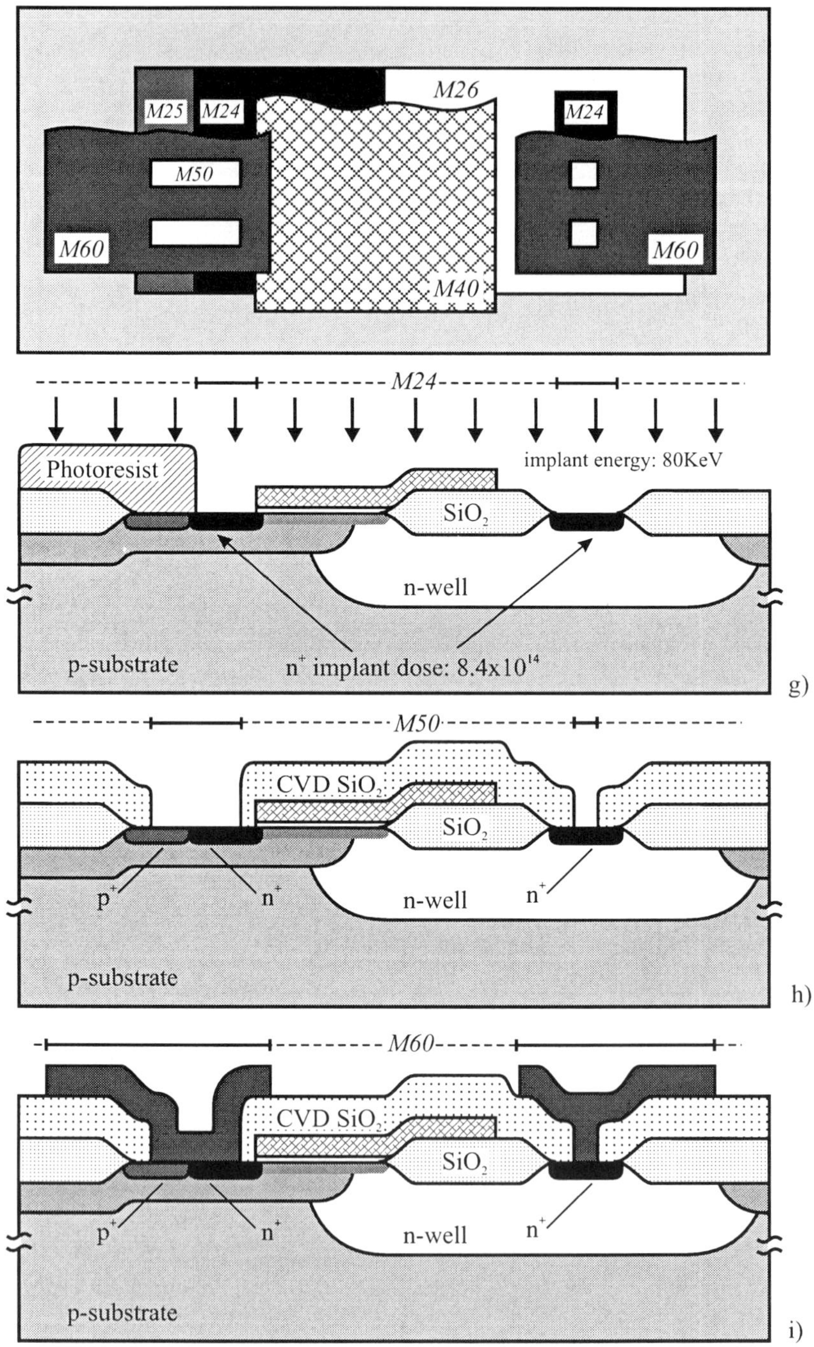

Figure 4.8: Processing steps of the high-voltage nMOS transistor (*continued*)

4.2.2 High-voltage pMOS transistor (HVPMOS)

While the use of the n-WELL layer as a lightly doped drain is straightforward in the case of the HVNMOS transistor, the lightly doped drain of the high-voltage pMOS transistor is more tricky to achieve. *Apel et al.* proposed a solution in [1] based on the lateral diffusion of two close n-WELL layers combined with the p-TUB implant. In fact, the p-TUB implant is used to form the lightly doped drain of the HVPMOS transistor. But, since no boron implant is possible in the n-WELL layer, except the one used for the p^+ diffusion, two close n-WELL layers must be used to overcome this limitation. Indeed, if they are placed at a distance at least twice the "over-sizing" constant, but closer than twice the n-WELL lateral diffusion, the p-TUB implant can be performed between the two layers to form the lightly doped drain of the high-voltage pMOS device. This technique is shown in Figure 4.9. Its main disadvantage is the maximum drift region length, which is limited in this case to less than twice the lateral diffusion of the n-WELL. The use of the multiple n-WELL technique, which consists of placing multiple layers close to each others, is a possible solution to overcome this limitation. However the thinning of the n-WELL, due to its lateral diffusion, remains a serious limitation of the junction avalanche breakdown voltage.

Another solution for implementing a p-type lightly doped drain using the p-TUB implant, consists of modifying the logic equation at the mask shop level. Care must be taken to avoid any altering of the low-voltage devices or the process flow steps when performing such modification. For instance, additional implantation steps and change in the implant energy levels or dose and in the duration of the processing steps are not allowed. Moreover, any modification of the logic equations when generating the masks must be transparent to the low-voltage devices. Therefore, to perform a p-TUB implant in the n-WELL layer, the modification of the logic equation consists of an additional variable delimiting the p-drift region of the high-voltage device. Thus, the new logic equation generating the mask M11 becomes typically

$$M11 = (CNW@1.5\mu m)*-pDRIFT \tag{4.2}$$

where *pDRIFT* is the additional variable for the lightly doped drain, "-" the logic "*not*" operator and "*" the logic "*and*" operator. Since M11 is a clear mask the p-TUB implant is performed at the *pDRIFT* location in the n-WELL layer, in addition to the locations defined by the low-voltage process flow. Actually, instead of using a new variable such as *pDRIFT*, it is preferable to use an existing one in the mask shop which is not used in the standard process flow. For instance, in the present technology the logic variable "*BLOK*" is typically used to define the routing channels in the gate arrays circuits of the same process. Therefore, it can be used to define the

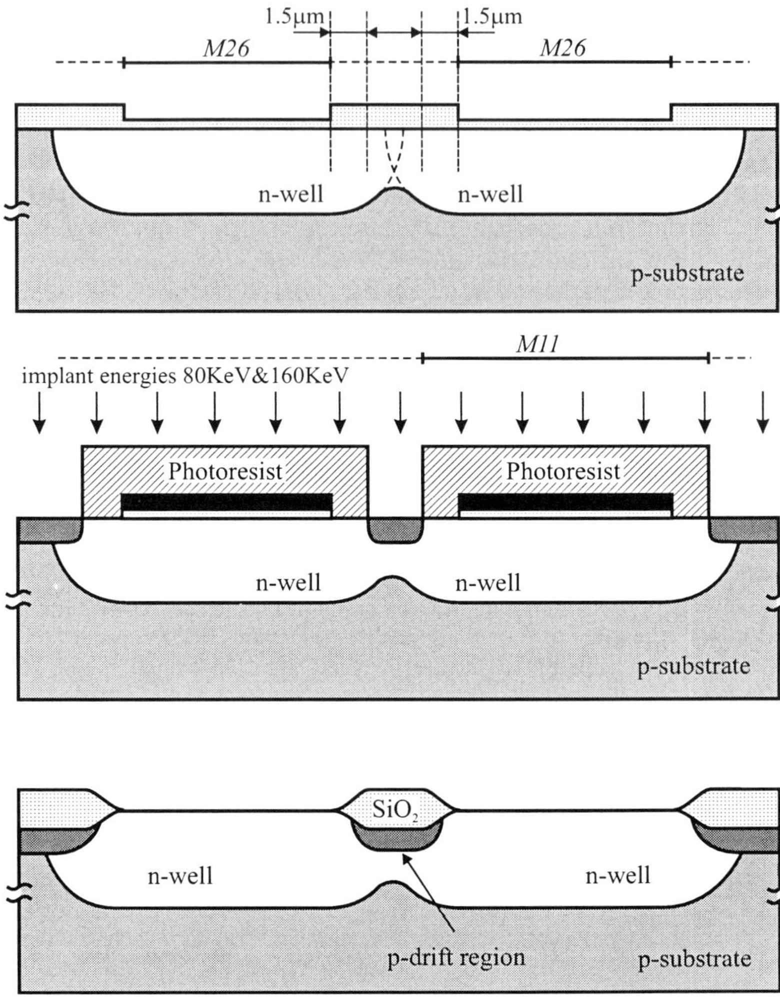

Figure 4.9: Double n-WELL technique combined with the p-TUB implant

lightly doped drain of the HVPMOS transistors. The main processing steps of such devices are shown in Figure 4.10, Figure 4.11 and Figure 4.12 and are discussed in the following paragraphs.

1) The n-WELL implant and drive-in (Figure 4.10a): In the high-voltage pMOS device the n-WELL is basically used to isolate the transistor from the substrate. This step is performed in the same conditions as for the low-voltage devices.

2) Active areas definition (Figure 4.10b): The active areas are covered by a nitride layer under which a thin oxide film was grown previously. The areas are defined by mask M10, which allows the stripping of the Si_3N_4 and SiO_2 outside these locations.

3) The lightly doped drain (Figure 4.10c): After the definition of the field doped mask M11 through the new logic equation given by (4.2), the "p-TUB" and "channel stop" layers are implanted at the lightly doped drain location, defined by the *BLOK* variable, and at the extremities of the n-WELL. Indeed, the photoresist is etched over these locations. Note that at the drain contact the "p-TUB" layer is not as deep as in the drift region because of the nitride layer. Hence, the lightly doped drain of the HVPMOS transistor is not as uniform as the HVNMOS one, leading to lower high-voltage capabilities. This represents the third consequence of keeping the process steps unchanged.

4) field-oxide growth (LOCOS) (Figure 4.11d): As in the case of the HVNMOS transistor, the "p-TUB" layer diffuses of about 1μm during this step. It results in an increase of the different radius of curvatures of the lightly doped drain junction, thus improving avalanche breakdown. Since the lateral diffusion of the "p-TUB" layer is almost as high as the vertical one, the channel length of the high-voltage device is reduced by the same amount. Therefore, the channel shortening must be taken into account when designing the masks of the HVPMOS transistor. Finally, during the field-oxide growth step the n-WELL layer diffuses of about 1μm as well.

5) Polysilicon gate deposition and gate oxide growth (Figure 4.11e): After the gate oxide growth the V_t adjust implant is performed at a depth of about 0.4μm as in the case of the HVNMOS device. Then the 3800Å polysilicon layer is deposited and the gate is defined by mask M40. The same smooth stepping of the gate is observed over the "bird's beak" of the field-oxide. Note that in contrast to the HVNMOS transistor case, the field-plate parameter L_{g1} can be fixed by overlapping the *BLOK* mask over the active area. But, it must be recalled that due to the presence of the nitride layer during the "p-TUB" implant, the depth of the drift region under the field-plate defined by L_{g1} is lower than the one under the thick field-oxide. This is the last consequence of keeping the process steps unchanged.

6) Source and drain contact implants (Figure 4.11f and Figure 4.12g): Once the polysilicon gate is defined, a boron dose of 8.4×10^{14} at/cm^2 is implanted in the "p-TUB" layer to form the lightly doped drain contact. The source contact is self-aligned with the polysilicon gate. Note that the drain contact is performed at the location where the nitride layer partially stopped the "p-TUB" implant, leading to a severe thinning of the drift region under the contact itself. After the drain and source implants, a phosphorus dose of 8.4×10^{14} at/cm^2 is performed to contact the n-WELL layer.

7) CVD SiO2 and metal1 deposition (Figure 4.12h and Figure 4.12i): The remaining steps, such as CVD SiO_2 deposition, metal1 and metal2 deposition and passivation of the wafer are achieved in the same manner as for the low-voltage devices.

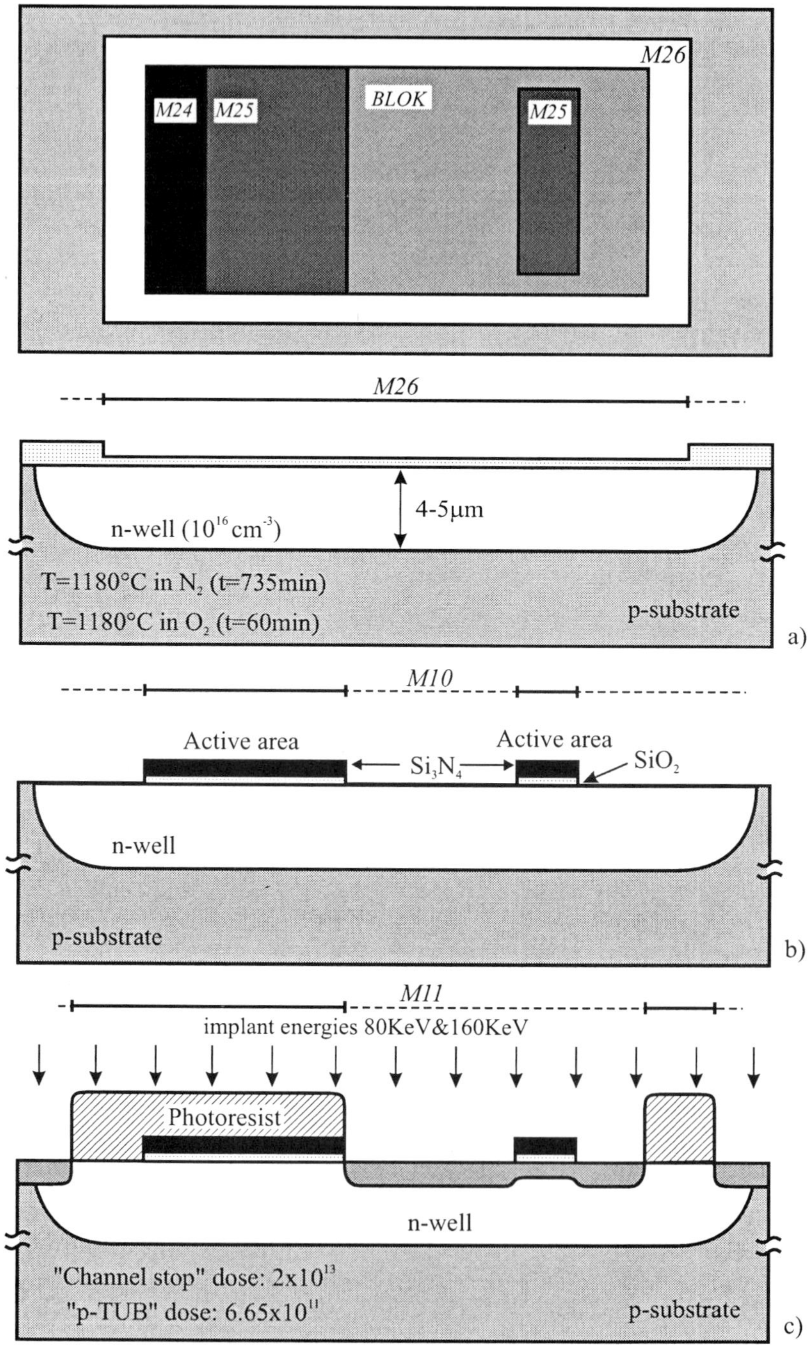

Figure 4.10: Processing steps of the high-voltage pMOS transistor

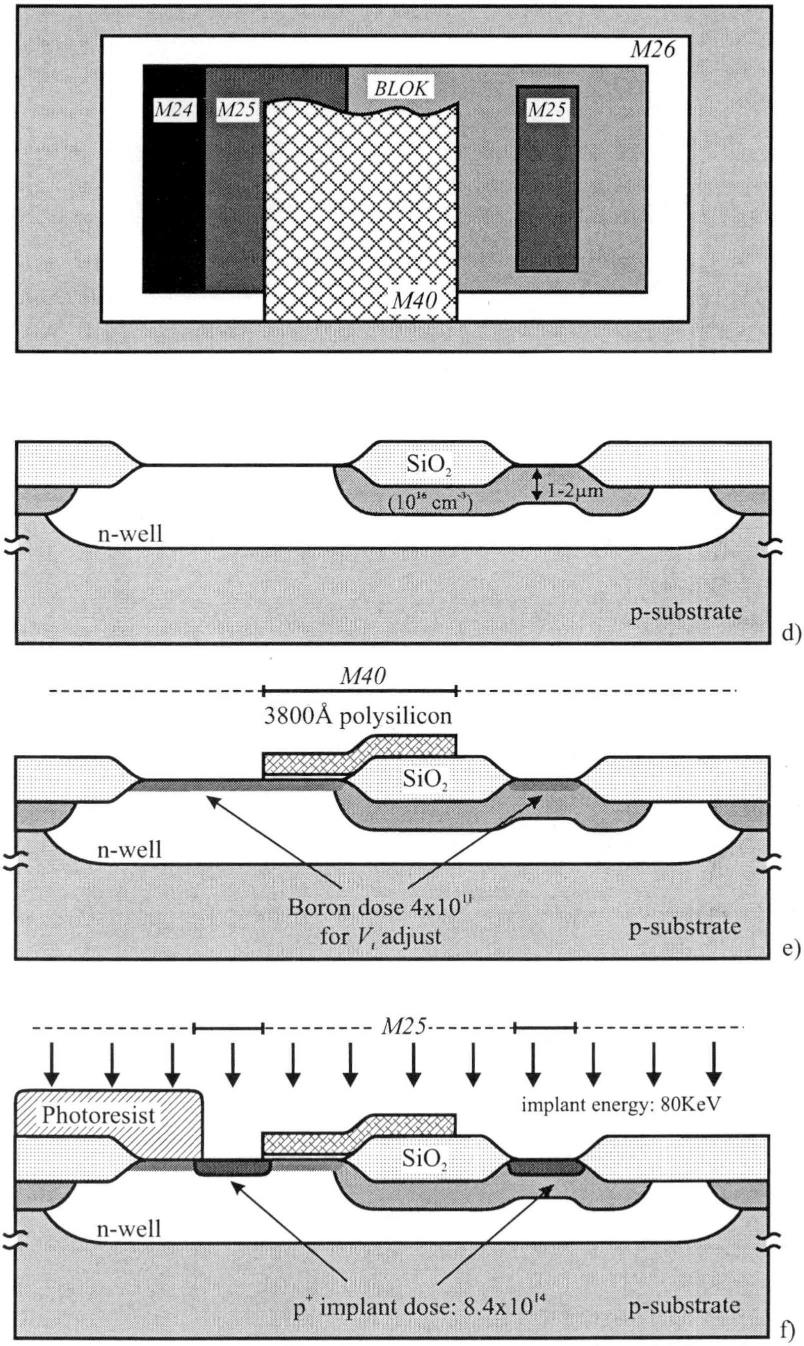

Figure 4.11: Processing steps of the high-voltage pMOS transistor (*continued*)

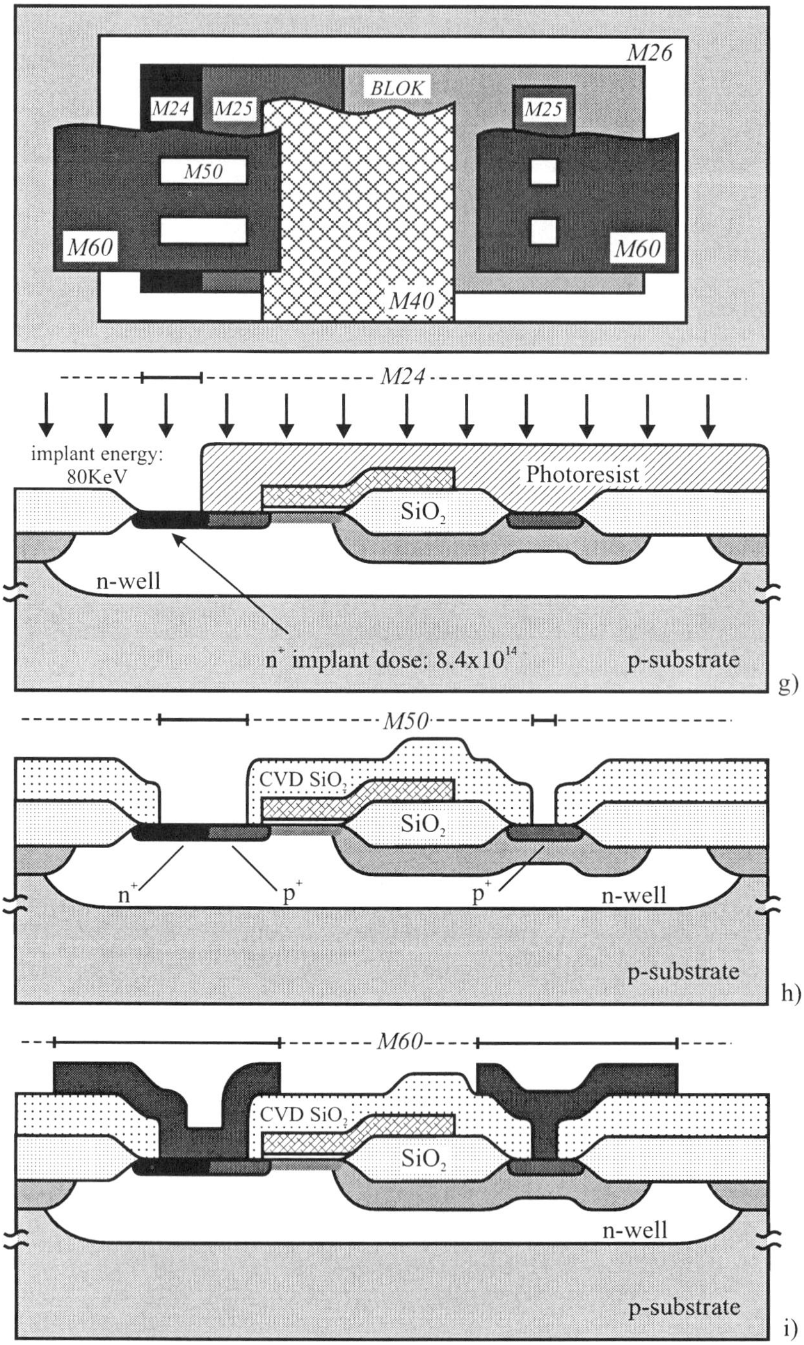

Figure 4.12: Processing steps of the high-voltage pMOS transistor (*continued*)

4.3 Scaling of the high-voltage devices

In Chapter 3 section 3.1.2, a first order theory was given to scale the different parameters of a lateral high-voltage double diffused MOS device. It is obvious that because of the two dimensional nature of the electric field in the lightly doped drain, an exact analytical scaling is not a simple job. Actually, computer simulation programs, using numerical algorithms to solve the 2-D Poisson's equation, are generally employed to optimise the scaling of the different parameters. Nevertheless, the rough estimations obtained using the first order theory are usually a good start-up for computer simulations and a simple understanding of the complex mechanisms occurring in this region. The scaling of the high-voltage devices designed with the SVX technique is similar to the DMOS one, except that the channel is not linearly graded and the lightly doped drain is not made with an epitaxial layer. The field-plate theory applies in the same conditions except that the field-plate parameter L_{g1} is zero leading to $L_g=L_{g2}$. The scaling of these devices is discussed in the following sections.

4.3.1 High-voltage nMOS transistor (HVNMOS)

The design of the high-voltage nMOS transistor starts with an estimation of the avalanche breakdown voltage that can be achieved with the structure discussed in section 4.2.1.

1) Avalanche breakdown voltage: In order to facilitate the scaling of the breakdown voltage the n-WELL layer is considered to be free of etching under the channel. It is obvious that the actual breakdown voltage would be probably reduced by the small radius of curvature of the etched area. Nevertheless, the previous assumption will fix a theoretical upper limit of this data, and a more precise value can be obtained from computer simulations. Assuming that spherical junctions are avoided by rounding the sharp corners of the n-WELL diffusion window, avalanche breakdown would mainly occur at the cylindrical termination of the lightly doped drain junction.

Considering the n WELL CMOS process presented in section 4.1, the substrate doping concentration is $7x10^{14}$ cm^{-3} and the n-WELL layer depth is about 5-6μm with a surface concentration $N_s=1x10^{16}$ cm^{-3}. Thus, the corresponding parallel plane breakdown voltage BV_{PP}=380V according to Figure 2.18 in Chapter **2** and the depletion layer width in the substrate at breakdown is W_{CPP}=24μm. As a consequence, the worst case normalised radius of curvature r_j/W_{CPP} is about 0.2, leading to a cylindrical breakdown voltage of BV_{CY}=171V, which is 45% of the parallel plane case. Hence, the upper limit of avalanche breakdown of the HVNMOS device is given by the cylindrical breakdown voltage of the n-WELL to substrate junction diode

$$BV_{\max} = 171V \tag{4.3}$$

2) Minimum channel length $L_{c,min}$: The minimum channel length $L_{c,min}$, in the case of the high-voltage nMOS devices, is fixed by the punchthrough condition corresponding to the merging of the drain and source depletion layers. In order to estimate the width of these layers as a function of the drain voltage, the lateral diffusion of the n-WELL layer under the channel and the etching caused by the "p-TUB" implant must be considered, as shown in the cross section of Figure 4.13. But, given the Gaussian doping profiles of the n-WELL and "p-TUB" layers, shown in Figure 4.14, the estimation of the depletion layer widths on both sides of the drain junction is a complex task. Moreover, a stepped approximation would lead to completely false minimum channel length values, given the actual doping profile of the "p-TUB"/n-WELL junction shown in Figure 4.15. In addition, this profile changes as a function of the position in the channel, since the lateral diffusion of the n-WELL layer itself is also Gaussian (compare the doping concentration levels between the doping profiles along the vertical axis x=1.5µm and x=3µm shown in Figure 4.14 and Figure 4.15 respectively). Therefore, two dimensional computer simulations are required here. The example shown in Figure 4.13, uses T-SUPREM4 and MEDICI [2] to perform technological and electrical simulations respectively. In this simulation, the depletion layer edge is shown just before punchthrough where L_c=5µm, L_g=3µm, L_{dd}=5µm, V_g=5V and V_d=50V.

3) field-plate length L_g: In section 4.2.1 the main processing steps of the high-voltage nMOS transistor were discussed. It was pointed out that the parameter L_{g1} of the field-plate was forced to zero by the etching of the n-WELL layer. Therefore, the unique field-plate length remaining is L_{g2} which is in this particular case equal to L_g. It is obvious that the field-plate efficiency is not as optimised as the one with a portion L_{g1} over thin gate oxide and a portion L_{g2} over thicker oxide. Considering the ideal high-voltage nMOS structure shown in Figure 4.5 with L_{g1}=0, the optimal field-plate length can be estimated from the geometrical first order consideration shown in Figure 4.16. Assuming a cylindrical approximation at the edge of the field-plate and for the termination of the drain junction at the channel end, the optimal field-plate length is considered when, at maximum reverse drain voltage, the cylindrical portion of the depletion layer created by the field-plate, intersects the one of the junction curvature. In this way, the electric field lines at the surface are spread along the resulting depletion layer, as shown by the hatched area in Figure 4.16. Note that the consideration of an ideal high-voltage nMOS structure with L_{g1}=0 and no etching of the n-WELL layer was made to simplify the estimation of the optimal field-plate length.

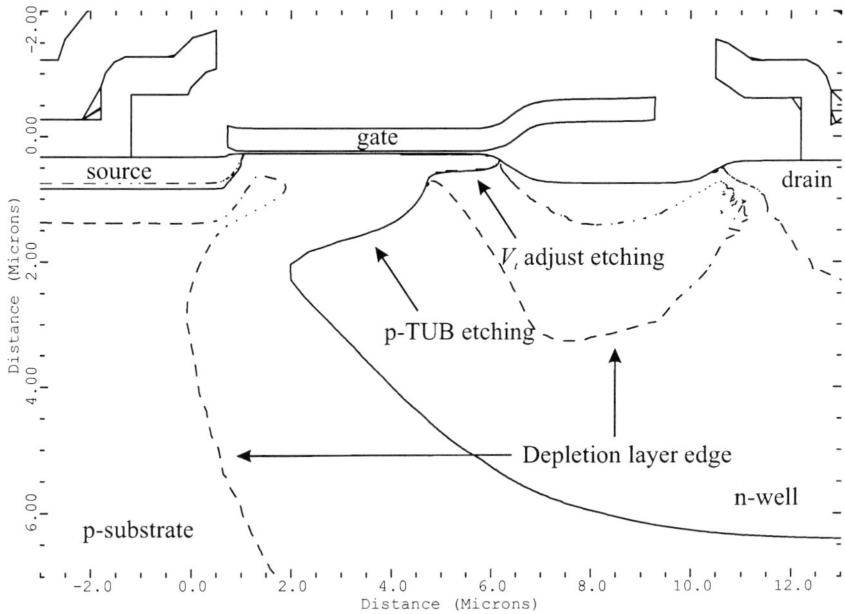

Figure 4.13: Cross section of a high-voltage nMOS device with L_c=5μm, L_g=3μm, L_{dd}=5μm, V_d=50V and V_g=5V, using T-SUPREM4® and MEDICI®

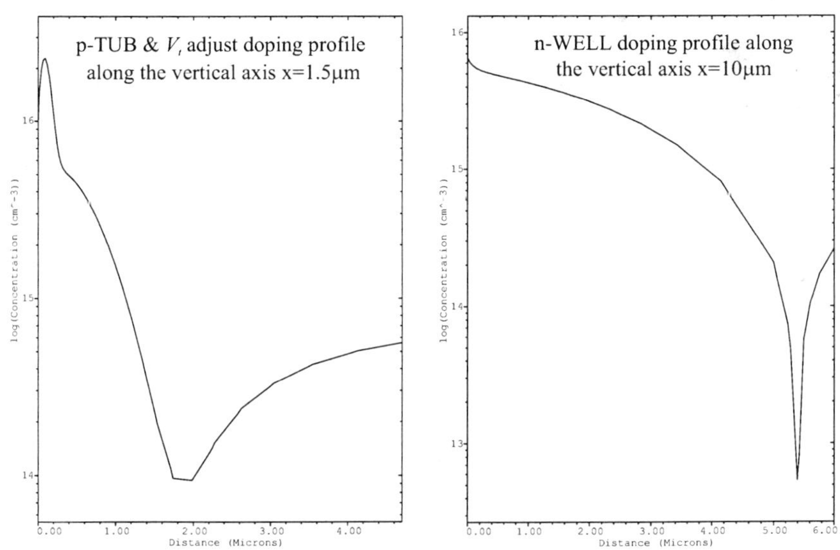

Figure 4.14: Gaussian doping profiles of the "p-TUB" and n-WELL layers

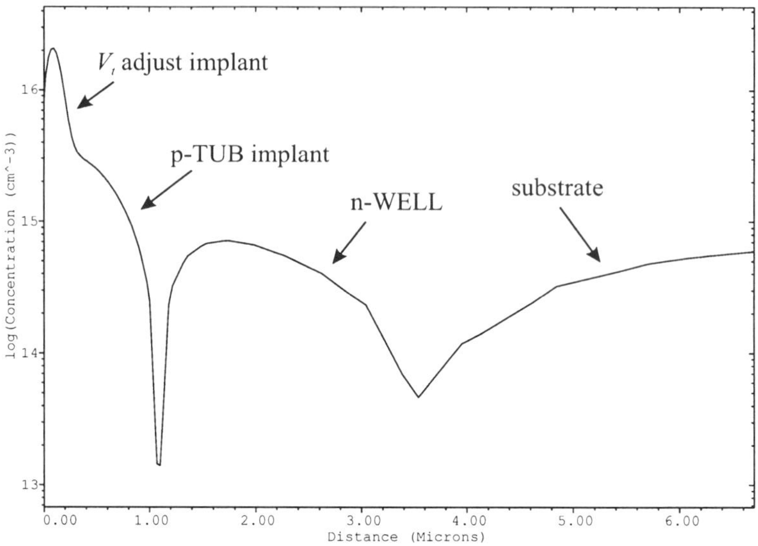

Figure 4.15: Actual doping profile of the "p-TUB"/n-WELL junction under the channel along the vertical axis x=3μm

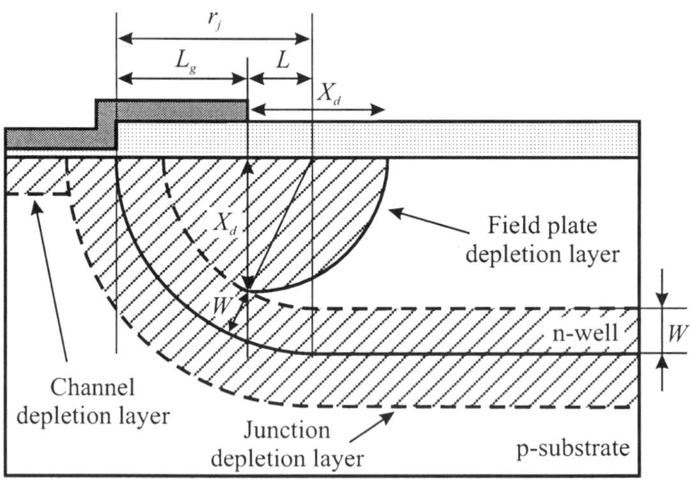

Figure 4.16: Geometrical first order consideration of the optimal field-plate length in the case of the HVNMOS device

More accurate results can be obtained from two dimensional computer simulations of the electric field and potential distributions in the field-plate region. Hence, the maximum efficiency of the field-plate effect is obtained when

$$L_g = L_{g2} = r_j - L = r_j - \sqrt{\left(r_j - W\right)^2 - X_d^2} \tag{4.4}$$

where W is the parallel plane depletion layer width of the drain junction and r_j is the radius of curvature of the drain junction. The depletion layer width X_d created by the field-plate cannot be easily estimated unless a constant doping concentration of the n-WELL layer is considered for the MOS structure formed by the field-plate, the thick oxide and the n-WELL layer. Under such conditions, X_d is calculated using the same approach as for the lateral DMOSFET discussed in section 3.1.2 of chapter 3. Thus, the width of the depletion layer created by the field-plate is given by

$$X_d = \sqrt{\left(\frac{\varepsilon_s}{\varepsilon_{ox}} t_{ox}\right)^2 + \frac{2\varepsilon_s}{qN_D}\left(V_d - V_g'\right)} - \frac{\varepsilon_s}{\varepsilon_{ox}} t_{ox} \tag{4.5}$$

where $V_g'=V_g-V_{FB}$, t_{ox} the thickness of the oxide layer and N_D the doping concentration of the n-WELL layer. Assuming a step approximation of the n-WELL-to-substrate junction, the doping concentration N_D resulting from the actual doping profile is $3.3x10^{15}$ cm^{-3} for a 5.5μm junction depth and an implant dose of $2.2x10^{12}$ at/cm^2. This value is calculated on the bases of the approximation shown in Figure 4.17.

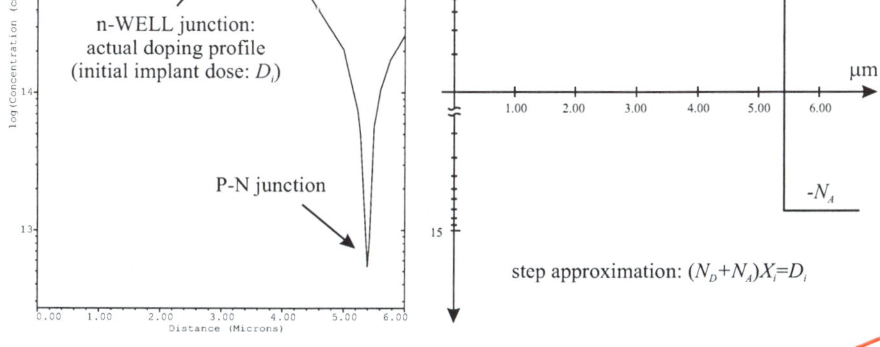

Figure 4.17: Step approximation of the Gaussian doping profile of the n-WELL to substrate junction diode

Replacing X_d and W in Equation (4.4) by their respective expressions, the resulting minimum field-plate length can be expressed as

$$L_g = r_j - \sqrt{\left(r_j - \sqrt{\frac{2\varepsilon_s N_A}{qN_D\left(N_D + N_A\right)}V_d}\right)^2 - \left(\sqrt{\left(\frac{\varepsilon_s}{\varepsilon_{ox}}t_{ox}\right)^2 + \frac{2\varepsilon_s}{qN_D}\left(V_d - V_g'\right)} - \frac{\varepsilon_s}{\varepsilon_{ox}}t_{ox}\right)^2}$$

(4.6)

Since the n-WELL is a deep junction with low surface concentration, the depletion layer width W in the n-WELL side is given by the expression of a step junction diode using two close doping concentration levels. The minimum field-plate length given by Equation **(3.21)** is represented in Figure 4.18 as a function of the reverse drain voltage ranging from 0V to the maximum theoretical avalanche breakdown voltage.

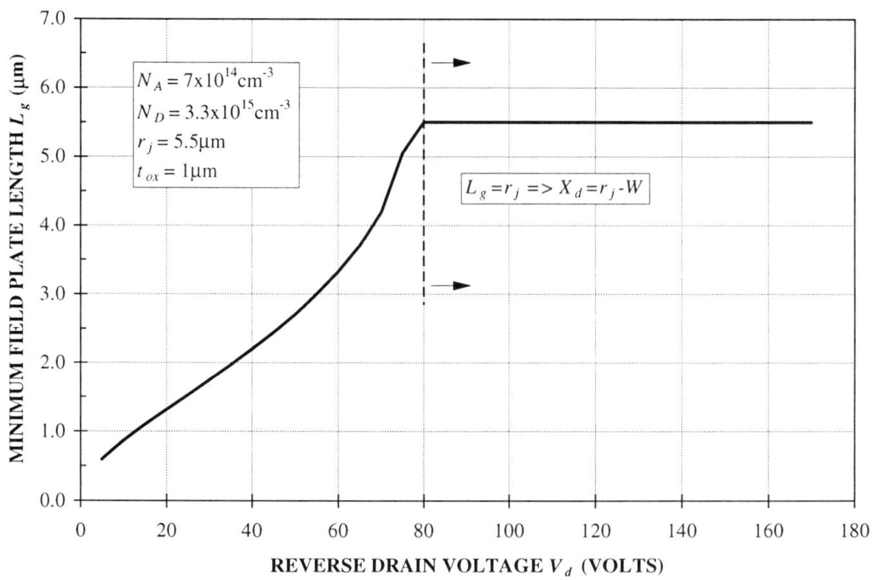

Figure 4.18: Minimum field-plate length L_g as a function of the drain voltage

Note that the minimum field-plate length saturates beyond a drain voltage V_d of 80V, since the depletion layer width created by the field-plate intersects the one of the parallel plane part of the junction. Therefore, the saturation value of the minimum field-plate length L_g is the radius of curvature r_j of the cylindrical junction termination.

4) Lightly doped drain extension L_{dd}: As shown in Figure 4.5, the lightly doped drain extension L_{dd} is given by the sum of the field-plate length L_g and the distance L_d' from the edge of the field-plate to n$^+$ drain contact. Hence,

$$L_{dd} = L_g + L_d' \tag{4.7}$$

where L_g is given by Equation (4.6). Concerning the distance L_d', the first order approximation made in the case of the lateral DMOS devices can be applied here as well. It consists of placing the n$^+$ contact just at the edge of the depletion layer created by the field-plate. Under such a condition, L_d' is simply given by the depletion layer width X_d expressed in Equation (4.5). The resulting value of L_d' and consequently of L_{dd}, are shown in Figure 4.19 as a function of the drain voltage.

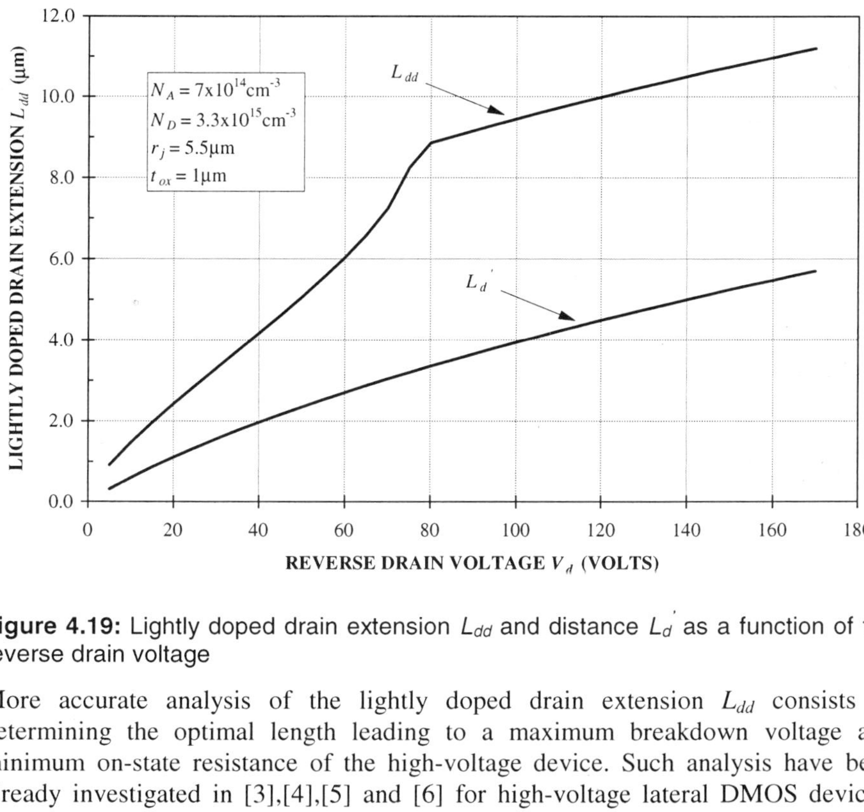

Figure 4.19: Lightly doped drain extension L_{dd} and distance L_d' as a function of the reverse drain voltage

More accurate analysis of the lightly doped drain extension L_{dd} consists of determining the optimal length leading to a maximum breakdown voltage and minimum on-state resistance of the high-voltage device. Such analysis have been already investigated in [3],[4],[5] and [6] for high-voltage lateral DMOS devices. They generally consider a uniform epitaxial layer as a lightly doped drain to estimate the corresponding parasitic resistance, and a two dimensional computer simulation to optimise the breakdown voltage. Such approach can be used to optimise the drift region of the present SVX devices by considering the step approximation of the n-WELL to substrate junction shown in Figure 4.17. Such an optimising procedure of the parameters of high-voltage devices is not discussed here as it is not the main objective of this book.

4.3.2 High-voltage pMOS transistor (HVPMOS)

The design of the high-voltage pMOS transistor uses the same steps discussed in section 4.3.1 for the HVNMOS transistor. Since the lightly doped drain of the pMOS device is made with the "p-TUB" and "channel-stop" layers implanted in the n-WELL, the resulting doping profiles lead to complex Poisson's equation solutions, as in the case of the minimum channel length of the HVNMOS transistor. Therefore, a step approximation of the Gaussian doping profiles will be used to scale the different high-voltage parameters of the transistor, when a first order theory can be applied.

1) Avalanche breakdown voltage: In order to estimate the maximum theoretical avalanche breakdown voltage of the HVPMOS transistor, a step approximation of the n-WELL doping profile is considered. Under such a condition, the parallel plane avalanche breakdown corresponding to a background doping level of $3.3x10^{15}$ cm^{-3} (n-WELL) and a planar diffused junction depth of 2μm with a surface concentration N_s=$1x10^{17}$ cm^{-3} ("p-TUB" and "channel stop" implants) is about BV_{PP}=170V, according to Figure 2.18 in Chapter **2**. The corresponding depletion layer width in the n-WELL side at breakdown is W_{CPP}=6μm, which is higher than the n-WELL layer depth under the lightly doped drain (*LDD*). Assuming that the n-WELL is deep enough, however, the worst case normalised radius of curvature r_j/W_{CPP} is about 0.3, leading to a cylindrical breakdown voltage of BV_{CY}=85V, which is 50% of the parallel plane case. Consequently, the theoretical upper limit of the maximum achievable breakdown voltage BV_{max} of the high-voltage pMOS device is

$$BV_{\max} = 85V \tag{4.8}$$

Actually, punchthrough breakdown occurs before avalanche, due to the DIBL mechanism (section 2.3 of 0) between the *LDD*/n-WELL and n-WELL/substrate junctions. In fact, since the remaining n-WELL diffusion depth under the *LDD* is about 4μm, the merging of the depletion layers of the *LDD*/n-WELL and n-WELL/substrate junctions occurs at relatively low V_{ds} values, leading to a start-up of "punchthrough" condition. But, in contrast to the punchthrough mechanism of the low-voltage devices, where the source-to-substrate potential barrier is lowered as soon as the depletion layers of the source and drain junctions merge, the potential barrier of the n-WELL to substrate junction is as high as the high-voltage power supply. Therefore, the drain of the HVPMOS transistor can reach relatively high values (high V_{ds}) before lowering the potential barrier of the n-WELL to substrate junction. A detailed analysis of this mechanism is discussed further in section 4.4.2 where two dimensional computer simulations are performed. In fact, considering the actual doping profile of the *LDD*/n-WELL and n-WELL/substrate junctions shown in Figure 4.21, the estimation of their depletion layer width and potential distribution is a complex job requiring two dimensional computer simulations. The cross section of Figure 4.20 shows a high-voltage pMOS transistor with L_c=5μm, L_g=3μm, L_{dd}=5μm, V_{ds}=0.1V, V_g=5V and V_s=V_b=50V, where the depletion layers of the different junctions are indicated as well.

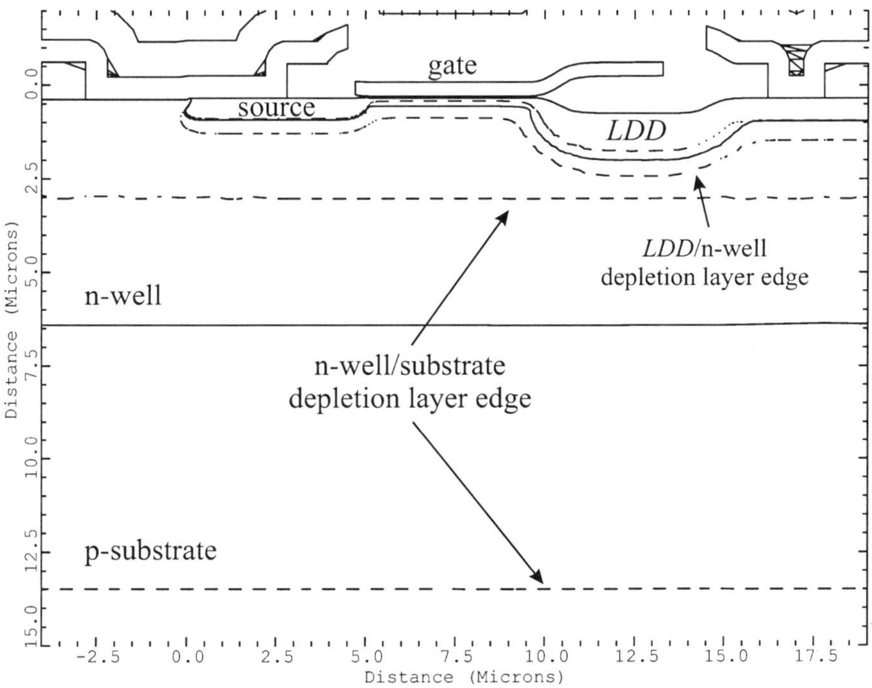

Figure 4.20: Cross section of a high-voltage pMOS device with L_c=5μm, L_g=3μm, L_{dd}=5μm, V_{ds}=0.1V, V_g=5V and V_s=V_b=50V

2) Minimum channel length $L_{c,min}$: The minimum channel length calculation of the HVPMOS transistor is as complex as the HVNMOS one due to doping profiles shown in Figure 4.21. Moreover, first order approximations cannot be considered as well since they would lead to false $L_{c,min}$ values. Therefore, the same approach as for the HVNMOS transistor, which consists of performing two dimensional computer simulations, must be used to estimate the minimum channel length as a function of the drain voltage.

3) Minimum field-plate and lightly doped drain lengths (L_g, L_{dd}): In contrast to the case of the HVNMOS device, since no possible step approximation can be made for the doping profile shown in Figure 4.21, the estimation of the remaining high-voltage parameters, such as minimum field-plate and lightly doped drain lengths, cannot be performed without using computer simulations as well. These are presented and discussed in the next section for both transistor types and for a specific size of the high-voltage devices.

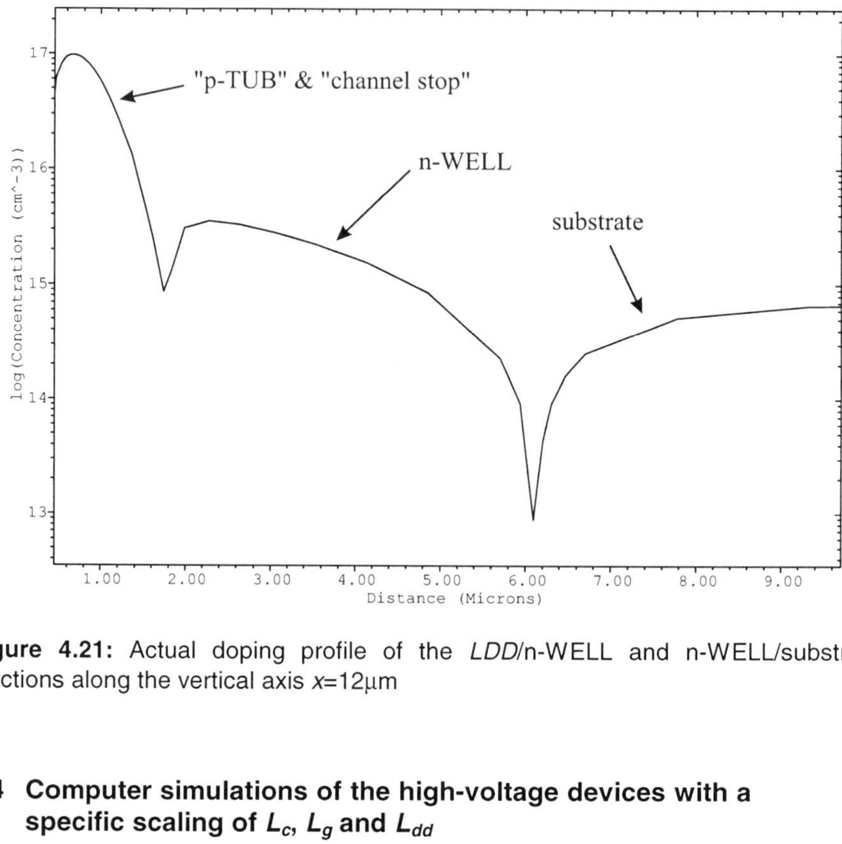

Figure 4.21: Actual doping profile of the *LDD*/n-WELL and n-WELL/substrate junctions along the vertical axis x=12µm

4.4 Computer simulations of the high-voltage devices with a specific scaling of L_c, L_g and L_{dd}

As already mentioned, the optimisation of the different geometrical parameters of the high-voltage devices, as a function of the drain voltage, is not the main scope of this book. Actually, this book deals in the same time with the design of such devices and their use in mixed analog and digital circuits oriented for high-voltage applications. Therefore, a specific size only of the HVNMOS and HVPMOS transistors is optimised for all the high-voltage circuits discussed in the following chapters. These circuits have been used in different industrial applications requiring high-voltage power supplies ranging between 12V and 24V.

In this section two dimensional computer simulations are performed on the high-voltage nMOS and pMOS transistors using a power supply value of 50V. The corresponding geometrical parameters such as minimum channel length $L_{c,min}$, field-plate length L_g and lightly doped drain length L_{dd} are also presented. Note that the high-voltage capabilities of the SVX devices are much higher than 50V when another set of geometrical parameters is used, as presented in section 5.1.

4.4.1 High-voltage nMOS transistor with L_c=5μm, L_g=3μm, L_{dd}=5μm

The geometrical parameters L_g=3μm and L_{dd}=5μm are extracted from Figure 4.18 and Figure 4.19 respectively, assuming a maximum power supply voltage of 50V. The minimum channel length is fixed a priori to 5μm considering the actual lateral diffusion of the n-WELL layer (80% of the diffusion depth). These values were chosen as initial conditions for computer optimisation. They happen to already be the optimum parameters in terms of high voltage capabilities, as will be shown in the following paragraphs.

The most important simulation data, when applying the maximum drain voltage, is the electric field and potential distributions. The edges of the different depletion layers must be checked as well to avoid punchthrough effects. Hence, In Figure 4.22 the electric field distribution is shown in the cross section of the high-voltage nMOS transistor using the geometrical parameters calculated with the first order theory of section 4.3.1. The regions of constant electric field magnitude are represented by the electric field lines ranging between 10V/cm and ~$3x10^5$V/cm. Note that the maximum ionising electric field is located at the field-plate edge in the thick oxide layer. The electric field magnitude in the lightly doped drain (n-WELL layer) and channel end region (thin gate oxide) ranges between $1x10^4$V/cm and ~$3x10^4$V/cm, which is lower than the typical value of the critical electric field E_{sat} (~$4x10^4$V/cm) for velocity saturation of electrons at the drain end of low-voltage devices (see chapter 0 section 2.1.1). Consequently, the electric field distribution in the channel end region of the high-voltage device, with V_{ds}=50V and V_g =5V, is very close to the one present in the channel end region of a low-voltage device entering the saturation mode. Hence, the thin gate oxide of such HVNMOS transistor is under the same stressing conditions as the low-voltage one operating in the saturation mode. Therefore, no hot-carrier effects should be expected up to 50V at least for this particular scaling of the high-voltage nMOS transistors.

The previous observation can be confirmed with the simulation shown in Figure 4.23 where the equipotential lines are plotted under the same static bias conditions. The first line in the n-WELL layer is at the same potential as the gate electrode, i.e. 5V. Considering a step of 5V between the equipotential lines, the second one is at 10V and is already pushed by the field-plate into the thick oxide layer. Note the crowding of the lines at the field-plate edge between the gate electrode and the drain one. It explains the high electric field concentration in this region and at the surface of the n-WELL layer, as shown in Figure 4.22. The smaller the distance L_d' is between the gate electrode and the drain one, the higher is the crowding of the electric field and equipotential lines in these regions, leading to avalanche breakdown at the surface of the n-WELL layer, as already observed by *Declercq et*

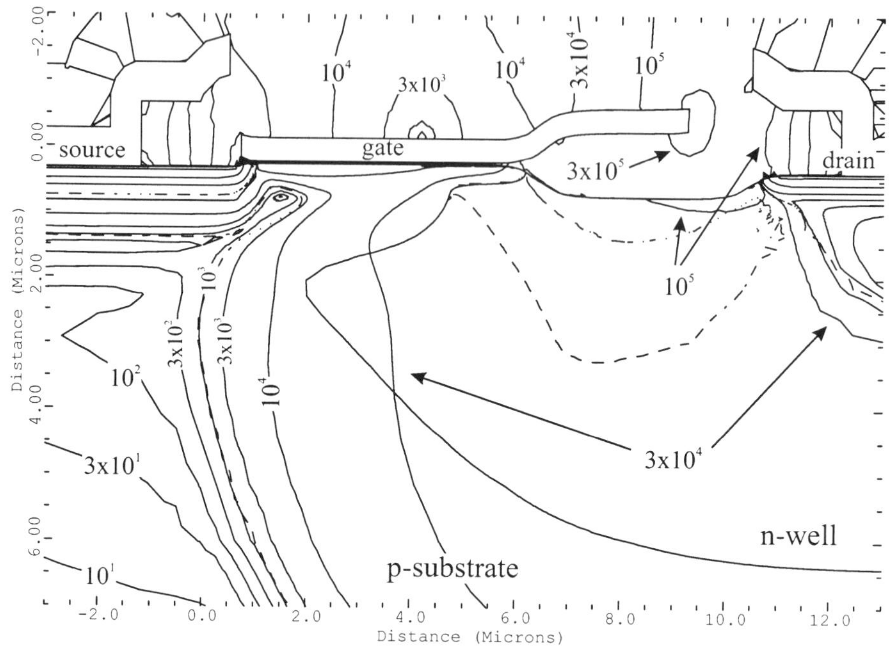

Figure 4.22: Constant electric field lines in V/cm when V_g=5V, V_d=50V, V_{sb}=0V, L_c=5µm, L_g=3µm and L_{dd}=5µm

al. in [7]. In fact, the cross section of Figure 4.24 shows the constant lines of electron-hole pair generation. Although it remains negligible with L_d'=2µm, the highest peak density of electron-hole pair generation due to impact ionisation, is located along the vertical axis x=10µm at about 0.6µm depth ($3x10^{25}$ pair/cm^3/s). Actually, all the regions at the surface of the n-WELL, where the magnitude of the electric field is 10^5V/cm, exhibit a high density of electron-hole pair generation. The total current calculated by the simulator and corresponding to this region is about 57nA/µm, which is in fact negligible.

The last interesting simulation results are the boundaries of the depletion layers combined with the current flow lines, as shown in Figure 4.25. Such simulation helps to verify punchthrough conditions by monitoring the possible merging of the depletion layers and the corresponding current flow lines. Figure 4.25 shows that L_c=5µm is the minimum channel length that must be used for a 50V power supply. In fact, beyond 50V the source and drain depletion layers merge, leading to a punchthrough current that increases drastically with the drain voltage. The current flow lines of Figure 4.25 confirm that no significant subsurface current is flowing between source and drain.

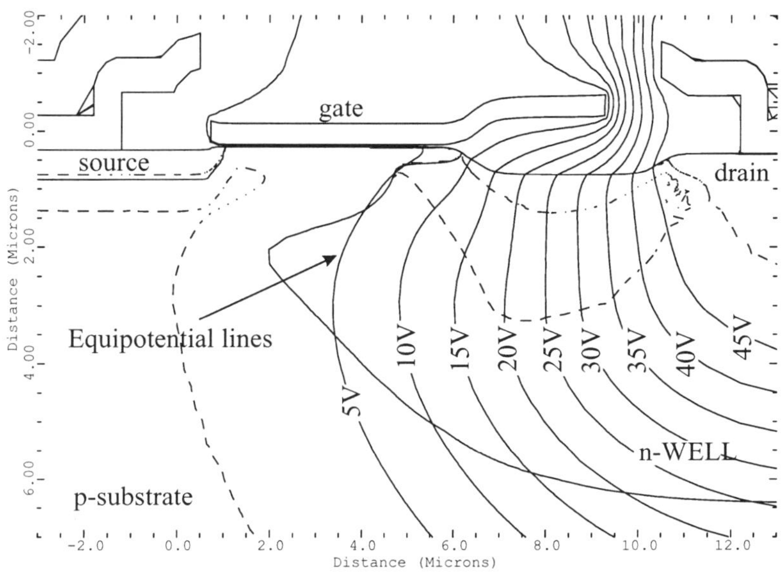

Figure 4.23: Equipotential lines when V_g=5V, V_d=50V, V_{sb}=0V, L_c=5µm, L_g=3µm and L_{dd}=5µm

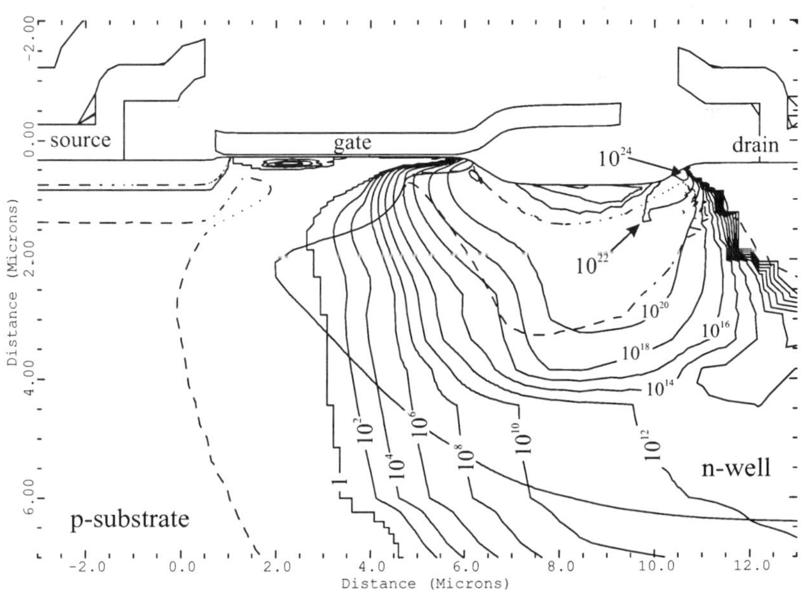

Figure 4.24: Constant lines of electron-hole pair generation density due to impact ionisation in pair/cm^3/s using the same device and under the same bias conditions

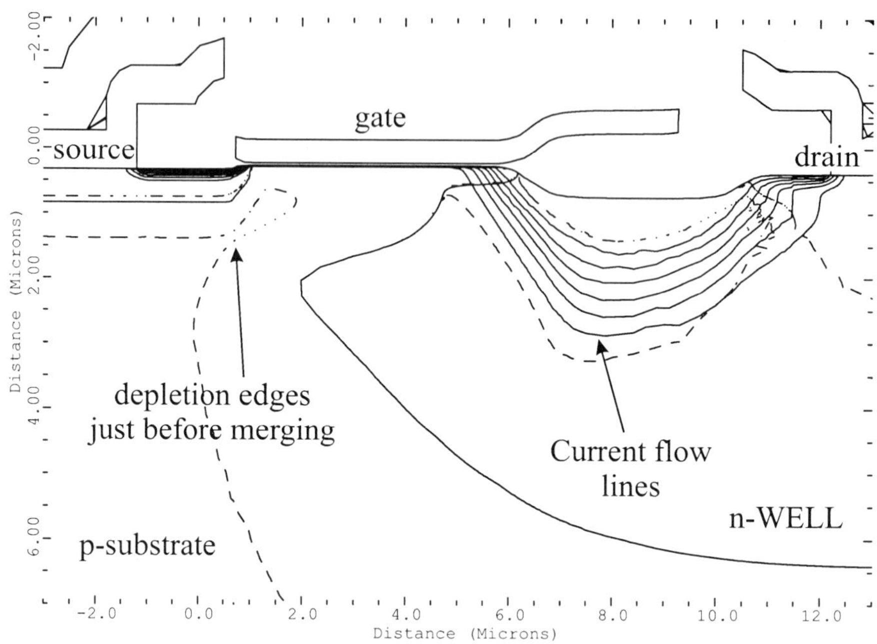

Figure 4.25: Depletion layers boundaries and current flow lines using the same device and under the same bias conditions

4.4.2 *High-voltage pMOS transistor with* $L_c=5\mu m$, $L_g=3\mu m$, $L_{dd}=5\mu m$

The same simulation approach as for the HVNMOS case is used for the high-voltage pMOS devices. Namely, The bulk is fixed at 50V and the high-voltage parameters are assumed to be identical. Moreover, The electric field and potential distribution as well as hot carriers generation are also simulated for the maximum drain-to-source voltage V_{dsmax}. But, before presenting such simulation results, the "punchthrough" condition between the *LDD*/n-WELL and n-WELL/substrate junctions is first analysed.

As was stated in section 4.3.2, the high-voltage capabilities of the pMOS device are limited by the punchthrough breakdown occurring between the *LDD*/n-WELL and n-WELL/substrate junctions. Actually, the punchthrough condition, corresponding with the merging of the depletion layers, occurs at low V_{ds} values (about -4.5V). But, since the n-WELL/substrate potential barrier is as high as the power supply voltage, punchthrough breakdown resulting from the drain induced barrier lowering, occurs at V_{ds} values as high as -70V. In fact, in the cross sections of Figure 4.26 and Figure 4.27, the potential distribution is shown before and after punchthrough respectively. In the first case, the applied bulk voltage is V_b=V_s=70V and the drain-to-source voltage V_{ds} is -60V. Under such bias conditions, the potential barrier of the n-WELL/substrate junction is about 20V. In the second case, the bulk voltage is kept constant but V_{ds} is now raised to -70V. The result is a complete lowering of the potential barrier, leading to a punchthrough breakdown which appears in the measured static characteristics shown in Figure 5.2.

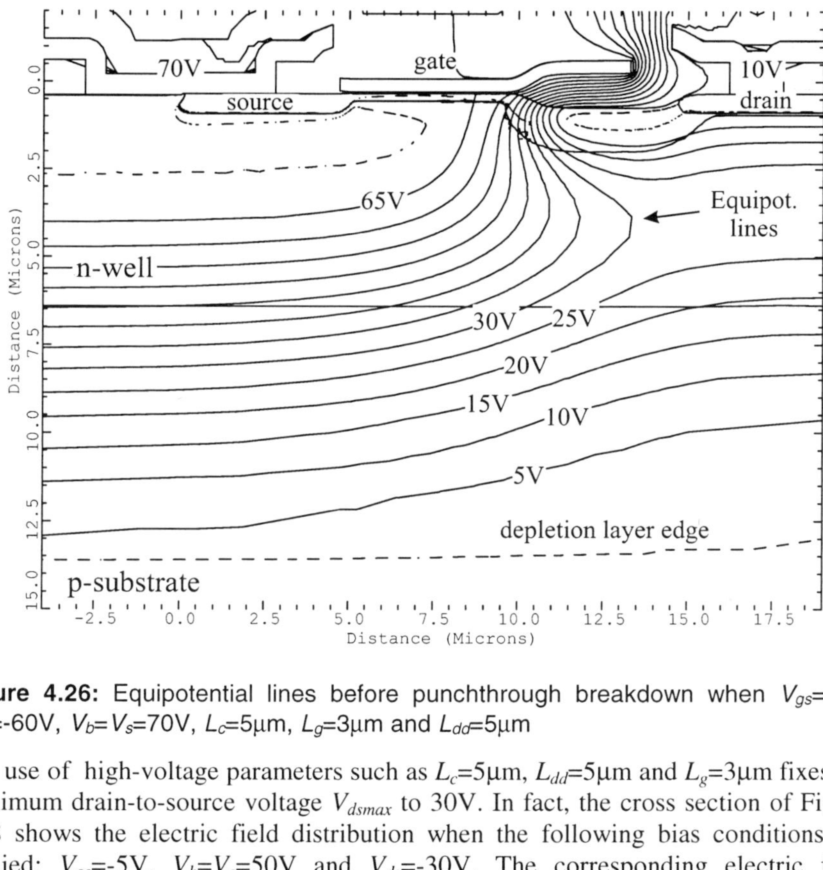

Figure 4.26: Equipotential lines before punchthrough breakdown when V_{gs}=-5V, V_{ds}=-60V, V_b= V_s=70V, L_c=5μm, L_g=3μm and L_{dd}=5μm

The use of high-voltage parameters such as L_c=5μm, L_{dd}=5μm and L_g=3μm fixes the maximum drain-to-source voltage V_{dsmax} to 30V. In fact, the cross section of Figure 4.28 shows the electric field distribution when the following bias conditions are applied: V_{gs}=-5V, V_b=V_s=50V and V_{ds}=-30V. The corresponding electric field magnitude at the channel end ranges between $3x10^4$V/cm and ~$1x10^5$V/cm, which is lower than the typical value of the critical electric field E_{sat} (~$1.2x10^5$V/cm) for velocity saturation of holes at the drain end of low-voltage pMOS devices. Consequently, the electric field distribution in the channel region of the high voltage device, with V_{ds}=-30V and V_{gs}=-5V is very close to the one present in the channel region of a low voltage device entering the saturation mode. Hence, the thin gate oxide of such a HVPMOS transistor is under the same stressing conditions as the low-voltage one operating in the saturation mode. Therefore, no hot-carrier effects should be expected up to 30V for this particular scaling of the high-voltage pMOS transistors. Beyond this limit, the constant electric field line of $3x10^5$V/cm spreads along the surface of the silicon under the thin gate oxide, resulting in a significant increase of hot-carriers generation.

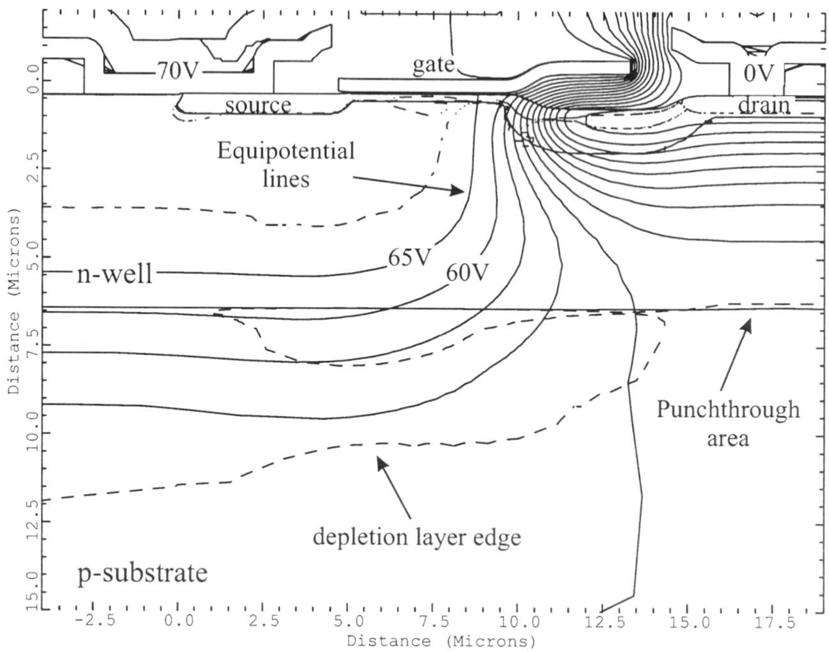

Figure 4.27: Equipotential lines before punchthrough breakdown when V_{gs}=-5V, V_{ds}=-70V, V_b=V_s=70V, L_c=5μm, L_g=3μm and L_{dd}=5μm

This is pointed out further in the measurement section, where no significant gate current I_g is observed as far as $|V_{ds}| \leq 30$V. I_g starts increasing significantly at V_{ds}=-40V. Such limitation is easily overcome by changing the high-voltage parameters scaling. On the one hand, the use of a stepped field-plate over thin gate oxide (L_{g1}≠0) increases its efficiency by pushing the electric field lines away from the surface of the silicon, reducing thus the electric field magnitude at the channel end. As already stated in section 4.2.2 the field-plate parameter L_{g1} can be fixed by overlapping the *BLOK* mask over the active area. On the other hand, the lightly doped drain length L_{dd} can be increased as well, in order to reduce the crowding of the equipotential lines in this region. However, since all the circuits discussed in the next chapters use power supplies ranging between 12V and 24V, such optimising techniques as a function of the drain voltage are not performed here. Therefore, the above high-voltage parameters are still valid and are used for the HVPMOS transistors of these circuits.

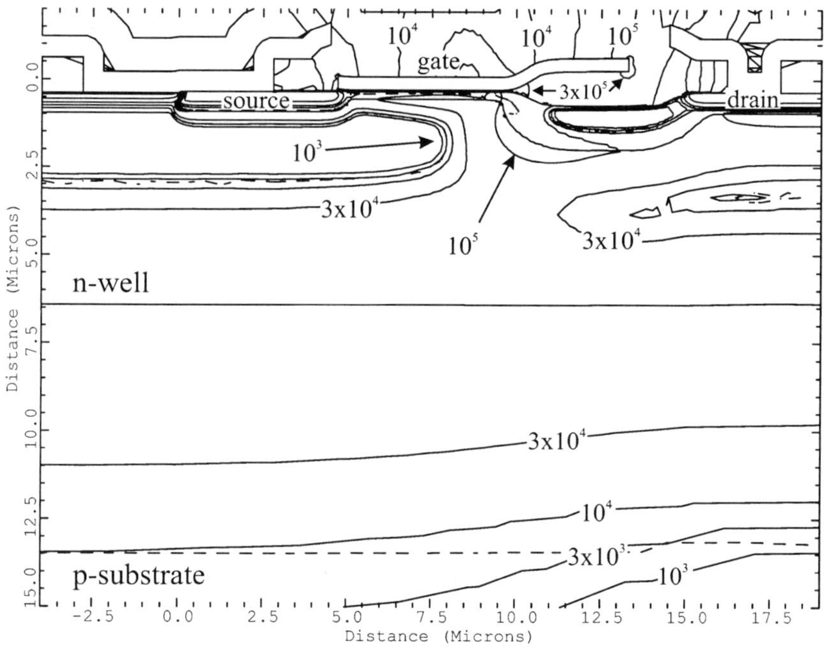

Figure 4.28: Constant electric field lines in V/cm when V_{gs}=-5V, V_{ds}=-30V, V_b= V_s=50V, L_c=5μm, L_g=3μm and L_{dd}=5μm

The simulation of the potential distribution, under the same bias conditions as for the electric field one, is shown in Figure 4.29. The first line is at the same potential as the gate electrode, i.e. 45V assuming that the source is at 50V. As in the case of the HVNMOS transistor, the line at 40V is already pushed into the thick oxide layer but with less efficiency, which confirms that V_{dsmax}=-30V is the upper limit for this particular scaling of the HVPMOS transistors. Indeed, beyond 30V the efficiency of the field-plate decreases to the extent that the equipotential lines higher than 5V with respect to the source penetrate into the thin gate oxide region. For instance, note the difference in the potential distribution under the field-plate between the previous simulations performed at V_{ds}=-60V and V_{ds}=-70V (Figure 4.26 and Figure 4.27) and the one performed at V_{ds}=-30V. In the former case the crowding of the equipotential lines is such that the line at 60V penetrates into the thin gate oxide layer. Hence, the corresponding electric field magnitude at the channel end is much higher than the critical electric field for velocity saturation of holes, leading to hot carriers generation. For V_{ds}=-30V the field-plate remains efficient and the line at 40V is pushed into the thick oxide layer.

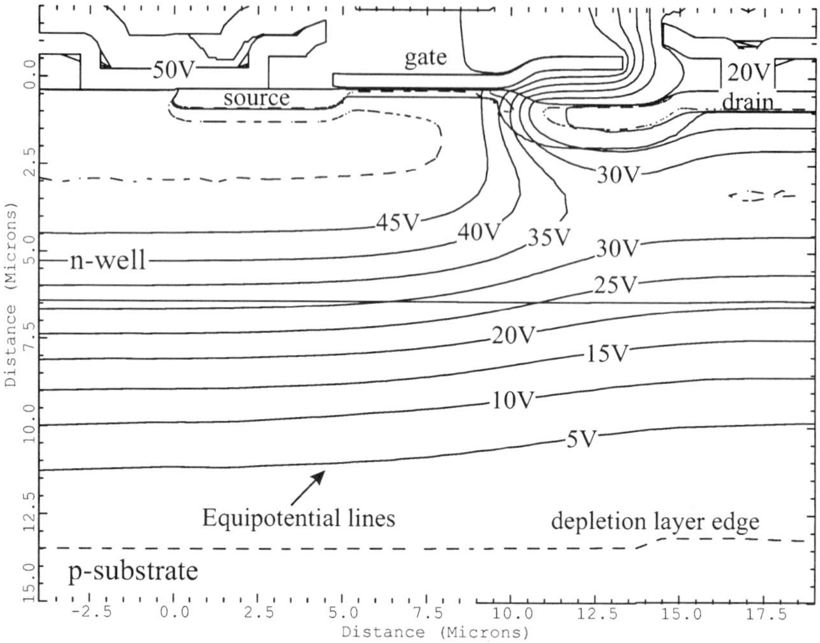

Figure 4.29: Equipotential lines when V_{gs}=-5V, V_{ds}=-30V, V_b=V_s=50V, L_c=5μm, L_g=3μm and L_{dd}=5μm

A direct quantitative monitor of hot carriers generation is the density of electron-hole pair generation due to impact ionisation. The cross section of Figure 4.30 shows the constant lines of electron-hole pair generation density. In contrast to the HVNMOS device, the region of high electron-hole pair generation density is not located at the surface of the silicon and at the edge of the field-plate, but under its smooth stepping over the "birds beak". Actually, the highest peak is about $3\text{x}10^{24}$ pair/cm^3/s and is located along the vertical axis x=10μm at about 0.9μm depth. The total current calculated by the simulator and corresponding to the region of high generation of electron-hole pair is about 49nA/μm width, which is as high as the HVNMOS one. Finally, the current flow lines shown in Figure 4.31, confirm that no significant current is flowing under the surface of the device and that no punchthrough current between the *LDD*/n-WELL and n-WELL/substrate is flowing to the substrate.

As a conclusion of the technological and electrical simulations performed on both devices, the use of the same scaling for the high-voltage parameters, where L_g=3μm, L_{dd}=5μm and L_c=5μm, results in lower high-voltage capabilities for the HVPMOS transistor. While the HVNMOS transistor can be used with power supplies as high as 50V for this particular scaling, the HVPMOS transistor must be supplied below 30V due to hot carrier effects. As already mentioned, since the high-voltage circuits discussed in the next chapters are used in high-voltage applications with power

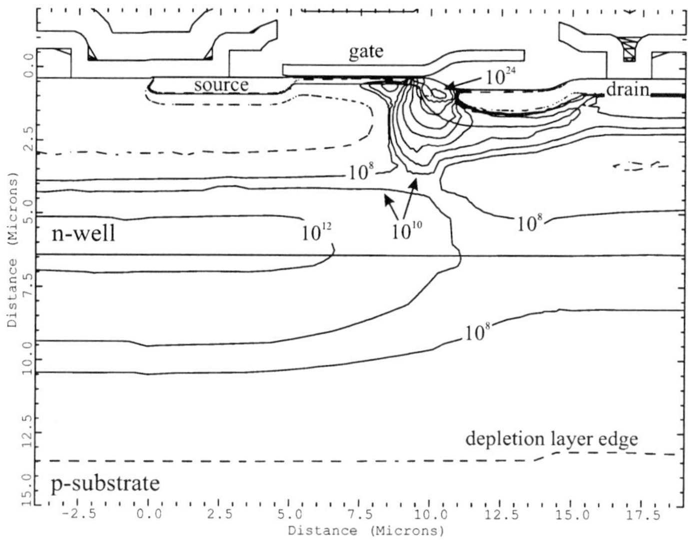

Figure 4.30: Constant lines of electron-hole pair generation density due to impact ionisation in pair/cm^3/s using the same device and under the same bias conditions.

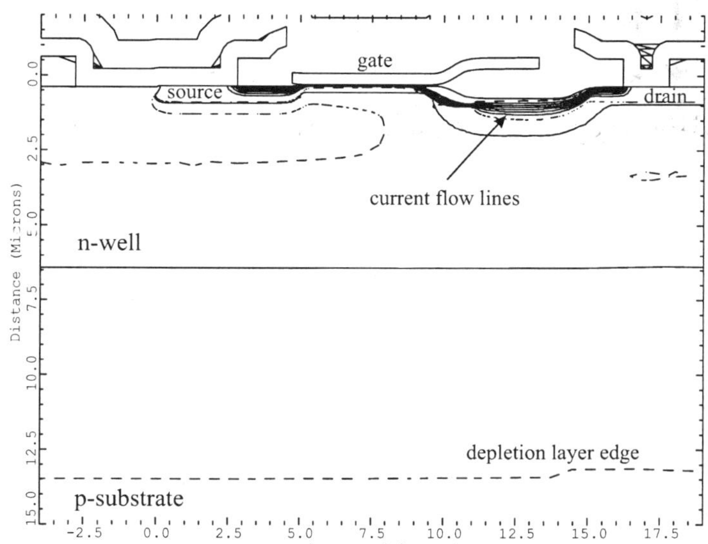

Figure 4.31: Current flow lines using the same device and under the same bias conditions

supplies ranging between 12V and 24V, the simulated devices are the best trade-off between high-voltage performance and surface. But first, the simulated performances must be approved by measurement data, which is the scope of the next chapter.

4.5 References

[1] U. Apel, E. Habekotte, B. Hofflinger, "High-voltage CMOS Transistors in a Low-Voltage CMOS Process", *Proceedings of the International Symposium on Signals, Systems and Electronics,* University of Erlangen, pp. 52-55, URSI, 1989

[2] Technology Modeling Associates, Inc. Paolo Alto, California

[3] S. Colak, "Effects of Drift Region Parameters on the Static Properties of Power LDMOST", *IEEE Trans. Elec. Dev.*, Vol. ED-28, NO. 12, December 1981, pp. 1455-1466

[4] S. C. Sun, J. D. Plummer, " Modeling of the on-resistance of LDMOS, VDMOS, and VMOS power transistors", *IEEE Trans. Elec. Dev.*, Vol. ED-27, NO. 12, 1980, pp. 356

[5] M. D. Pocha, R. W. Dutton, "A Computer-Aided design Model for High-voltage double diffused MOS (DMOS) transistor", *IEEE J. Solid State Circuits*, Vol. SC-11, 1976, pp. 713

[6] X. B. Chen, Z. Q. Song, Z. J. Li, "Optimization of the Drift Region of Power MOSFET's with Lateral Structures and Deep Junctions", *IEEE Trans. Elec. Dev.*, Vol. ED-34, NO. 11, November 1987, pp. 2344-2349

[7] M. J. Declercq, J. D. Plummer "Avalanche breakdown in High-Voltage D-MOS Devices", *IEEE Trans. Elec. Dev.*, Vol. ED-23, January 1976, pp. 1-4

5 MEASUREMENT AND MODELLING OF THE HIGH-VOLTAGE DEVICES

Abstract:
The measured static and dynamic characteristics of the SVX devices are presented first, where the high-voltage capabilities and the reliability aspects are pointed out through the measurement data. On the basis of these results, a first order modelling is then performed for both devices in order to allow further the simulation of high-voltage integrated circuits. It consists of two macro-models in which discrete devices are combined with a low-voltage transistor to model the behaviour of the lightly doped drains. Finally, the handling of high-voltage power supplies in a standard low-voltage CMOS process, necessitates the use of special layout precautions against parasitic effects. These precautions are discussed in the last part of this chapter.

Various high-voltage transistors with different sets of parameters L_c, L_g and L_{dd} were integrated and tested. In sections 5.1.1, 5.1.2, 5.1.3 and 5.1.4 the static and dynamic measurement results of the devices optimised in the previous simulations are presented in detail. In section 5.2, a first order model is introduced for both transistor types based on the measurement data. Finally, because of the high voltage power supplies handled on the chip special precautions must be taken to avoid parasitic effects such as latchup or field MOSFETs. This is the scope of section 5.3. But first, since the high-voltage capabilities of the SVX transistors are higher than 50V, a rough presentation of the upper limit of their static characteristics is introduced.

5.1 Measurement results of the high-voltage devices

Starting with the high-voltage nMOS transistor, the measured I_d-V_d characteristics are shown in Figure 5.1 for a device with L_c=10μm, L_g=5μm and L_{dd}=10μm. The maximum achieved breakdown voltage is about 120V. It can be shown with two dimensional computer simulations that the breakdown mechanism occurring in this transistor is due to the avalanche process taking place under the field-plate edge [1]. Therefore, increasing L_{dd} to 15μm while keeping L_g unchanged, i.e. increasing L_d from 5μm to 10μm, pushes the breakdown voltage up to the theoretical limit of 170V, caused by the cylindrical avalanche breakdown of the n-WELL diffusion. Unfortunately such measurement data is difficult to reproduce when no field-plate is used at the boundaries of the n-WELL to substrate junction, helping to cancel surface effects. Therefore, the characteristics shown in Figure 5.1 are the best trade-off between the surface occupied by the transistor and a reproducible breakdown voltage. To perform a higher and reproducible breakdown voltage floating field rings [2] and field-plates (chapter 2 section 2.2.2) can be used at the cost of occupying much higher surface.

The high-voltage characteristics of the HVPMOS transistor using the same set of parameters as the HVNMOS ones, are shown in Figure 5.2. As already discussed in section 4.4.2, the breakdown mechanism occurring in the pMOS transistor is a punchthrough breakdown between the lightly doped drain junction and the n-WELL to substrate junction. Therefore, and in contrast to the high-voltage nMOS case, the breakdown voltage shown in Figure 5.2 is the highest voltage that can be achieved with the pMOS devices, even with the use of field-plates or floating field rings.

5.1.1 Static characteristics of the HVNMOS transistor with L_c=5μm, L_g=3μm, L_{dd}=5μm

The static characteristics of the previously simulated and optimised high-voltage nMOS transistor are discussed in this section. Measurement data such as substrate current I_{sub}, gate current I_g, subthreshold slope ΔS and threshold voltage V_t are of particular interest to monitor if hot carrier effects are taking place in the device. Moreover, when the comparison can be made, the characteristics of a low-voltage

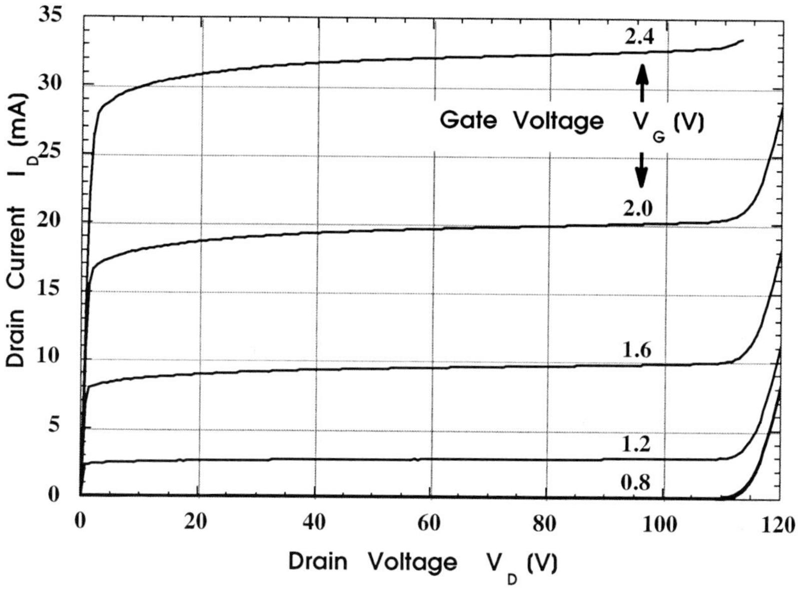

Figure 5.1: Measured I_d-V_d characteristics of the HVNMOS transistor with L_c=10μm, L_g=5μm and L_{dd}=10μm

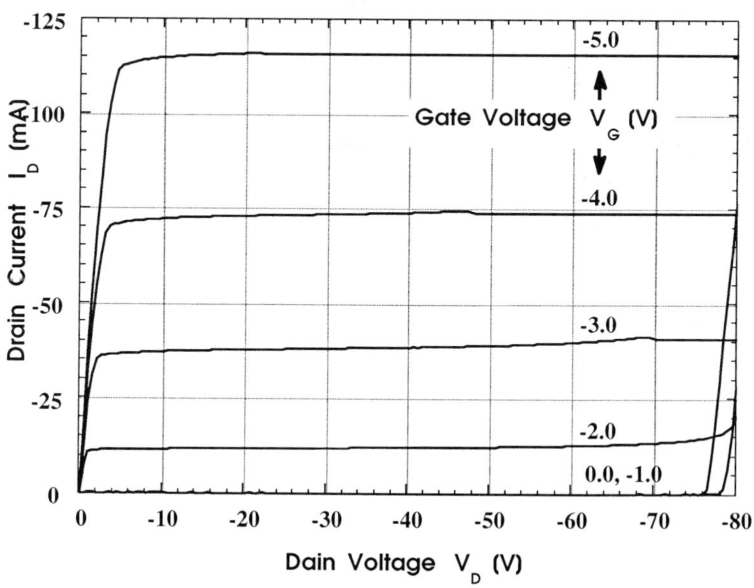

Figure 5.2: Measured I_d-V_d characteristics of the HVPMOS transistor with L_c=10μm, L_g=5μm and L_{dd}=10μm

transistor in the same technology and having the same drawn length and width, are considered as reference data. For instance, the I_d-V_d curves plotted in Figure 5.3 for various gate voltages V_g, show an increase of the saturation current in comparison to the low-voltage transistor. The increase ranges between 300μA for V_g=2V and 500μA for V_g=5V. Such behaviour is related to the channel shortening caused by the etched n-WELL layer under the channel forcing the injection point of electrons to be located before the channel end. This phenomenon appears clearly in Figure 4.25. Since the shortening length ΔL depends strongly on the drain and gate bias conditions as well as the shape of the drain diffusion, an expression of ΔL as a function of these parameters would require an accurate modelling of the electric field and potential distributions in the channel end region, which is not performed in this book.

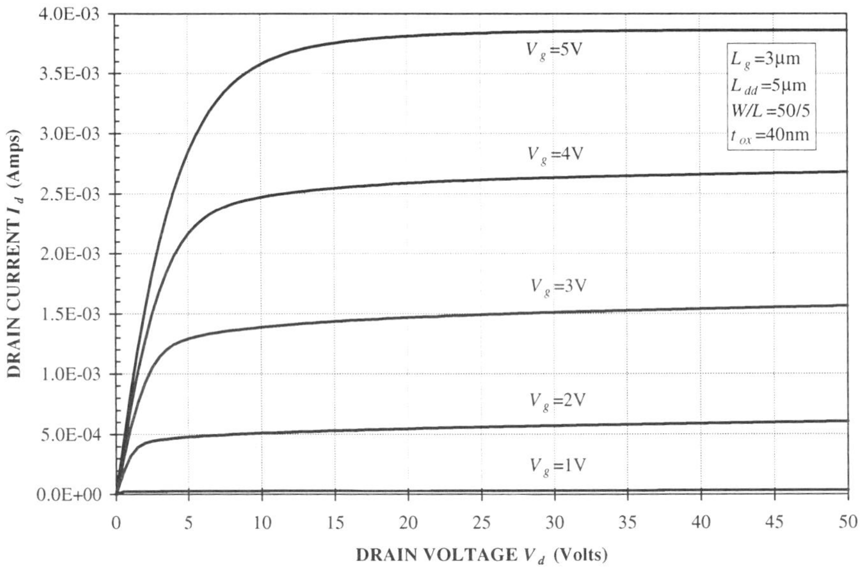

Figure 5.3: Static I_d-V_d characteristics of the HVNMOS transistor

Another interesting information arising from Figure 5.3 is the saturation voltage V_{dsat}. It ranges between 2.5V and 15V for V_g=2V and V_g=5V respectively. This is due to the current flow path between the depletion layers of the lightly doped drain, the one created by the field-plate and the one created by the n-WELL/substrate junction (Figure 4.25). Since the width of the depletion layers is a function of the drain and gate voltage, the current flows through a variable resistance. Such behaviour alters significantly the characteristics in the conduction region and the value of the saturation voltage. This will be discussed more in detail in section 5.2.

The substrate current as a function of the gate voltage is a typical example where the comparison with the low-voltage device cannot be performed, because the origin of the current is different. In fact, the ionisation process occurring in the low-voltage device is mainly located at the channel end under the thin gate oxide and, generally, a high substrate current results from a high electric field magnitude. The relationship between them was discussed in chapter 0 section 2.1.1, where a peak substrate current of 150nA was measured for V_{ds}=5V, W=50µm, L=5µm, t_{ox}=40nm and X_j=0.4µm. According to Equation **(2.3)**, the corresponding peak of the electric field magnitude is about $E_m=8x10^4$V/cm. A higher electric field value will generate more substrate current leading to hot carrier effects generation. In the HVNMOS transistor, the ionisation process is shifted into the lightly doped drain by the field-plate. As already presented in the simulation section, it is located under the field-plate edge at the surface of the n-WELL layer where the oxide thickness is 10000Å. Therefore the relationship between I_{sub} and E_m is no more valid and the substrate current value, corresponding to a peak electric field beyond which hot carrier effects must be considered, is completely different.

In order to estimate an analytical expression of I_{sub} as a function of the electric field magnitude, which is useful for device modelling in VLSI circuit simulators, a pseudo 2D-model of the electric field in the lightly doped drain should be developed. This approach would be similar to the one used by *HU* in [3] for LDD devices in submicron MOSFETs except that the field-plate effect must be considered and the lightly doped drain is deeper. Such modelling is not performed here. The substrate current of 57nA/µm calculated in the two dimensional simulation is in general agreement with the measurement data shown in Figure 5.4, where $I_{sub}\approx$1.5µA for V_{ds}=50V and V_g=5V. The corresponding electric field magnitude ($E=1x10^5$V/cm) at the surface of the n-WELL layer is comparable to the one present at the channel end of the low-voltage device. Hence, for almost equal electric field magnitudes, the substrate current is one decade higher in the high-voltage device where the drain voltage is also 10 times higher, which confirms that the pseudo 2D-modelling approach would be a correct starting point. Finally, since I_{sub} cannot be used as an indirect monitor for hot-carriers generation, as far as no substrate current model is available, the gate current remains the best direct monitor of hot carriers injection into the gate oxide. For the high-voltage nMOS transistor presented in this book, no significant gate current was measured for a drain voltage as high as 50V. Consequently, no hot carriers effects are occurring in the device and no long term degradation of the device characteristics should be observed.

The I_d-V_g characteristics of Figure 5.5 show the method used to define the threshold voltage. It is the most common method of determining the threshold voltage V_t and it consists in extrapolating the linear portion of the I_d-V_g characteristics, from the point of maximum slope. The latter is the point where the transconductance g_m reaches its maximum. The cross point with the V_g axis is at V_t+0.5V_{ds} due to the expression of the drain current in the linear region. When the same method is applied on the low-voltage transistor the resulting threshold voltage V_t is 0.78V, while in the case of the high-voltage device with the same drawn length (L=5µm) V_{tl}=0.67V. The difference ΔV_t is almost 100mV. Considering such V_t shift, one could conclude for device degradation due to hot carrier effects. It is obviously not the case since channel hot carriers injection and trapping into the gate oxide of

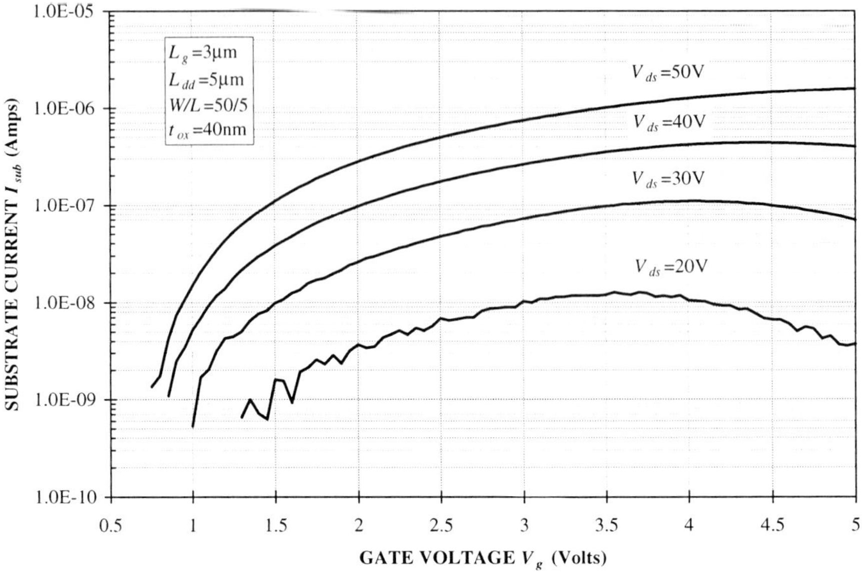

Figure 5.4: Substrate current of the HVNMOS transistor as a function of the gate voltage and for various drain voltages

an nMOS transistor, would result in a positive shift of the threshold voltage. The main reason of this negative shift is the lateral diffusion of the n-WELL layer under the channel, which has its maximum effect for L=5μm. More precisely, for L=5μm, the lateral diffusion of the n-WELL layer becomes as high as the channel length. Therefore, the channel region is formed by a p-TUB implant in the n-WELL layer, leading to a lower equivalent substrate doping concentration level. Consequently, the threshold voltage is the lowest for this channel length, and gradually approaches the low-voltage transistor one when the lateral diffusion of the n-WELL layer becomes negligible. This is confirmed by the measured I_d-V_g characteristics of a 10μm length device shown in Figure 5.5, where the threshold voltage increases of about 40mV. An analytical expression of the threshold voltage shift as a function of the channel length is not performed here, given the actual doping profile of the channel. The situation is similar to the one discussed in section 4.3.1 for the minimum channel length estimation.

The subthreshold current characteristic of the high-voltage transistor is shown in Figure 5.6. Such measurement data is necessary for hot carrier effects, surface DIBL and punchthrough monitoring. Since the measured subthreshold slope is identical to the low-voltage transistor one (ΔS=99.5mV/decade), it is an additional confirmation that no hot carriers effects are occurring in the device, and that the threshold voltage shift is effectively caused by the lateral diffusion of n-WELL layer. The slight shift in the characteristics for V_{ds}=1V and V_{ds}=50V is caused by a negligible surface DIBL. In fact, it was mentioned in section 2.3 of chapter 0 that an acceptable value of the gate voltage shift ΔV_{gs} normalised to the drain voltage increase ΔV_{ds} is

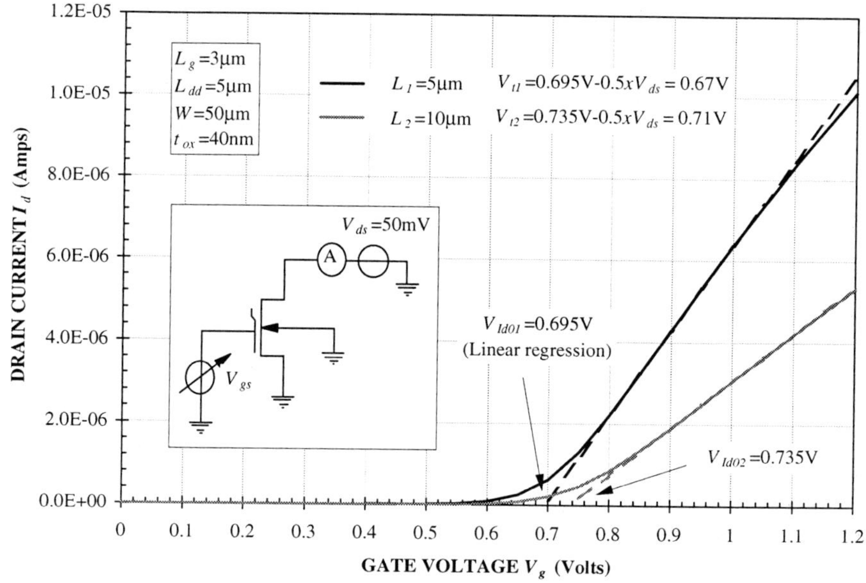

Figure 5.5: Threshold voltage of the HVNMOS transistor extracted for two channel lengths L_1=5µm and L_2=10µm

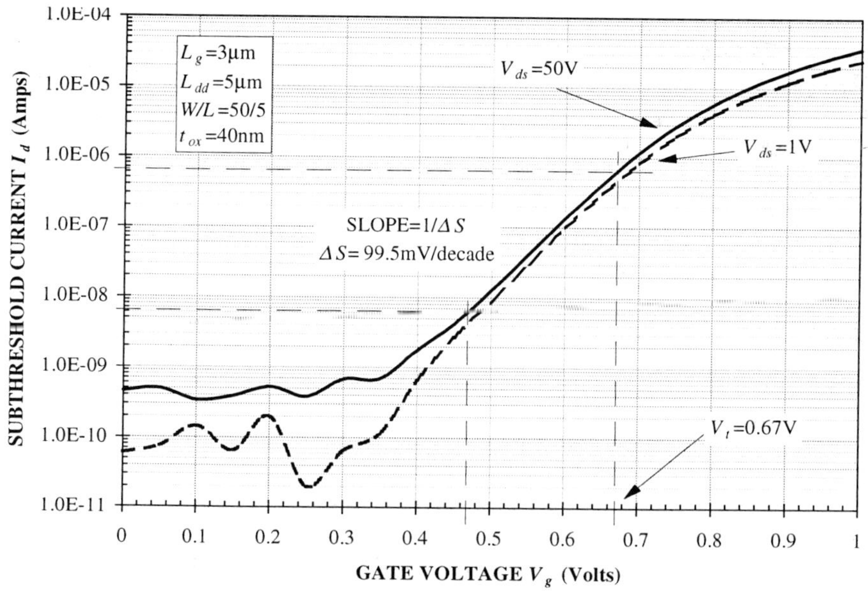

Figure 5.6: Subthreshold slope of the HVNMOS transistor with V_{ds}=1V and V_{ds}=50V

approximately 25mV/V. Figure 5.6 shows that this value is observed for ΔV_{ds}=50V. Moreover, since no increase of the subthreshold slope is observed for an increase of the drain voltage up to 50V, no punchthrough current is flowing under the channel region. Finally, The lower limit of the subthreshold current, measured at V_{ds}=50V, corresponds to the leakage current of the lightly doped drain junction. The one measured at V_{ds}=1V cannot be considered since the actual current value is lower than the noise floor of the measurement setup.

For switching applications where the high-voltage device is used as a simple switch, it is interesting to evaluate the total on state resistance R_{dson} . In Figure 5.7, R_{dson} is shown as a function of the gate voltage and for some typical values of the drain-to-source voltage suited for switched "power" applications. The influence of the lightly doped drain resistance R_{ldd} is clearly seen when these characteristics are compared to the low-voltage ones. In fact, assuming that both transistors are operating in the linear region, R_{ldd} can be calculated by subtracting R_{dson} from the on state resistance R_{on} of the low-voltage device. It can be seen that R_{ldd} is constant, about 32.5KΩ per micron width, as far as the low voltage transistor is in the linear region. Consequently, for V_{ds} ranging between 0V and about 2V the high voltage transistor can be considered as a low-voltage transistor in series with a constant resistor. Note that R_{ldd} is not plotted for V_{ds} higher than 0.5V as the low voltage transistor enters the conduction region for all V_g and the calculation of R_{ldd} is no more valid.

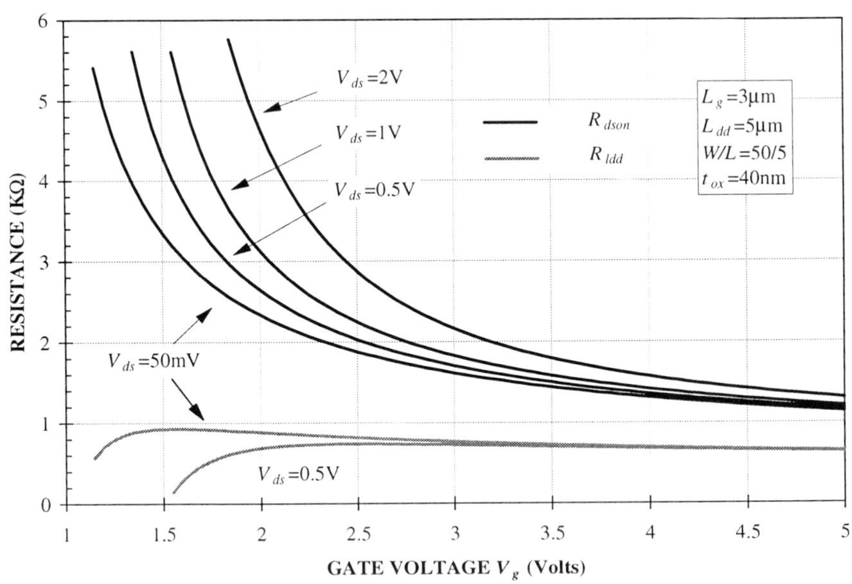

Figure 5.7: Total on state resistance and parasitic lightly doped drain resistance of the HVNMOS transistor

The last static measurement data essential for switched power applications is the forward current characteristic of the drain-to-bulk junction. The parameters required to model this diode are the saturation current I_s, the ideality factor η and the parasitic series resistance r_s. They are essential to evaluate the forward current and the voltage drop across r_s when the body diode of the transistor is used as a "free wheel" one. This is typically the case when the load connected to the drain of the high-voltage device is inductive, as the case is in the class-D audio amplifier discussed in chapter 7. Figure 5.8 and Figure 5.9 show two common measurement procedures used to extract such parameters [4]. These procedures are recalled here for convenience. In the first case, $\log(I_d)$ is plotted versus V_d, and the corresponding least-square linear regression curve is extrapolated to V_d=0V, determining thus the saturation current (I_s=1.6x10^{-13}A). The ideality factor η is determined from the slope m of the least-square linear regression curve using

$$\eta = \frac{1}{2.3U_t} \cdot \frac{1}{m} \tag{5.1}$$

since $\log(I_d)$ is given by Equation (5.2) when the voltage drop across r_s is negligible in comparison to the voltage drop across the intrinsic diode (V_d>>r_sI_d and V_d>>U_t).

$$\log(I_d) = \frac{V_d}{2.3\eta U_t} + \log(I_s) \tag{5.2}$$

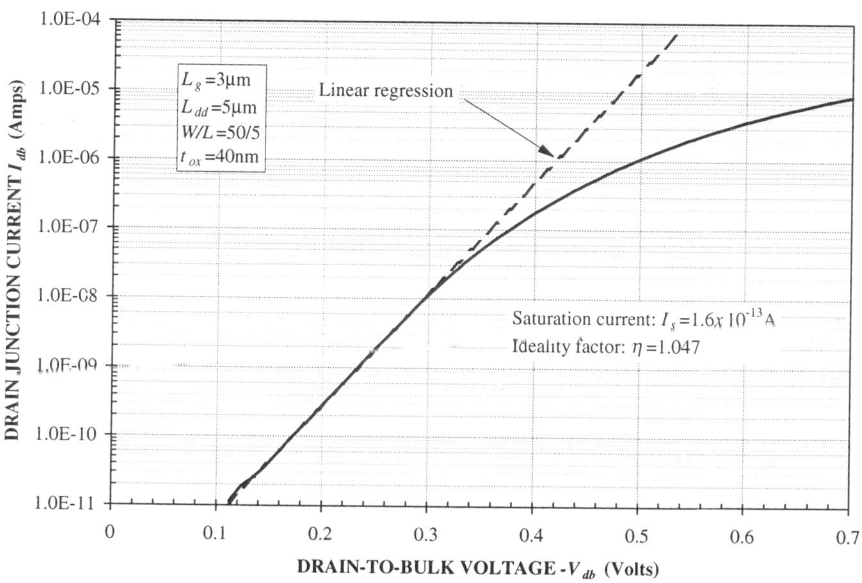

Figure 5.8: Saturation current and ideality factor of the HVNMOS Drain-to-substrate junction diode

The series resistance r_s becomes significant when the diode current is large enough so that the previous assumption is no more valid. Under such operating conditions the diode current is given by

$$I_d = I_s \left[\exp\left(\frac{V_d - I_d r_s}{\eta U_t} \right) - 1 \right] \tag{5.3}$$

rearranging Equation (5.3) in terms of V_d ($V_d - r_s I_d >> U_t$)

$$V_d = \eta U_t \ln\left(1 + \frac{I_d}{I_s} \right) + r_s I_d \tag{5.4}$$

Assuming that $I_d/I_s >> 1$ and differentiating with respect to I_d results in

$$I_d \frac{dV_d}{dI_d} = \eta U_t + r_s I_d \tag{5.5}$$

The plot of $I_d*(dV_d/dI_d)$ as a function of the diode current results in a straight line, the slope of which gives the series resistance r_s.

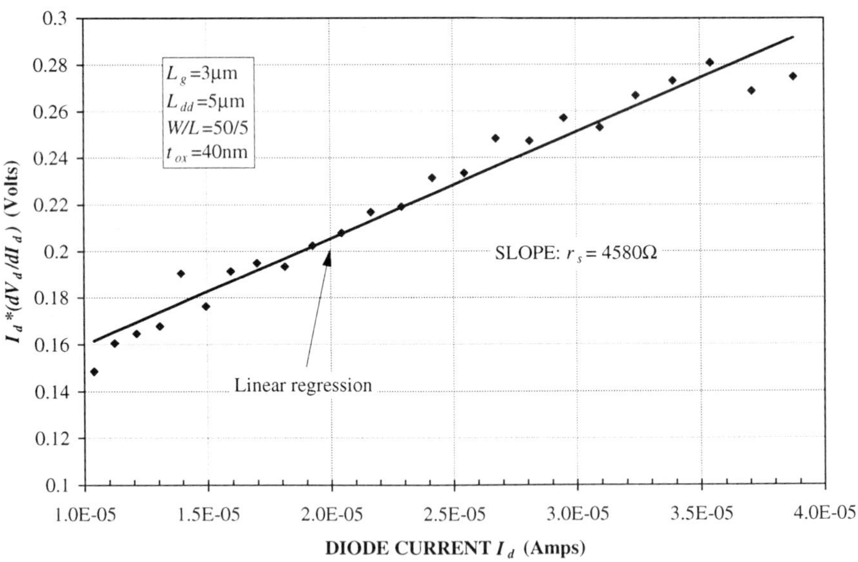

Figure 5.9: Series resistance of the HVNMOS drain-to-substrate junction diode

5.1.2 Dynamic characteristics of the HVNMOS transistor with $L_c=5\mu m$, $L_g=3\mu m$, $L_{dd}=5\mu m$

The dynamic behaviour of a MOSFET is due to the device capacitive effects, which in turn are the results of the charges stored in the device. In this section, the capacitive characteristics of the high-voltage nMOS transistor are discussed. These characteristics are the sum of the intrinsic and extrinsic capacitances present in the device. Given the difference between the drains of the high-voltage and low-voltage transistors, the capacitive characteristics are obviously not the same. But before presenting such characteristics, small signal parameters such as transconductance g_m and output conductance g_{ds} are discussed with respect to the low-voltage ones. Figure 5.10 shows the transconductance as a function of the gate voltage and for different drain voltages. The behaviour of g_m is very similar to the low-voltage one for V_{ds} values ranging between 0 and 2V, except that the magnitude is smaller. This is due to the parasitic series resistance of the lightly doped drain, which makes the effective drain-to-source voltage V_{ds}' smaller than in low-voltage case. In fact, in section 5.1.1 it was shown that for such V_{ds} range, the high-voltage transistor can be considered as a low-voltage transistor in series with the lightly doped drain resistance, R_{ldd}, measured in the linear region.

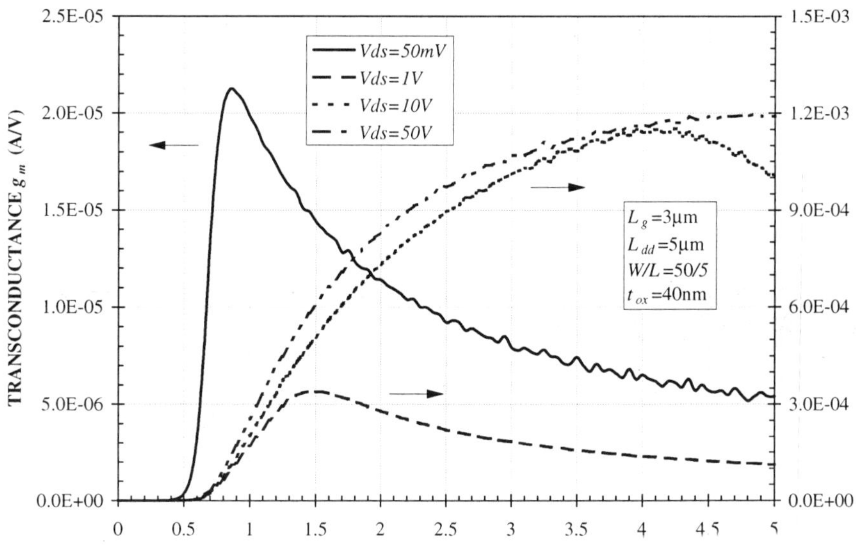

Figure 5.10: Transconductance g_m of the HVNMOS transistor for different V_{ds} values

For higher V_{ds} values the previous assumption is no more valid as R_{ldd} becomes dependent on V_g and V_d, leading to the behaviour shown by the curves where V_{ds}=10V and V_{ds}=50V. Moreover, the degradation of the mobility becomes a function of the lateral electric field as well. These considerations are discussed more in detail in section 5.2 concerning the modelling of the high-voltage devices.

The output conductance of the device is shown in Figure 5.11 as a function of the drain voltage and for different gate voltage values. The curves behave as the low-voltage transistor one except for some differences related to the lightly doped drain resistance. For instance, the absolute value of all the g_{ds} curves is smaller in the case of the high-voltage transistor, which is obvious given the higher effective channel resistance. Moreover, for low V_{ds} values the same behaviour as in the static I_d-V_d characteristics appears here as well. That is, as far as the transistor is operating in the linear region the output conductance results from the sum of the channel resistance and the lightly doped drain resistance R_{ldd}. When the transistor enters the conduction mode, the output conductance is given by the sum of the channel resistance and a variable resistance caused by the current flow path between the depletion layers of the field-plate and the n-Well-to-substrate junction. In the saturation region, g_{ds} stops increasing for V_g higher than 2V. It is even smaller for V_g=4V than for V_g=3V. Such behaviour is caused by the self heating of the device for high current values, resulting in a reduction of the output conductance. If the drain current is higher, g_{ds} becomes negative appearing as an increase in the logarithmic scale if the absolute value $|g_{ds}|$ is plotted.

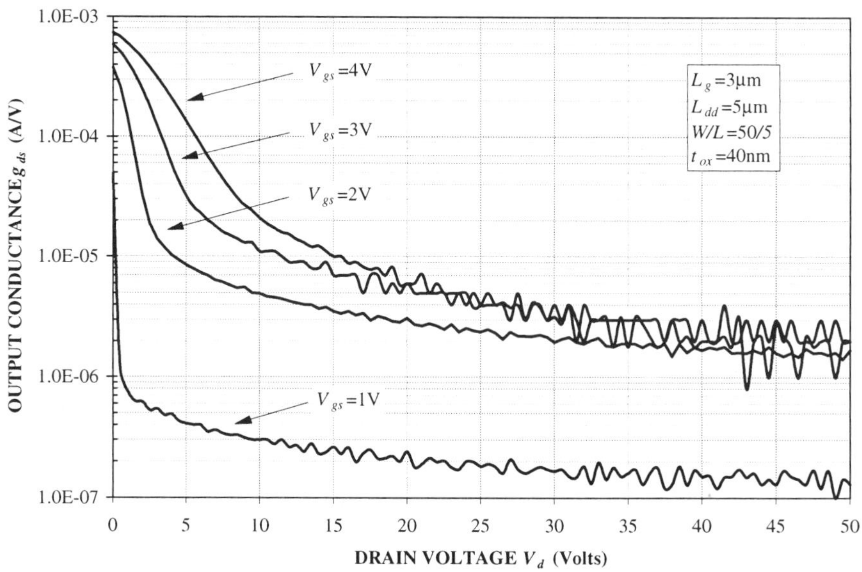

Figure 5.11: Output conductance g_{ds} of the HVNMOS transistor for different V_{gs} values

The first capacitive characteristic of the device is the gate-to-bulk capacitance C_{gb} shown in Figure 5.12 as a function of the gate voltage and for different V_{ds} values. Since the only difference between the high-voltage and low-voltage devices is the drain, the same capacitance should be measured for both devices. Actually, some differences appear between the characteristics, particularly in the accumulation and inversion regions. This could be explained by the effects of the n-WELL depletion layer capacitance under the channel region and the field-plate one. In fact, the measured capacitance in the accumulation region is about 250fF against 200fF in the low-voltage case. Such difference is due to the field-plate capacitance C_{fp} over the n-WELL layer, which is in parallel with the thin gate oxide capacitance C_{ox}. Since the extra capacitance of 50fF is caused by field-plate, it should be observed for higher channel length values. Indeed, the measured C_{gb} capacitance for a high-voltage transistor with twice the previous channel length is about 450fF against 400fF in the low voltage case.

In the inversion region, the behaviour of C_{gb} is more complex due to the influence of the channel depletion layer, when the device is operating in the saturation mode, and the depletion layer of the lightly doped drain created by the field-plate. Hence, when the transistor is in the linear mode C_{gb} depends on the field-plate capacitance, while in the saturation mode C_{gb} depends on the channel depletion layer as well as the field-plate one. More accurate analysis is required at this level to model such behaviour of C_{gb}. Particularly, a charge-based capacitance model must be constructed taking into account the charge of the lightly doped drain depletion layer in addition to the charge of the channel and the bulk. Then the derivative of the total gate charge Q_g must be performed with respect to the gate-to-bulk voltage V_{gb}, in order to express C_{gb} as a function of V_{gb}. Such approach is also valid for the estimation of C_{gs} and C_{gd} as a function of V_{gb}, which is not performed in this book. A qualitative analysis of the remaining capacitive characteristics of the device is discussed in the following paragraphs, and the modelling will be limited to a first order one, as will be pointed out in the next section.

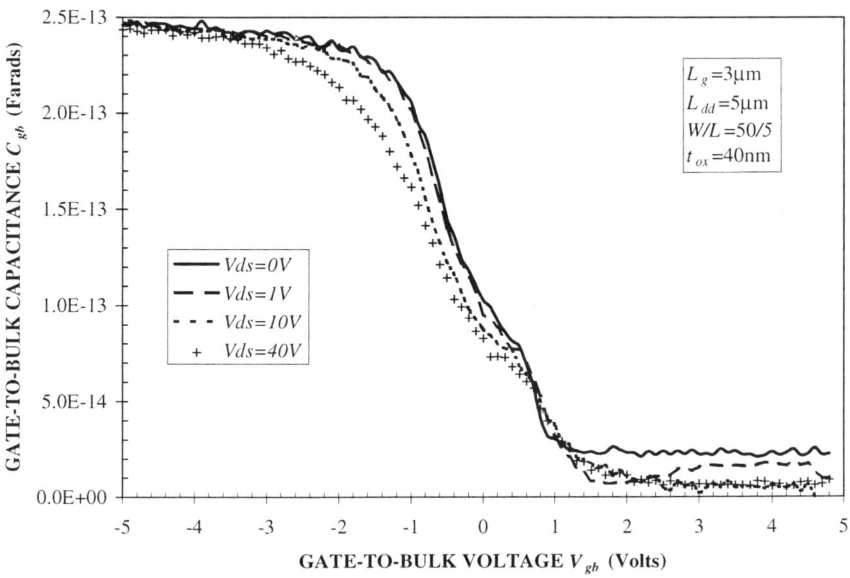

Figure 5.12: Gate-to-bulk capacitance of the HVNMOS transistor as a function of the gate voltage and for different V_{ds} values

The behaviour of the gate-to-source capacitance as a function of the gate voltage is shown in Figure 5.13 for different V_{ds} values. When these characteristics are compared to the low-voltage ones, the differences can be explained as follows. In the linear mode and considering the low-voltage case, C_{gs} remains constant at about 120fF as soon as an inversion layer is formed in the channel. In the high-voltage case, for V_{ds} ranging between 0V and about 2V, C_{gs} increases gradually up to 200fF as a function of the gate voltage. This is related to the gradual increase of the accumulation layer under the field-plate, resulting in a longer effective channel length. In the saturation mode and for V_{ds} ranging between about 2V and 10V, C_{gs} steps to a smaller value then increases almost linearly versus V_{gb}. This is due to the channel shortening caused by the injection of electrons before the channel end. As the gate voltage is increased, the injection point is shifted towards the channel end until the device enters the linear region, where the behaviour of C_{gs} becomes similar to the previous case . Finally, for V_{ds} higher than 10V all the C_{gs} curves are superposed in the saturation mode because the channel shortening of the device does not increase anymore.

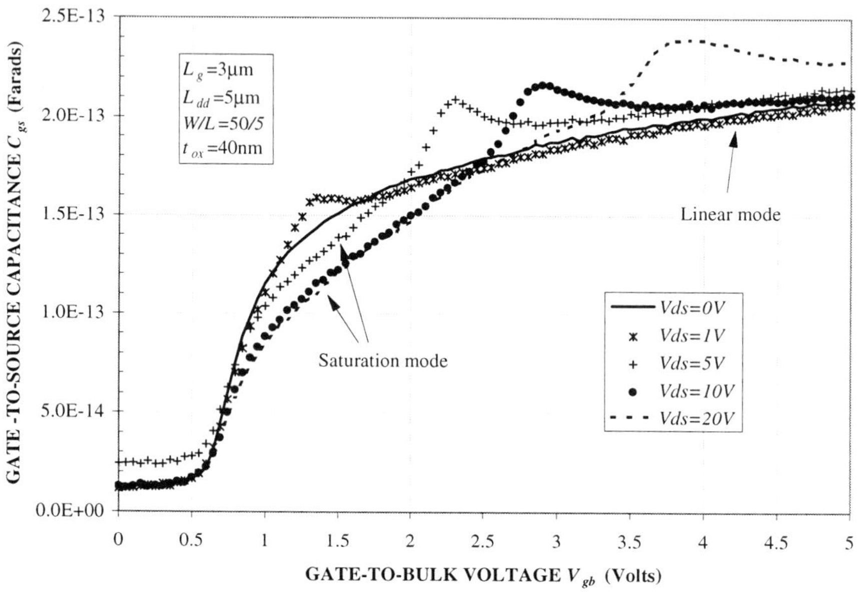

Figure 5.13: Gate-to-source capacitance of the HVNMOS transistor as a function of the gate voltage and for different V_{ds} values

The gate-to-drain capacitive characteristics shown in Figure 5.14 behave typically as the characteristics observed for low-voltage LDD devices in submicron technologies. The gate-to-drain capacitance C_{gd} steps to a constant value as soon as an inversion layer appears at the channel end, i.e. when the device enters the linear mode. The starting value of the step is different than zero, since in the absence of an inversion layer in the saturation mode, C_{gd} is given by the field-plate capacitance C_{fp}. The latter decreases as a function of the drain voltage because the depth of the depletion layer created by the field-plate increases with V_{ds}. Finally, the only difference between the characteristics of the high-voltage transistor and the low-voltage LDD devices is the decrease of the capacitance as a function of the gate voltage in the linear region. Such behaviour can be explained as follows. Considering V_{ds}=0V, the step of the capacitance is caused by the appearance of an inversion layer in the channel. Hence, the gate-to-drain capacitance is given by the portion of the charge in the channel controlled by the drain voltage. When V_{gb} is increased further, an accumulation layer is formed under the field-plate increasing the effective channel length. Consequently, while the portion of the charge in the channel controlled by the source is increased, which results in an increase of C_{gs}, the portion of the charge in the channel controlled by the drain is decreased by the same amount. Since the contribution of the charge under the field-plate to C_{gd} is negligible in comparison to the loss of the charge in the channel, the capacitance decreases as a function of the gate voltage.

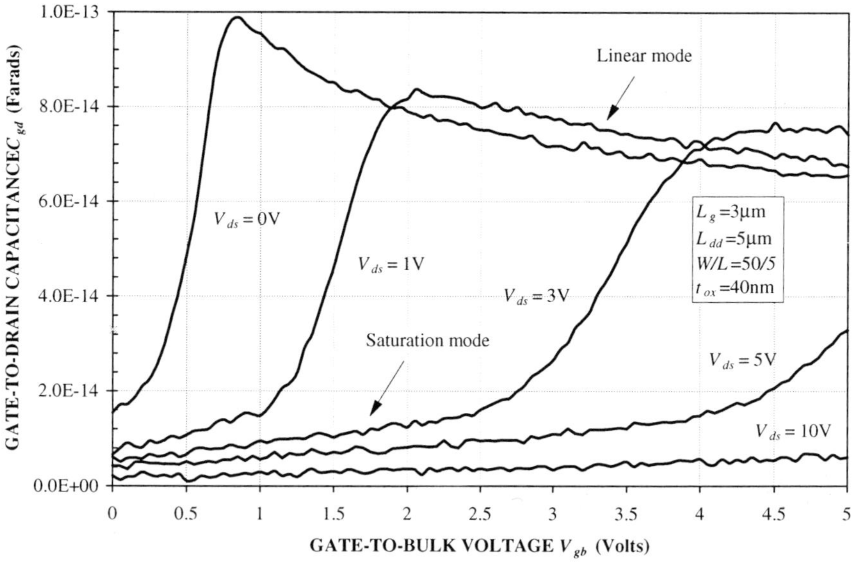

Figure 5.14: Gate-to-drain capacitance of the HVNMOS transistor as a function of the gate voltage and for different V_{ds} values

The last dynamic measurement data of the HVNMOS transistor is shown in Figure 5.15, where the capacitive characteristic of the drain-to-bulk junction is plotted versus the drain voltage. The behaviour of the capacitance is typical to biased junctions, where in the reverse mode it decreases as a function of the applied voltage due to the increase of the depletion layer thickness. In the forward mode the capacitance increases to infinity as the voltage reaches the built-in potential V_{bi}. As in the case of the static characteristics, such measurement is important to implement a correct model of the body diode for switched power applications. In fact, it will be used in the modelling section to help the fitting of the theoretical curve with the measured one by adjusting the basic parameters of the SPICE diode model.

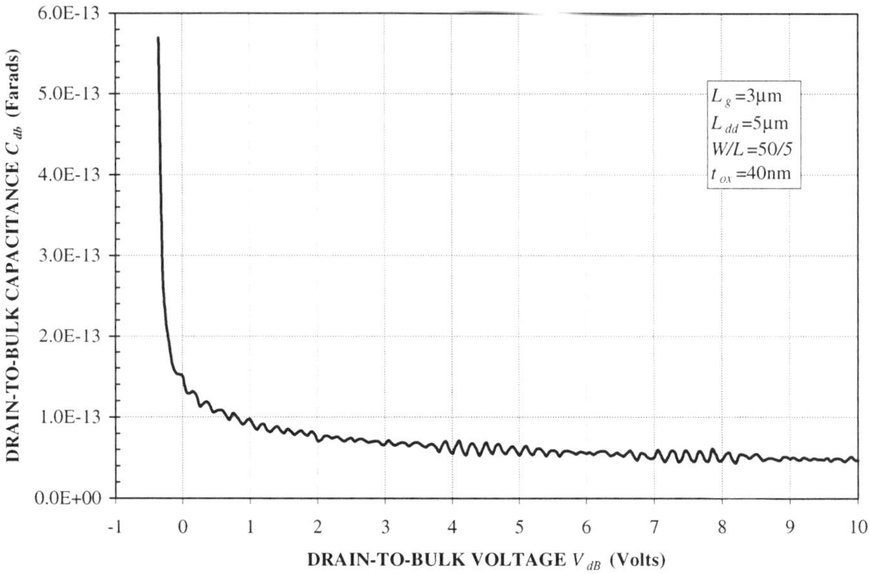

Figure 5.15: Drain-to-bulk capacitance of the HVNMOS transistor as a function of the drain voltage

5.1.3 Static characteristics of the HVPMOS transistor with L_c=5µm, L_g=3µm, L_{dd}=5µm

The same static measurement data as for the case of the HVNMOS transistor are presented in this section for the high-voltage pMOS device. The consideration concerning the comparison of the low-voltage characteristics with the high-voltage ones applies in this case as well. Starting with the drain current characteristics shown in Figure 5.16, the channel shortening is not as important as in the case of the HVNMOS transistor. In fact, the increase of the saturation current ranges between 0µA V_g=-2V and 100µA for V_g=-5V, assuming that the device is entering the saturation mode. The main reason is the absence of etching of the lateral diffusion of the lightly doped drain. Beyond the saturation limit, the current rises gradually as a function of the drain voltage due to the output conductance of the devices with L=5µm. Note that such output conductance does not appear in the HVNMOS transistor for high V_g values. It is due to the self heating of the device since the measurements were performed with a continuous DC voltage source (Hewlett-Packard HP4145), while in the case of the HVPMOS transistor a pulsed DC voltage source was used (HP4142).

In contrast to the HVNMOS case, the saturation voltage is smaller. It ranges between -2V and -6V for V_g=-2V and V_g=-5V respectively. This is related to the difference between the shape of the depletion layers in the lightly doped drains. In the nMOS transistor, the depth as well as the etching of the n-WELL layer is such

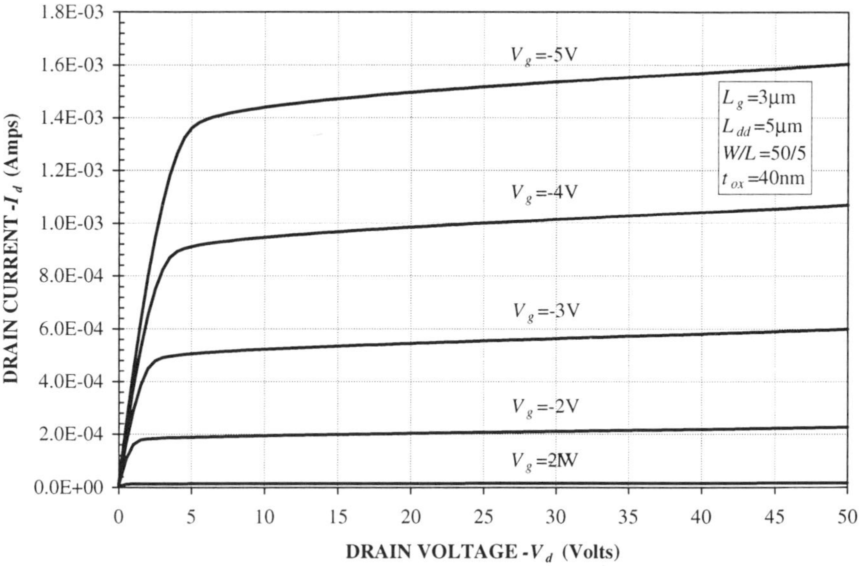

Figure 5.16: Static drain current characteristics i_d-V_d of the HVPMOS transistor

that the depletion layers created by the field-plate and the junction do not merge for V_d as high as 50V. Therefore, the potential and the electric field are spread along the entire length, leading to a distributed voltage drop across the variable resistance (Figure 4.22 and Figure 4.23). Consequently, a higher drain voltage is required to saturate the velocity of electrons. In the pMOS transistor, the depth of the lightly doped drain is approximately three times less. The depletion layers created by the field-plate and the junction merge along a portion of the lightly doped drain for lower V_d values. Thus, most of the potential and electric field lines are confined within this short region (Figure 4.28 and Figure 4.29), leading to an early saturation of the velocity of holes as a function of the drain voltage.

A peak bulk current I_{bulk} of 10pA was measured for the low-voltage pMOS transistor as a function of the gate voltage and with V_{ds}=-5V. Such low value, in comparison to the substrate current of the low-voltage nMOS transistor (150nA), is due to the lower energy of hot-holes with respect to electrons, assuming the same bias conditions. Thus, the impact ionisation rate is much lower leading to a smaller bulk current. In the case of the HVPMOS transistor, however, the same discussion made for the substrate current of the HVNMOS transistor applies as well, except for the location of the impact ionisation process. In fact, it was shown in chapter 4 section 4.4.2 that the peak of electron-hole pair generation density is located under the "bird's beak" of the field-plate. The corresponding bulk current calculated by the simulator, assuming V_{ds}=-30V, was about 50nA/μm width. The measured peak bulk current of the HVPMOS transistor with V_{ds}=-30V is about 45nA, which is almost two decades lower than the simulated one. This is due to the exponential dependence of the ionisation coefficient α_p as a function of the inverse of the electric field. An

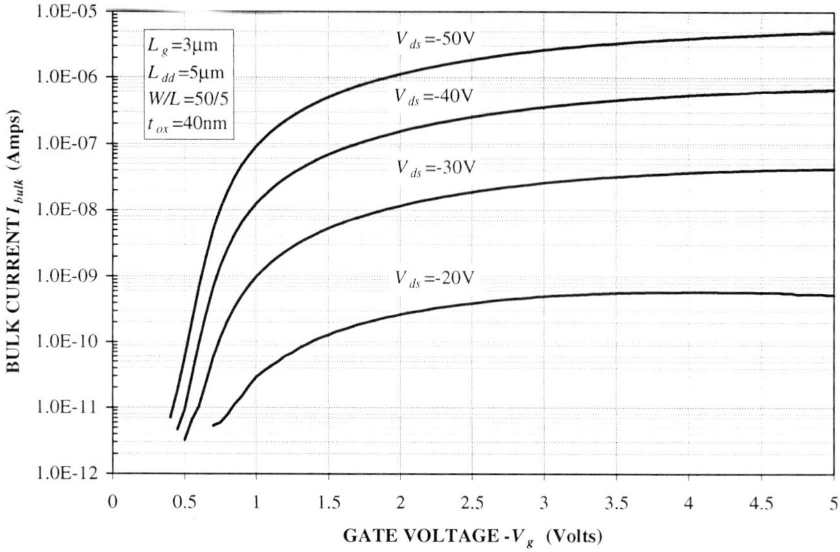

Figure 5.17: Bulk current of the HVPMOS transistor as a function of the gate voltage and for various drain voltages

electric field ranging between $2x10^5$V/cm and $3x10^5$V/cm would result in a change of the ionisation rate of about two decades [5]. Although no accurate modelling of the bulk current is yet available for these devices, I_{bulk} is related to the ionisation coefficient through the multiplication factor and the drain current, as discussed in section 2.1.1. Finally, since the simulated current value is higher than the measured one, the actual electric field distribution is apparently lower.

In the two dimensional simulation concerning the electric field distribution, it was stated that the maximum drain-to-source voltage over which hot-carrier effects must be considered is about -30V. In fact, Figure 5.18 shows a significant increase of the gate current for higher V_{ds} values. Particularly, the peak gate current of 300pA measured at V_{ds}=-50V, is equivalent to a bias condition of V_{ds}=-13V and V_g=-3V applied to the low-voltage pMOS transistor. This was shown in Figure 2.9. When such bias condition is applied to the device for about 6 minutes, it results in a degradation of the electrical characteristics.

When the drain voltage is restricted to -30V no hot carriers effect are observed in the characteristics of the device, as it appears in the extraction of the threshold voltage shown in Figure 5.19 and in the subthreshold slope shown in Figure 5.20. In contrast to the HVNMOS case, the threshold voltage is identical to the one of the low-voltage transistor. Moreover, the extracted value is reliably reproducible. This applies for the subthreshold slope ΔS as well, where no surface DIBL or punchthrough effects are observed. The lower limit of the subthreshold current cannot be considered as the leakage current of the drain junction, since the actual value reaches the noise floor limit of the measurement setup. Note that the difference between the noise floors of the nMOS and pMOS is due to the fact that the HP4142 voltage source has a higher current meter resolution.

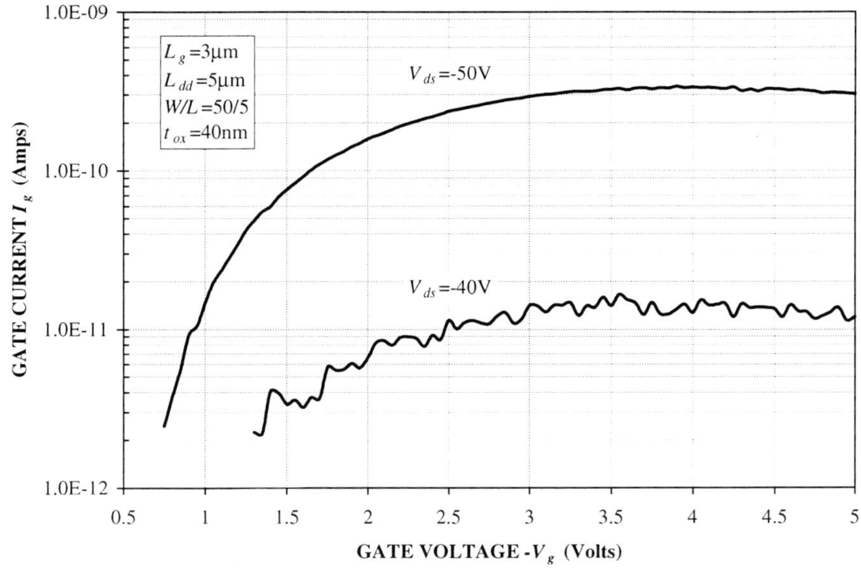

Figure 5.18: Gate current of the HVPMOS transistor, as a function of V_g, increasing significantly for V_{ds} higher than -30V

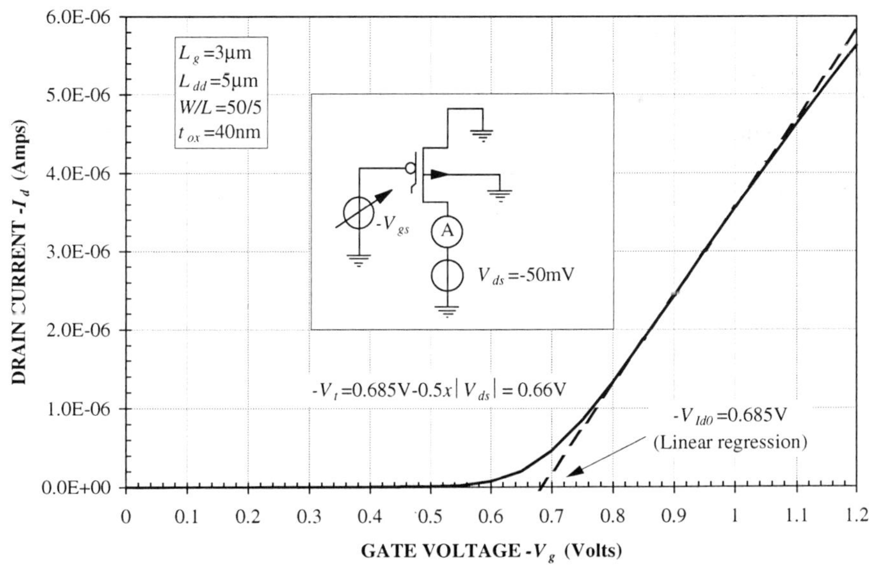

Figure 5.19: Threshold voltage extraction procedure of the HVPMOS transistor

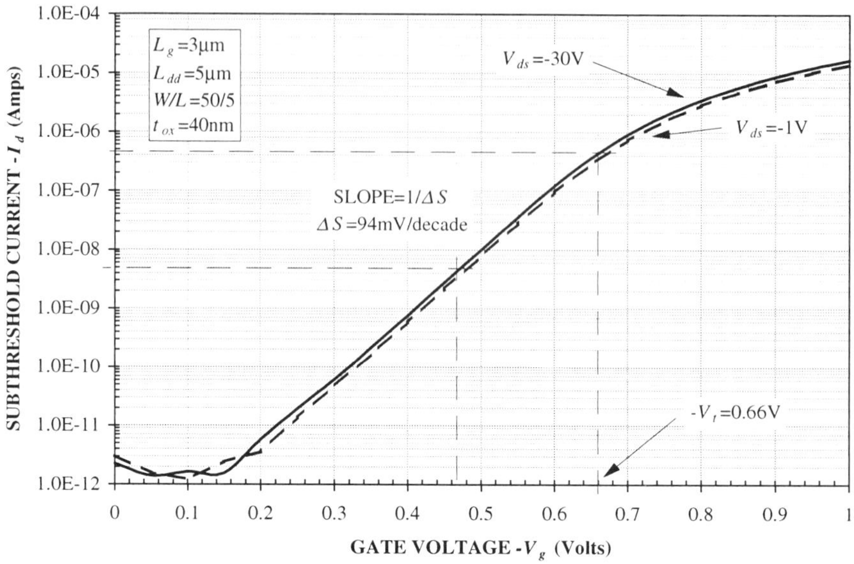

Figure 5.20: Subthreshold slope of the HVPMOS transistor with V_{ds}=-1V and V_{ds}=-30V

In Figure 5.21, the total on state resistance R_{dson} of the HVPMOS transistor is shown with its parasitic lightly doped drain resistance R_{ldd}. The latter is computed as for nMOS case. Assuming $|V_g|$=5V, the total on state resistance of the HVPMOS transistor is about 2.2KΩ, against 1.2KΩ for the HVNMOS transistor. The resulting R_{dson} ratio is about 1.8 only instead of 3 measured for the low-voltage devices having the same channel length and width. The main reason of such difference is related to the parasitic resistance of the lightly doped drains, which are almost equal in both cases. In fact, the calculated value for the HVPMOS transistor is about 28.5KΩ per micron width against 32.5KΩ.

The drain-to-bulk junction diode is characterised by using the same measurement procedure as for the HVNMOS transistor. Thus, when the least square linear regression of the log(I_d) versus V_d is extrapolated to V_d=0, the resulting saturation current I_s is about $1.84x10^{-14}$A. The ideality factor η=1.06 results from the slope m of the linear regression. Such measurement data is shown in Figure 5.22 and the one concerning the series resistance r_s of the junction diode is shown in Figure 5.23. Note the difference between the series resistance of the HVPMOS transistor and the HVNMOS one, 60Ω against 4580Ω. It is caused by the difference between the depth of the lightly doped drains as well as the doping concentration levels of the junction layers.

In fact, the lightly doped drain of the HVNMOS transistor is formed by an n^-p^- diode. Hence, the forward current flows through a total series resistance formed by the 6µm depth n-WELL layer ($N_s \approx 1x10^{16}$cm^{-3}) and the portion of the substrate ($N_A \approx 7x10^{15}$cm^{-3}) between the metallurgical junction and the substrate contact. While in the HVPMOS case, the forward current flows through a total series resistance formed by the 2μm depth *LDD* layer ($N_s \approx 1x10^{17} cm^{-3}$) and the portion of the n-WELL between the metallurgical junction and the bulk contact.

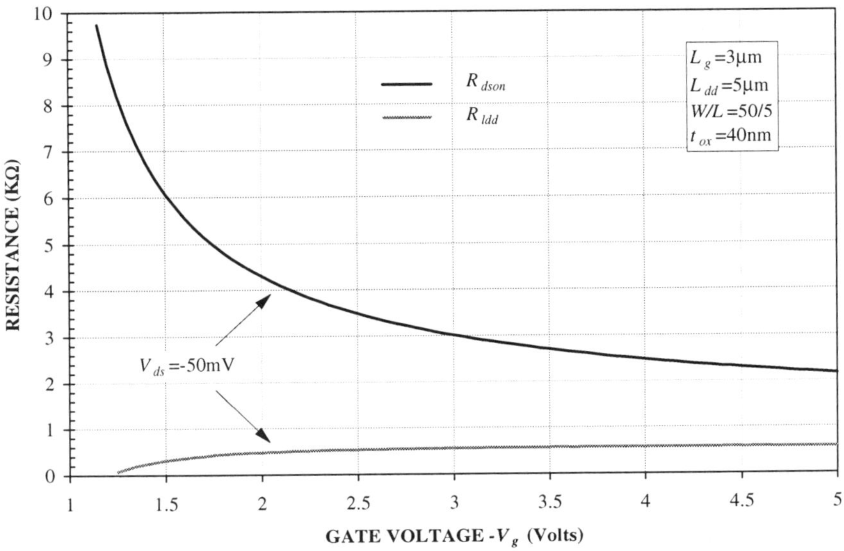

Figure 5.21: Total on state resistance and parasitic lightly doped drain resistance of the HVPMOS transistor

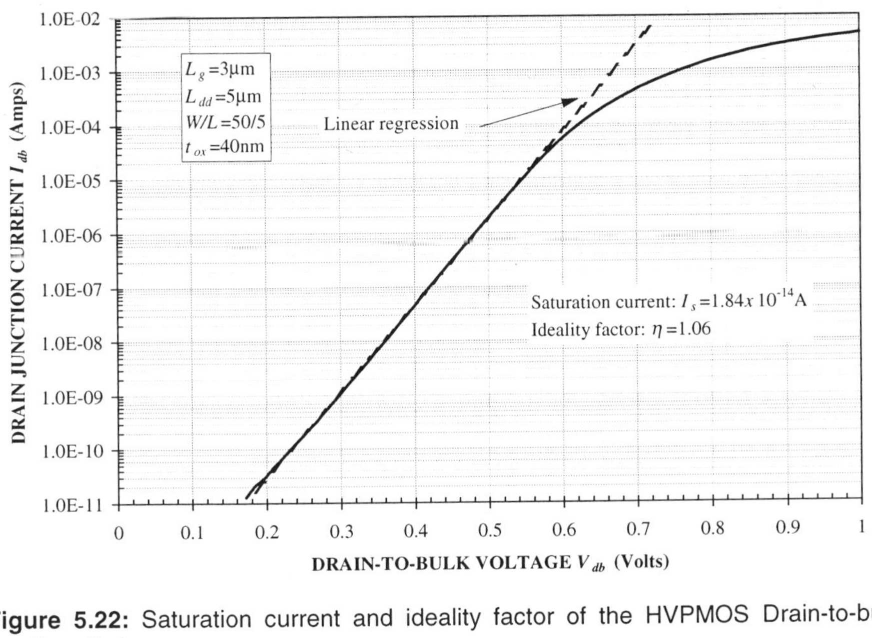

Figure 5.22: Saturation current and ideality factor of the HVPMOS Drain-to-bulk junction diode

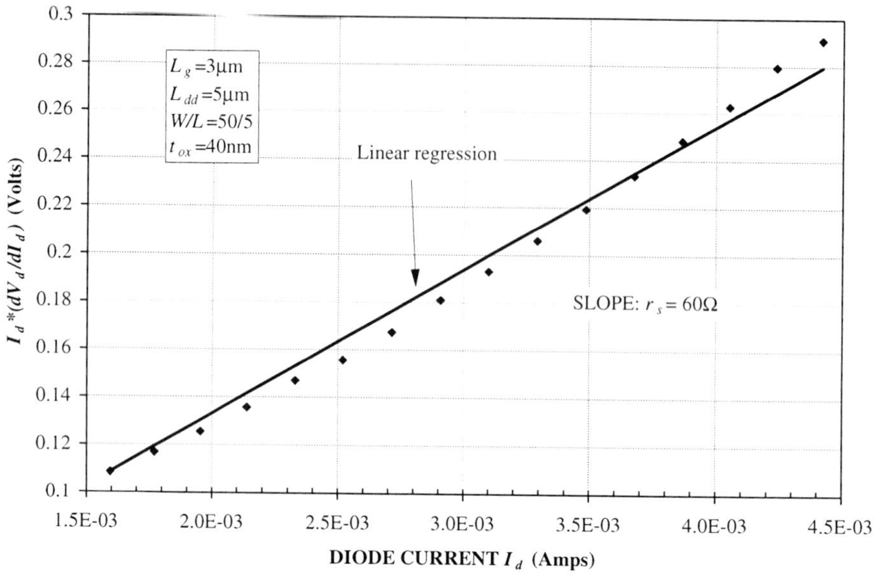

Figure 5.23: Series resistance of the HVPMOS Drain-to-bulk junction diode

5.1.4 Dynamic characteristics of the HVPMOS transistor with L_c=5μm, L_g=3μm, L_{dd}=5μm

In this section the dynamic characteristics of the HVPMOS transistor are discussed for drain voltage values ranging between 0V and -30V. The general behaviour of these characteristics is similar to high-voltage nMOS case discussed in section 5.1.2. The slight differences in the capacitive characteristics are mainly related to the depletion layer of the lightly doped drains. In the small signal characteristics, the differences are caused by the behaviour of the parasitic series resistance as a function of V_g and V_d, which is different in each case.

Hence, for V_{ds} ranging between 0V and about 2V the transconductance g_m shown in Figure 5.24 can be obtained by modelling the high-voltage transistor with a low-voltage one in series with the lightly doped drain resistance, R_{ldd}, measured in the linear region. For higher drain voltages, typically for V_{ds}=-10V and V_{ds}=-30V, the transconductance in the saturation mode is not linear anymore as a function of the gate voltage. This is caused by different effects occurring in the saturation region. On the one hand, R_{ldd} is no more constant since it depends on V_g and V_d. On the other hand, the degradation of the mobility depends on the vertical electric field resulting from the gate voltage as well as the lateral electric field resulting from the high drain voltage.

When the characteristics of the output conductance g_{ds} shown in Figure 5.25 are compared to the HVNMOS case, the effect of the variable resistance present in the n-WELL layer are more visible. In fact, in the high-voltage nMOS device, the

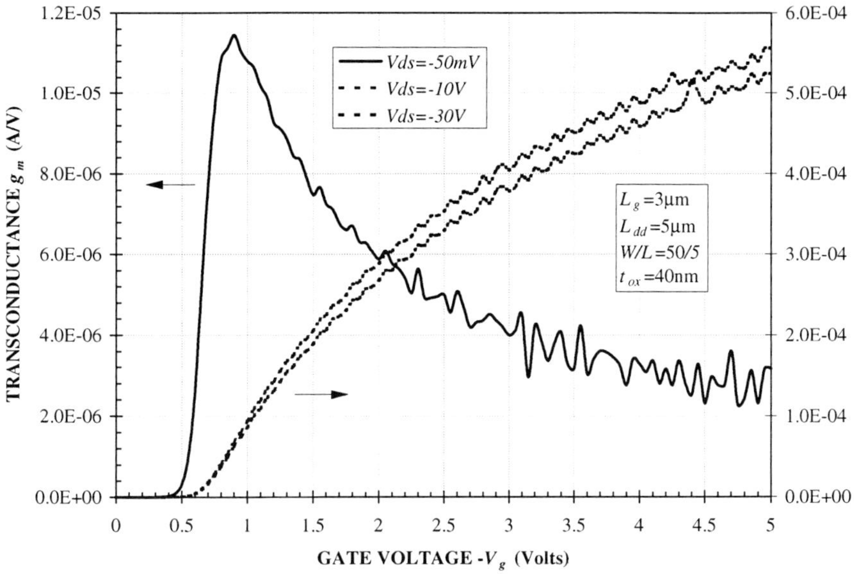

Figure 5.24: Transconductance of the HVPMOS transistor as a function of the gate voltage and for different V_{ds} values

electric field is distributed along the entire length of the lightly doped drain. Therefore, the velocity saturation of the carriers occurs in the variable resistance created by the current flow path between the depletion layers, and the device is resistive for a wider V_{ds} range (up to 15V). In the HVPMOS case, the merging of the depletion layers forces the electric field to be concentrated within the fully depleted region, leading thus to a faster saturation of the velocity of the carriers as a function of V_{ds}. It results in lower saturation voltages as in the former case (up to 5V). It should be mentioned at this level that, because of such differences, the modelling of the lightly doped drains should not be the same. Finally, and for high V_{gs} values, no self heating effects are visible on the characteristics of the output conductance.

The capacitive characteristics of the HVPMOS transistor are almost identical to the HVNMOS case, except for some slight differences caused by the lateral diffusion of the lightly doped drains and their depletion layers. It must be recalled that in the high-voltage pMOS device, the *LDD* layer diffuses laterally but is not etched. It results in a smaller effective channel length. In the HVNMOS case, although the lateral diffusion of the n-WELL layer is much higher, the etching caused by the "p-TUB" and V_t adjust implants makes the effective channel length higher than the pMOS one. Moreover, the depletion layers of the *LDD*/n-WELL and n-WELL/substrate junctions merge for a drain-to-source voltage higher than -5V. Consequently, the region under the lightly doped drain is fully depleted, as it appears in Figure 4.31 for instance, and as soon as it merges with the depletion layer created by the field-plate, the capacitive behaviour of the device is influenced. In the following paragraphs, a qualitative analysis only is performed for the capacitive characteristics of the device.

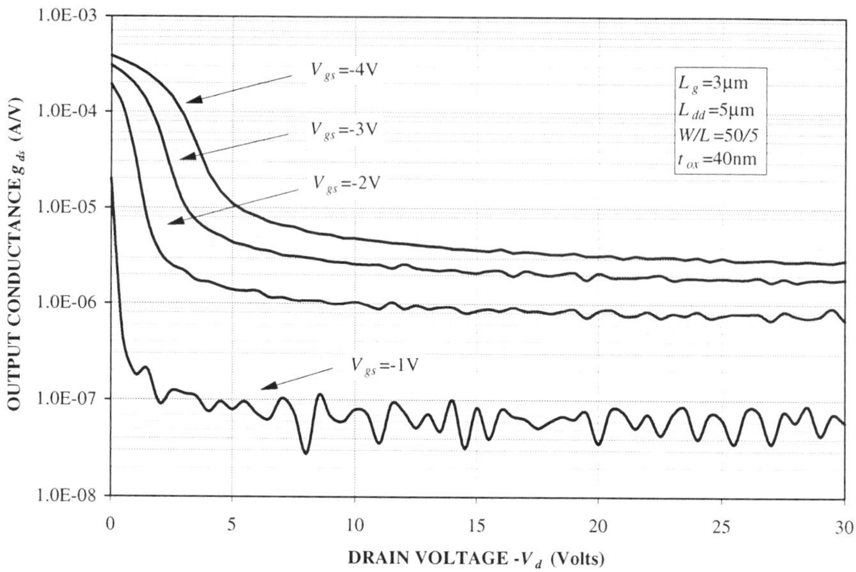

Figure 5.25: Output conductance of the HVPMOS transistor as a function of the drain voltage and for different V_{gs} values

In the accumulation region, the gate-to-bulk capacitance C_{gb} shown in Figure 5.26 is 20fF smaller than the HVNMOS one. This is due to the smaller effective channel length, leading to a smaller thin gate oxide capacitance C_{ox}. Since C_{gb} results from the sum of C_{ox} and the field-plate capacitance C_{fp}, however, it remains higher than the low-voltage case. As V_{ds} and V_{gs} increase, the transistor gradually leaves the accumulation mode, but the decrease of the capacitance value is stronger than the nMOS one for V_{ds} higher than -5V. Such behaviour is caused by the sudden increase of the width of the depletion layer under the field-plate, resulting from the merging discussed above. In the inversion region, when the device is operating in the saturation mode, the capacitance is higher than the linear mode, which is opposite to the behaviour observed for the HVNMOS transistor. This is also related to the wide depletion layer present at the channel end. More accurate analysis is required here as well to carry out a correct modelling of the charge present in the device.

The gate-to-source capacitance characteristics, shown in Figure 5.27, are also smaller than the HVNMOS ones for all V_g and V_d values. This is related to the lateral diffusion of the lightly doped drain layer leading to a smaller effective channel length. Moreover, in the saturation mode, the channel shortening of the devices causes some differences in the characteristics. In fact, in the HVNMOS transistor, it was stated that for V_{ds} higher than 10V the channel shortening does not increase anymore, which results in the superposition of the characteristics, as it appears in Figure 5.13 for V_{ds}=10V and V_{ds}=20V. In the HVPMOS transistor, the shortening of

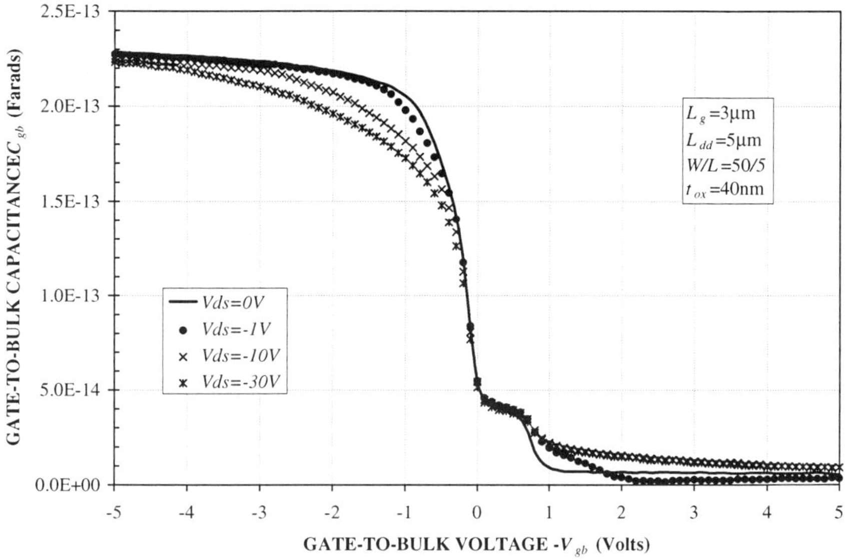

Figure 5.26: Gate-to-bulk capacitance of the HVPMOS transistor as a function of the gate voltage and for different V_{ds} values

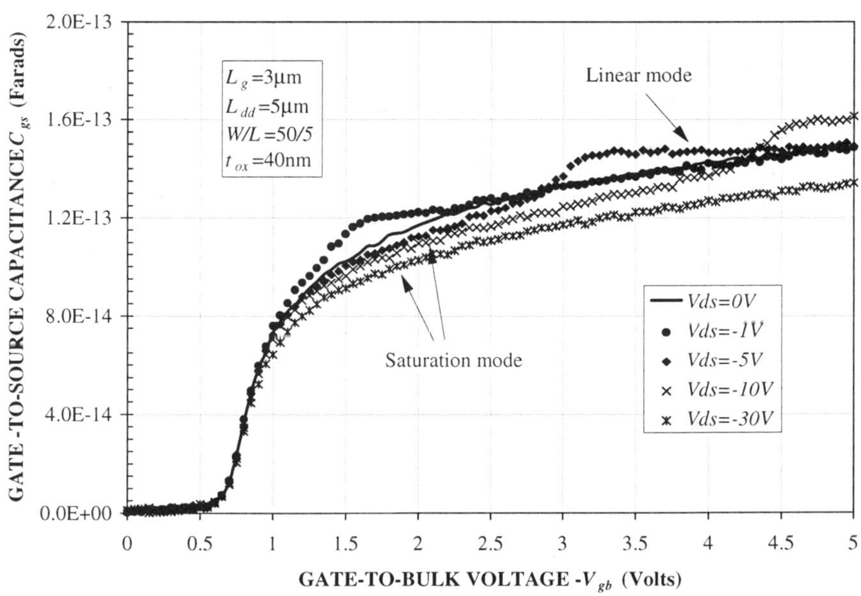

Figure 5.27: Gate-to-source capacitance of the HVPMOS transistor as a function of the gate voltage and for different V_{ds} values

the channel is not limited for V_{ds} up to -30V, resulting in a continuous decrease of C_{gs} as the drain voltage is increased.

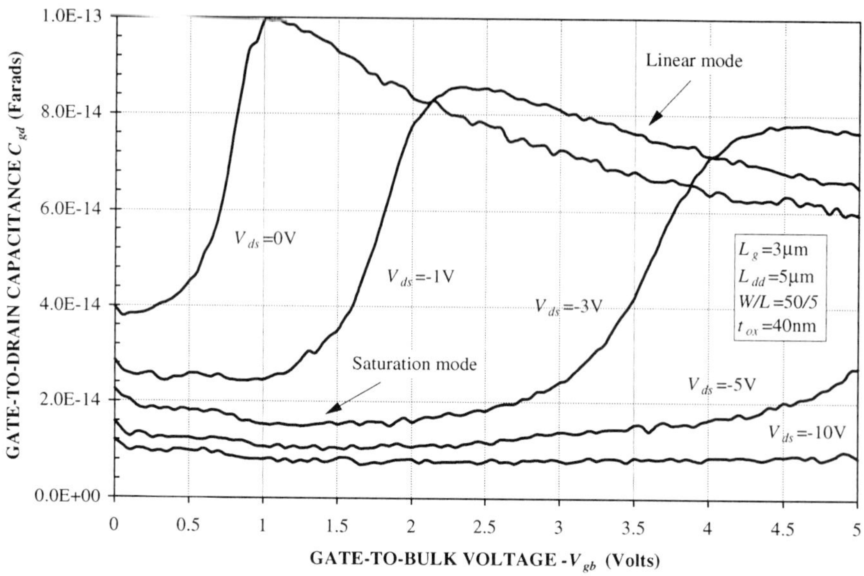

Figure 5.28: Gate-to-drain capacitance of the HVPMOS transistor as a function of the gate voltage and for different V_{ds} values

The gate-to-drain capacitive characteristics shown in Figure 5.28 are identical to the HVNMOS ones except for the parasitic field-plate capacitance C_{fp}. Once again, the lateral diffusion of the lightly doped drain is responsible of such behaviour. In the HVNMOS transistor, the parasitic field-plate capacitance is over the thick oxide layer only, since the etching of the n-WELL layer by the "p-TUB" and V_t adjust implants do not allow a field-plate over thin gate oxide. In the HVPMOS transistor, the lateral diffusion of the lightly doped drain is not etched, which results in an overlap of the field-plate over thin gate oxide. This explains the higher gate-to-drain capacitance value in the saturation mode and under the same drain bias conditions. Note that if the HVPMOS transistor is designed with a field-plate over thin gate oxide, by overlapping the *BLOK* mask and the active region such that $L_{g1}\neq 0$, C_{gd} will increase significantly in the saturation mode.

The drain-to-bulk capacitance as a function of the drain voltage is shown in Figure 5.29. As in the case of the HVNMOS transistor, it will be used to fit the theoretical characteristic implemented for the circuit simulator. It is interesting to note the step of the capacitance value in the reverse mode at V_{db}=-4.5V. It corresponds to the merging of the depletion layers of the *LDD*/n-WELL and n-WELL/substrate junctions, which leads to a decrease of the capacitance due to the sudden widening of the depletion layer thickness.

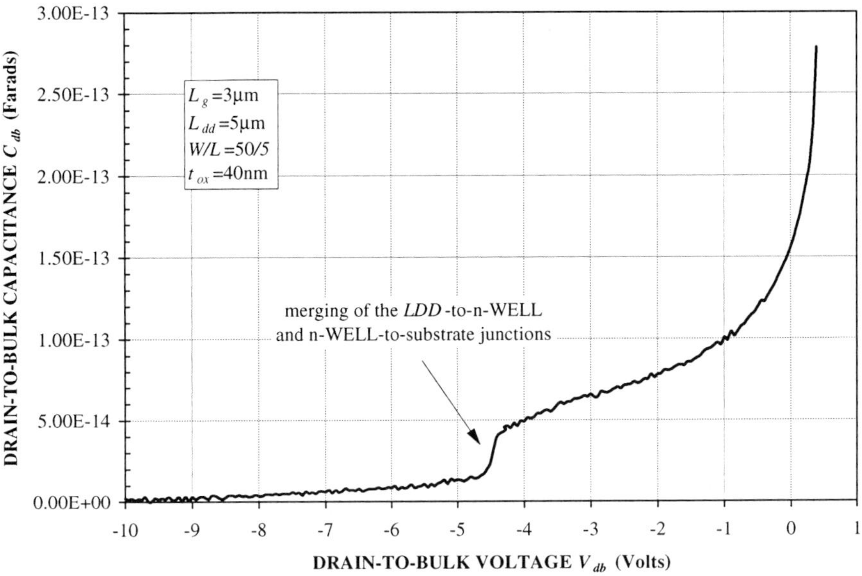

Figure 5.29: Drain-to-bulk capacitance of the HVPMOS transistor as a function of the drain voltage

5.2 Device modelling for circuit simulators

Modelling of the high-voltage transistors for circuit simulation is one of the most important aspects that must be considered for the design of high-voltage integrated circuits. It is obvious that the use of a two dimensional model based on numerical computing of Poisson's equation, is not dedicated for this topic. Although it is the best solution in terms of accuracy, it is time consuming for circuit simulations. Therefore, the use of simulators based on this approach is restricted to the simulation at a device level, as presented in chapter 4 for instance. For VLSI circuit simulation, less accuracy is often accepted against higher simulation speed. Thus, different modelling levels with increasing complexity can be used, depending on the required accuracy. Section 5.2.1 presents two first order models based on the observations made in the simulation and measurement sections. It is followed in section 5.2.2 by a comparison between the simulated and measured results. But first, let us define what is an accurate model of the high-voltage devices for VLSI circuit simulation.

The evolution of MOSFET modelling during the last few years was motivated by the continuous down scaling of the devices, leading to short channel effects. These effects are modelled in SPICE simulators by the use of the pseudo-2D modelling approach of the electric field at the channel end. This model was extended by *C. Hu et al.* [3] to LDD devices in MOSFET submicron technologies. It consists in replacing the expression of the electric field at the drain end of the channel, given by Equation (2.3), by the following one

$$E_m = \frac{V_{ds} - V_{dsat}}{l'} = \frac{V_{ds} - V_{dsat}}{0.22 t_{ox}^{1/3} X_j^{1/2} - L_n'} \tag{5.6}$$

where L_n' is the length of the lightly doped drain layer. Note that in the expression of the electric field given in Equation (5.29) no field-plate effect is considered, since LDD devices in submicron technologies use a lightly doped drain layer only to reduce the electric field magnitude. Thus, an accurate modelling of the SVX devices would first start with a pseudo-2D model of the lightly doped drain region taking into account the effect of the field-plate. The resulting electric field expression is then used to model accurately the channel length modulation, the substrate current generation and the degradation of the mobility due to the lateral electric field. Moreover, the effect of the series resistance of the lightly doped drain must be considered in the expression of the drain current in order to model the behaviour of the static characteristics in the linear and conduction regions. The low-voltage part of the device is modelled using the classic gradual channel approximation. To complete the resulting DC model, the temperature dependence of the drain current must be considered as well. That is, the effect of the temperature on the mobility of the carriers $\mu(T)$, on the threshold voltage $V_t(T)$ and on the carriers saturation velocity $v_{sat}(T)$. Finally, the second step of an accurate modelling approach consists of implementing a dynamic model for small and large signal variations. It is based on the estimation of the charges stored in the device and their variations as a function of the terminal voltages V_g, V_s, V_d and V_b, which results in the expressions of the different terminal capacitances.

The implementation of the modelling approach described above is a lengthy task that would result in a modelling oriented book. An intermediate solution consists of a macro-model formed by a low-voltage transistor, modelling the channel region, connected with other discrete devices, modelling the lightly doped drain. The macro-model of the HVPMOS transistor is shown in Figure 5.30a where the low-voltage transistor is connected in series with a simple resistor. The latter corresponds to the lightly doped drain resistance measured in the linear region. The diode placed between the drain and bulk access is used to model the forward characteristic of the drain junction in switched applications. In the HVNMOS case, an improved macro-model is required, as shown in Figure 5.30b. The series resistor is replaced by two non-linear conductance $G_{ld'}$ and G_{fp}, modelling the influence of the depletion layers created by the field-plate and the drain junction. The current flowing between these layers is a function of V_g and V_d, as in the case of a "JFET" transistor. A detailed analysis of both voltage controlled current sources is discussed in section 5.2.1, where the reason of the difference between the macro-models is pointed out. Since no charge-based capacitance model is available, and for the sake of modelling some important parasitic components, C_{fp} and C_{db} are added in both macro-models. They are used to simulate the behaviour of the field-plate and drain-to-bulk capacitances. Actually, they are not physical capacitors. C_{fp} for instance, is modelled by the SPICE parameter *CGDO* of the low-voltage transistor. It is set to the value of the gate-to-drain capacitance of the high-voltage devices measured in the saturation mode. Consequently, a rough modelling of C_{gd} is possible through the model of the low-voltage transistors by tuning the SPICE parameters. The drain-to-bulk capacitance is modelled more accurately with the junction capacitance of diode D_{ldd}. The SPICE parameters of the diode are tuned to fit the measured characteristic of C_{db}. Note that the step in the characteristic of the HVPMOS transistor cannot be reproduced by the use of this modelling approach. Finally, the channel length of the low-voltage transistor is 1μm shorter than the high-voltage transistor one. It models the channel shortening of the devices due to the lateral diffusion of the lightly doped drains, which

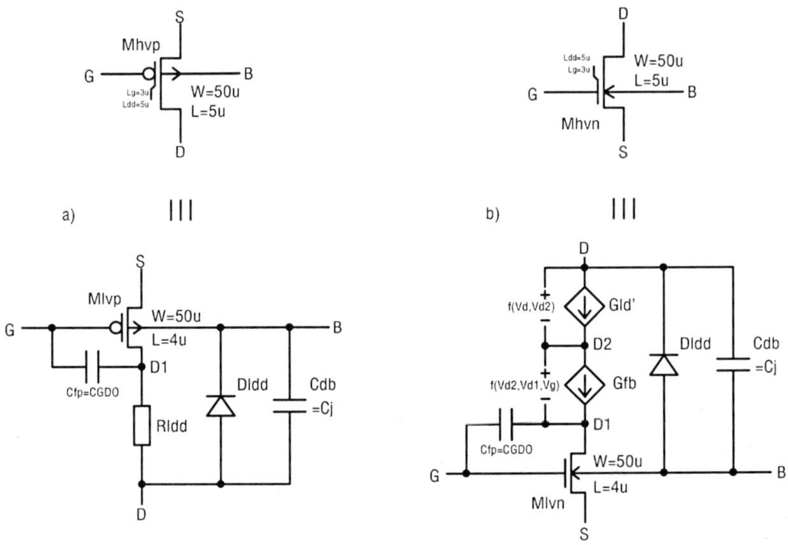

Figure 5.30: First order models used to simulate the high-voltage devices

corresponds to the effective channel length L_{eff} used in most of the SPICE models. The channel length modulation Δl occurring in the saturation mode is modelled by tuning the corresponding SPICE level parameter, *KAPPA* in the case of SPICE level 3 for instance. The fitting of the different SPICE parameters of the devices used in the macro-model are detailed in section 5.2.2.

5.2.1 Theoretical considerations of the macro-models

In the simulations performed at the device level in chapter 4, it was shown that the current of the HVNMOS transistor flows in the lightly doped drain between the depletion layer created by the field-plate and the one created by the drain junction. Since the width of these layers is a square root function of the drain and gate voltages, the current is non linear as a function of the terminal voltages. Thus, the section of the lightly doped drain covered by the field-plate behaves similarly to a

JFET transistor, except that the channel is controlled by two different gate electrodes. In the section located between the

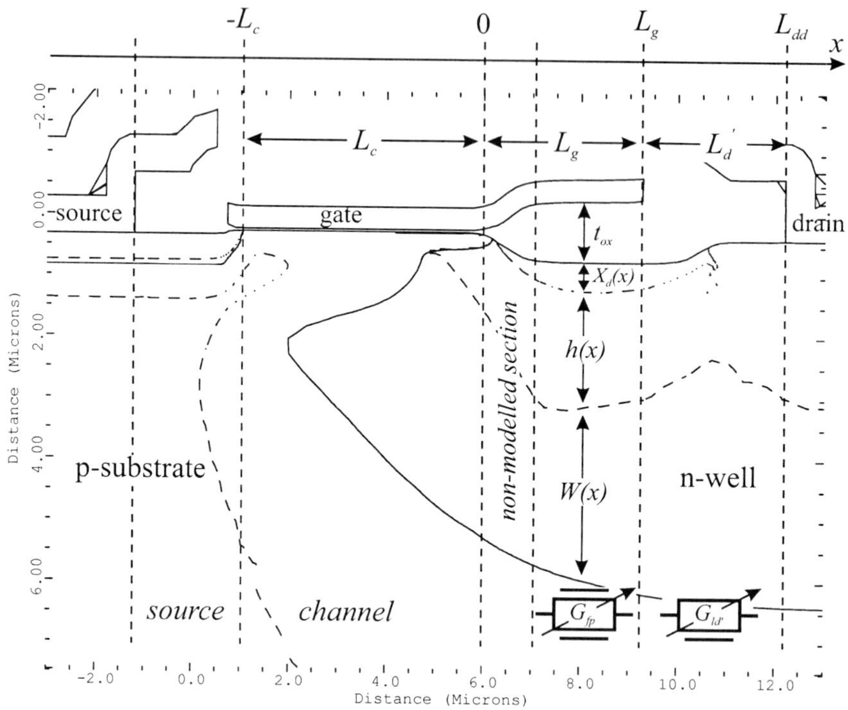

Figure 5.31: Cross section of the high-voltage transistor showing the different sections of the lightly doped drain (L_c=5µm, L_g=3µm, L_{dd}=5µm, V_g=5V, V_d=40V and V_{sb}=0V)

field-plate edge and the drain contact, the current is controlled by the depletion layer created by the drain junction only, leading to a non linear expression of the current as well. Actually, the current in both sections is forced by the low-voltage part of the transistor, and nodes D1 and D2 adapt their voltages according to the non linear relationships. The shape of these sections appears in the cross section of the HVNMOS transistor shown in Figure 5.31.

The current flowing in both sections is calculated using the same approach as for the JFET transistor [5], and lateral DMOSFETs [6],[7]. Thus, considering a step approximation of the n-WELL doping profile, the current flowing at a position x between the depletion layers of the section covered by the field-plate, is given by

$$I(x) = v(x) \cdot Q(x) \cdot W_{hv} = v(x) \cdot q \cdot N_D \cdot h(x) \cdot W_{hv} \quad (5.7)$$

where v is the velocity of the carriers, N_D the doping concentration level of the n-WELL layer, $h(x)$ the height of the current flow path at position x and W_{hv} the width of the device. The height $h(x)$ can be expressed by

$$h(x) = X_j - W(x) - X_d(x) \tag{5.8}$$

where $W(x)$ and $X_d(x)$ are the width of the depletion layers created by the drain junction and the field-plate respectively. Replacing $W(x)$ and $X_d(x)$ by their expressions, given by Equations **4.5** and **4.6**, $h(x)$ becomes

$$h(x) = \left(X_j + \frac{\varepsilon_s}{\varepsilon_{ox}} t_{ox} \right) - R_b \sqrt{V(x)} - R_{fp} \sqrt{A + V(x) - V_g'} \tag{5.9}$$

where all the voltages are defined with respect to the bulk, and

$$R_b = \sqrt{\frac{2\varepsilon_s N_A}{qN_D \left(N_D + N_A \right)}} \tag{5.10}$$

$$R_{fp} = \sqrt{\frac{2\varepsilon_s}{qN_D}} \tag{5.11}$$

$$A = \frac{q\varepsilon_s N_D}{2} \left(\frac{t_{ox}}{\varepsilon_{ox}} \right)^2 \tag{5.12}$$

Considering that the velocity-field relationship for electrons and holes is given by

$$v(x) = \frac{v_{sat} E_x}{E_{sat} + E_x} \tag{5.13}$$

where $v_{sat} = \mu E_{sat}$ is the velocity saturation of the carriers, and that the electric field in the current flow direction is $E_x = dV/dx$, the resulting current expression can be written as

$$I(x) = qN_D v_{sat} W_{hv} \frac{dV}{E_{sat} dx + dV} \cdot \left[\left(X_j + \frac{\varepsilon_s}{\varepsilon_{ox}} t_{ox} \right) - R_b \sqrt{V(x)} - R_{fp} \sqrt{A + V(x) - V_g^{'}} \right] \tag{5.14}$$

Considering that $I_{sat}=qN_D v_{sat} W_{hv}$ is the maximum saturation current per unit area, and integrating Equation (5.14) over the section length, assuming that $V(x=0) = V_{d1}$ and $V(x=L_g) = V_{d2}$, the current I_{fp} in the section covered by the field-plate can be expressed as

$$I_{fp} = \frac{I_{sat}}{E_{sat} L_g + V_{d2} - V_{d1}} \left[\left(X_j + \frac{\varepsilon_s}{\varepsilon_{ox}} t_{ox} \right) \left(V_{d2} - V_{d1} \right) - \frac{2}{3} R_b \left(V_{d2}^{3/2} - V_{d1}^{3/2} \right) - \frac{2}{3} R_{fp} \left[\left(A + V_{d2} - V_g^{'} \right)^{3/2} - \left(A + V_{d1} - V_g^{'} \right)^{3/2} \right] \right] \tag{5.15}$$

The current $I_{ld'}$ in the section between the field-plate edge and the drain contact can be computed using the same approach as for I_{fp}. The boundaries conditions are $V(x=L_g) = V_{d2}$ and $V(x=L_{dd}) = V_d$. Thus, the current in this section can be expressed as

$$I_{ld'} = \frac{I_{sat}}{E_{sat} L_d^{'} + V_d - V_{d2}} \left[X_j \left(V_d - V_{d2} \right) - \frac{2}{3} R_b \left(V_d^{3/2} - V_{d2}^{3/2} \right) \right] \tag{5.16}$$

I_{fp} and $I_{ld'}$ were carried out considering that the mobility is field dependent according to Equation (5.13). If they were computed considering that the mobility is field independent, i.e. $v=\mu E_x$, the validity domain of the currents in both sections would be given by the "pinch-off" of the current flow path. In fact, the width of the depletion layer created by the field-plate is maximum at the edge of the field-plate ($x=L_g$), and as far as $h(V_{d2})$ is positive, I_{fp} and $I_{ld'}$ are valid. When h=0 at $x=L_g$, the "pinch-off" point is reached beyond which I_{fp} becomes saturated, assuming V_{d1} and $V_g^{'}$ are fixed, and $I_{ld'}$ continues to increase as a function of the drain voltage V_d. Considering a step approximation of the n-WELL doping profile, the height h is shown in Figure 5.32 as a function of the voltage V_{d2}. It appears clearly that the depletion layers don't merge before the voltage at node D2 has reached 80V, assuming V_g'=5V. Actually, the simulation performed in Figure 4.23 shows that the merging occurs at about 45V. Such difference can be cancelled by adjusting the doping concentration level of the n-WELL layer from $3.3x10^{15}$ cm^{-3} to $2.5x10^{15}$cm^{-3}.

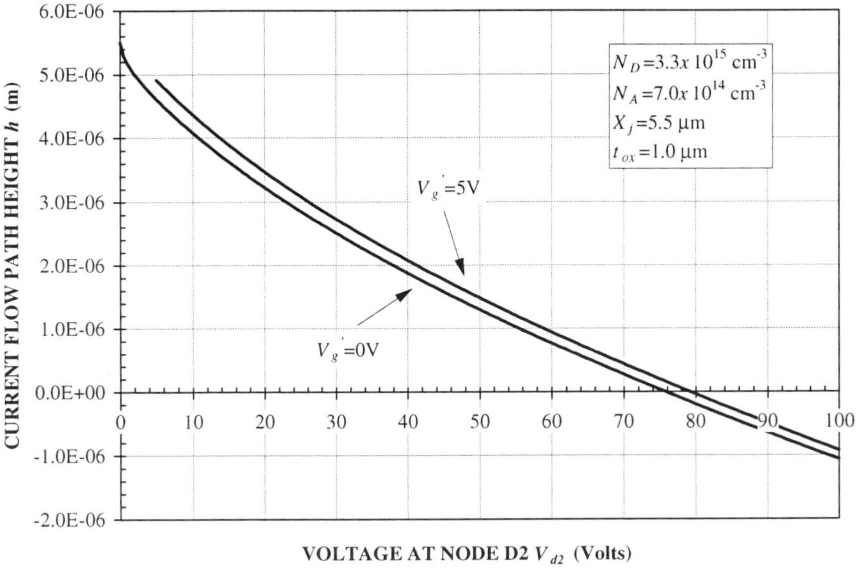

Figure 5.32: Current flow path height as a function of the voltage at the edge of the field-plate. (HVNMOS)

The field independent mobility approximation discussed above does not accurately model the behaviour of the currents in the lightly doped drain, due to the high concentration of the electric field in this region. In fact, the actual dependence of the velocity on the electric field is given by Equation (5.13), where the onset of velocity saturation occurs at $E_x=E_{sat}$. Consequently, the saturation does not correspond necessarily to the "pinching" of the current flow path at $x=L_g$, but to a finite height h_{sat}. The current I_{fp} is plotted in Figure 5.33 as a function of the voltage V_{d2}, for V_{d1}=0V and for different values of V_g', according to Equation (5.15). It increases initially with increasing V_{d2}, reaches a peak value and then begins to decrease with further increase of V_{d2}. Actually, the current must saturate once the peak value is reached. The reason why Equation (5.15) does not predict the saturation is that this equation becomes invalid once the peak is reached. In fact, the peak value is reached for $V_{d2}=V_{d2sat}$ corresponding to the saturation voltage at which $h=h_{sat}$, while in Equation (5.15), $h(V_{d2})$ continues to decrease and becomes even negative for V_{d2} higher than 80V, assuming V_g'=5V.

Therefore, when the saturation condition is detected V_{d2} in Equations (5.15) and (5.16) becomes V_{d2sat}. Discontinuities at the transition point from linear to saturation region are avoided by ensuring that the first derivative of the current is continuous, or using the following empirical function

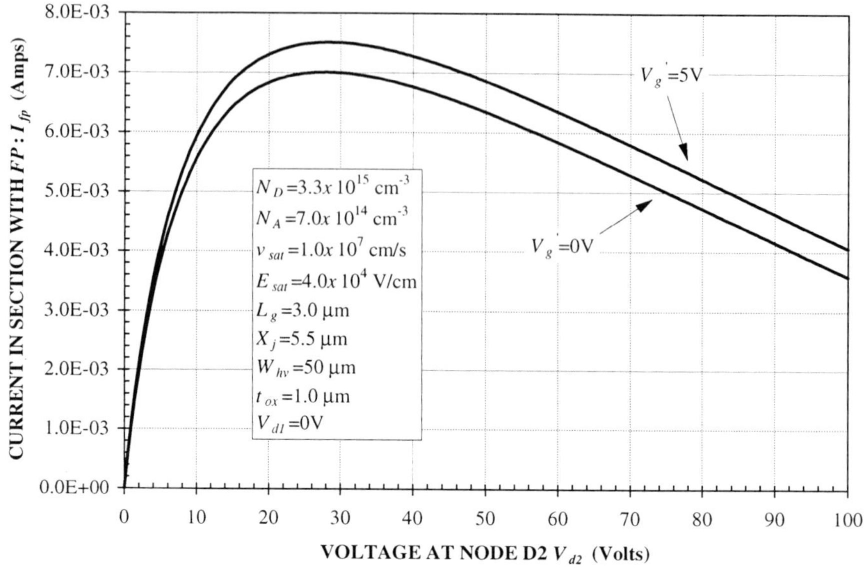

Figure 5.33: Current I_{fp} in the section covered by the field-plate (FP) as a function of the voltage V_{d2}. (HVNMOS)

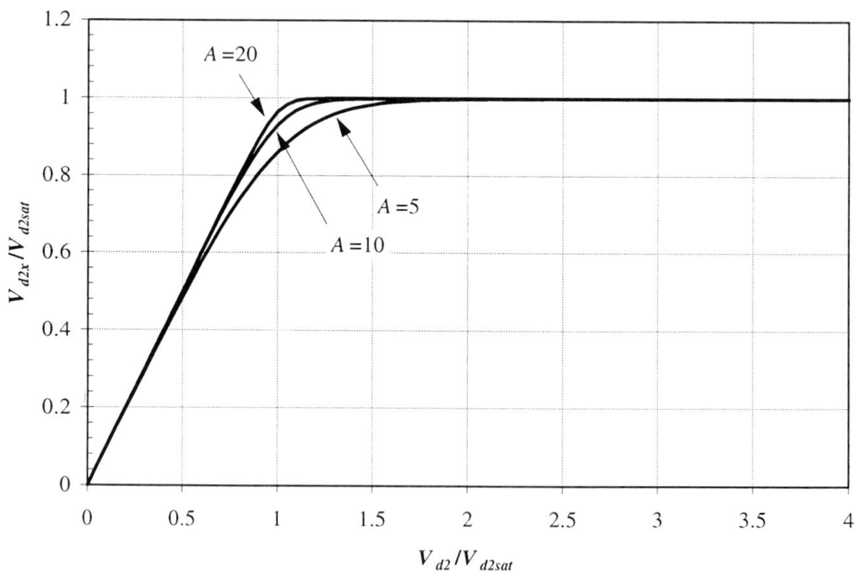

Figure 5.34: Smoothing function for the transition point from linear to saturation region. (HVNMOS)

$$V_{d2x} = V_{d2sat}\left\{1 - \frac{1}{\ln\left(1+e^{A}\right)}\left[1+e^{A\left(1-V_{d2}/V_{d2sat}\right)}\right]\right\} \quad (5.17)$$

where A is a parameter controlling the sharpness of the transition. Large values of A yields steep transitions while small values result in smooth transitions. The value of A=10 has been found to be a good choice [4]. Figure 5.34 shows the ratio of V_{d2x}/V_{d2sat} as a function of V_{d2}/V_{d2sat} and for different values of A.

The use of V_{d2x} in Equations (5.15) and (5.16) does not ensure smooth current only, but output conductance as well. It avoids the computing of the first and second derivatives of I_{fp} and their continuity at $V_{d2}=V_{d2sat}$. It also avoids the use of different current expressions for linear and saturation regions by reducing them to the following ones

$$I_{fp} = \frac{I_{sat}}{E_{sat}L_g + V_{d2x} - V_{d1}}\left[\left(X_j + \frac{\varepsilon_s}{\varepsilon_{ox}}t_{ox}\right)\left(V_{d2x} - V_{d1}\right) - \frac{2}{3}R_b\left(V_{d2x}^{3/2} - V_{d1}^{3/2}\right) - \frac{2}{3}R_{fp}\left[\left(A + V_{d2x} - V_g^{'}\right)^{3/2} - \left(A + V_{d1} - V_g^{'}\right)^{3/2}\right]\right] \quad \mathbf{(5.18)}$$

and

$$I_{ld'} = \frac{I_{sat}}{E_{sat}L_d^{'} + V_d - V_{d2x}}\left[X_j\left(V_d - V_{d2x}\right) - \frac{2}{3}R_b\left(V_d^{3/2} - V_{d2x}^{3/2}\right)\right] \quad \mathbf{(5.19)}$$

The saturation voltage V_{d2sat} is computed by solving numerically the first derivative of I_{fp} to zero, where V_{d2} is declared as a state variable. The resulting I_{fp} curve is shown in Figure 5.35 according to Equations (5.17) and (5.18). $I_{ld'}$ is not shown here as it increases gradually as function of V_d.

In the previous expressions the effect of a finite output conductance was not considered. The physical meaning of a finite output conductance can be understood as follows. In the field dependent mobility discussion, it was assumed that the saturation is reached when $E_x=E_{sat}$ leading to a finite height $h=h_{sat}$ at $x=L_g$. The corresponding voltage at node D2 is V_{d2sat} beyond which I_{fp} remains constant for further increase in V_{d2}. Actually, as V_{d2} is increased beyond V_{d2sat}, the position at

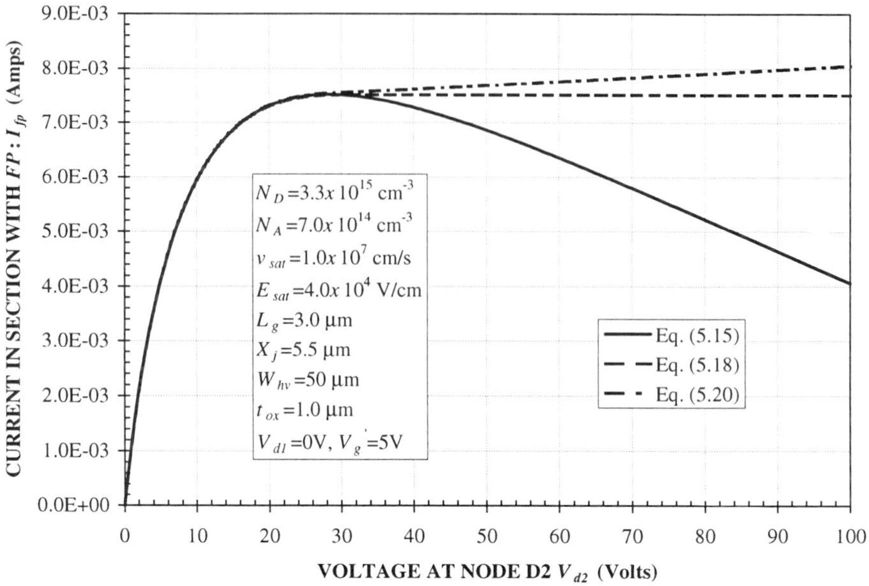

Figure 5.35: Current I_{fp} considering the linear and saturation modes. The effect of a finite output conductance is also shown according to (5.20). (HVNMOS)

which $E_x=E_{sat}$, moves from $x=L_g$ towards the channel end as the electric field distribution demands. This is shown schematically in Figure 5.36 for different drain bias conditions. Since the potential distribution as a function of the position decreases faster than the electric field one, the voltage at the new position $x=x_1$ for which $E_x=E_{sat}$ is lower than the previous case, resulting in a higher saturation height $h(x=x_1)=h_{1sat}$. Consequently, more current is injected into the velocity saturation region, according to Equation (5.7), leading to a finite output conductance as a function of V_{d2}.

Since no accurate modelling of the electric field distribution in the lightly doped drain is available yet, the output conductance is modelled by introducing an additional parameter λ_{hv} simulating the increase of I_{fp} as a function of V_{d2}. λ_{hv} will be used as a fitting parameter between the simulated and measured characteristics. Thus, Equation (5.18) can be rewritten as

$$I_{fp} = \frac{I_{sat}\left(1+\lambda_{hv}V_{gds}\right)}{E_{sat}L_g + V_{d2x} - V_{d1}}\left[\left(X_j + \frac{\varepsilon_s}{\varepsilon_{ox}}t_{ox}\right)\left(V_{d2x}-V_{d1}\right)\right.$$
$$\left. - \frac{2}{3}R_b\left(V_{d2x}^{3/2} - V_{d1}^{3/2}\right) - \frac{2}{3}R_{fp}\left[\left(A + V_{d2x} - V_g^{'}\right)^{3/2} - \left(A + V_{d1} - V_g^{'}\right)^{3/2}\right]\right]$$

(5.20)

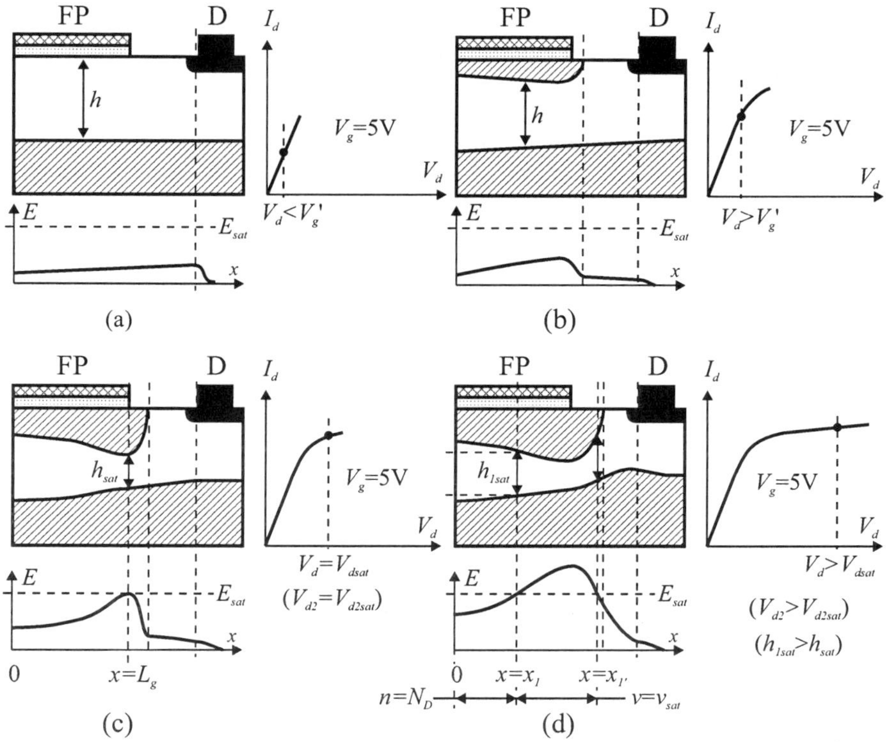

Figure 5.36: Depletion layer shapes and electric field profiles under various drain bias conditions

where

$$V_{gds} = 0.5 \cdot \left[\left(V_{d2} - V_{d2sat}\right) + \sqrt{\left(V_{d2} - V_{d2sat}\right)^2 + K^2} \right] \tag{5.21}$$

where K is a parameter controlling the sharpness of the transition between $V_{gds}=0$ and $V_{gds}=V_{d2}$ at $V_{d2}=V_{d2sat}$. Large values of K yields smooth transitions while small values result in steep transitions.

The current expressions developed for the HVNMOS transistor apply to the HVPMOS transistor as well. Equations (5.19) and (5.20) remain unchanged if the terminal voltages are defined with respect to the bulk, i.e. $V_{by}=V_b-V_y$. Since positive currents are defined as flowing from drain to source, the sign of the currents must be changed. Finally, a step approximation of the doping profile of the lightly doped drain must be performed here as well. Hence, the resulting current flow path height h is shown in Figure 5.37 as a function of the voltage at node D2 and for different V_{bg}' values. Given the high doping concentration level of the lightly doped drain layer, the variation of h is about 20% only of the total layer depth, for a voltage range of

100V. Therefore, the variations of the lightly doped drain resistance are expected to be negligible, resulting in a simple linear resistor in series with the low-voltage transistor.

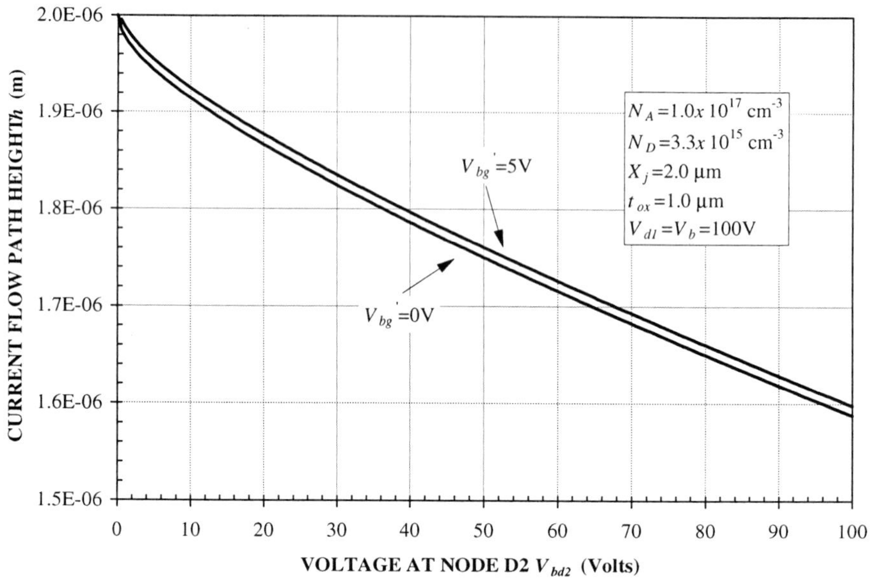

Figure 5.37: Current flow path height as a function of the voltage V_{d2} at node D2. (HVPMOS)

The previous observation can be confirmed by plotting the current I_{fp} according to Equation (5.20) and considering the actual current operating range, as shown in Figure 5.38. Note that the effect of the gate voltage on the current characteristic can be neglected. When the voltage at node D1 is fixed to the bulk, and V_{bd2} is swept from 0..100V, the current increases non linearly from 0..110mA. Since the actual current is forced by the low-voltage transistor, and given the low current range value (up to 1mA for V_{bg}=5V), the expression of I_{fp} can be linearised as a function of the voltage difference V_{bd2}-V_{bd1}. The linear approximation remains valid for current values as high as 20mA. It is based on the assumption that the variation of h is negligible for high doping concentration levels of the lightly doped drain. Thus, considering that the mobility is field independent, i.e. $v=\mu E_x$, the current can be written as

$$I(x) = v(x) \cdot q \cdot N_A \cdot h \cdot W_{hv} = q \cdot N_A \cdot X_j \cdot W_{hv} \cdot \mu \cdot E_x \qquad (5.22)$$

where $v_{sat}=\mu E_{sat}$ and $E_x=dV/dx$. Integrating Equation (5.22) over the field-plate length, where the boundaries conditions are $V(x=0)=V_{bd1}$ and $V(x=L_g)=V_{bd2}$, yields the expression of the linearised current I_{fp}

$$I_{fp} = G_{fp}\left(V_{bd2} - V_{bd1}\right) \tag{5.23}$$

where

$$G_{fp} = \frac{qN_A v_{sat} X_j W_{hv}}{E_{sat} L_g} \tag{5.24}$$

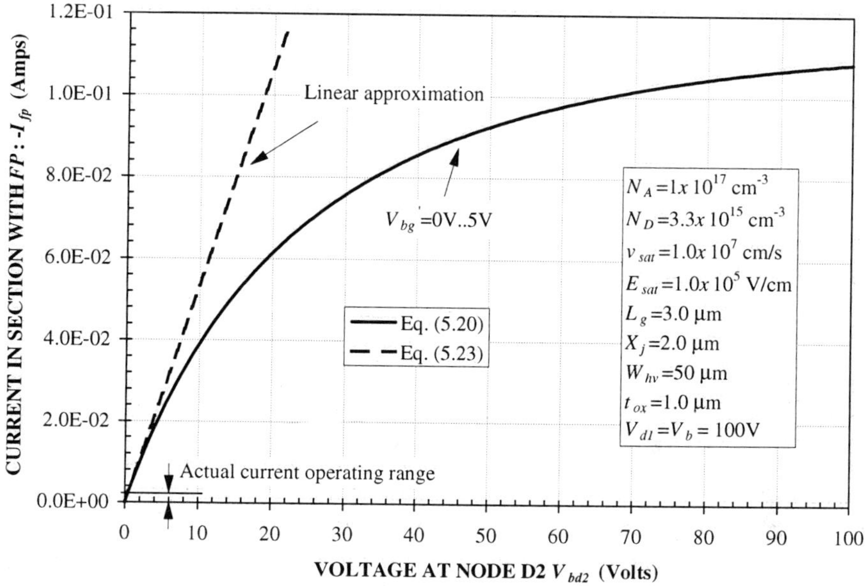

Figure 5.38: Current I_{fp} in the section covered by the field-plate (FP) as a function of the voltage V_{d2}. (HVPMOS)

The linear approximation of I_{fp} applies to $I_{ld'}$ as well, since the current range resulting from Equation (5.19) is almost identical, assuming the same voltage conditions. Hence, the linearised form of $I_{ld'}$ can be expressed as

$$I_{ld'} = G_{ld'}\left(V_{bd} - V_{bd2}\right) \tag{5.25}$$

where

$$G_{ld'} = \frac{qN_A v_{sat} X_j W_{hv}}{E_{sat} L_d'} \tag{5.26}$$

It is obvious that since the same current is flowing into the linear conductances G_{fp} and $G_{ld'}$, Equations (5.23) and (5.25) can reduce to a single current expression given by

$$I_{ldd} = G_{ldd}\left(V_{bd} - V_{bd1}\right) \tag{5.27}$$

where

$$G_{ldd} = \frac{1}{R_{ldd}} = \frac{qN_A v_{sat} X_j W_{hv}}{E_{sat} L_{dd}} \tag{5.28}$$

Thus, the lightly doped drain of the HVPMOS transistor is modelled with a simple linear resistor in series with the low-voltage part. The value of this resistor is given by Equation (5.28) representing the analytical form of the measured resistance in the linear mode. Actually, R_{ldd} is used as a fitting parameter since it depends on the doping concentration level N_A resulting from a step approximation of the actual doping profile.

It is interesting to consider at this level the physical meaning of the resulting model. According to Figure 5.38, and given the actual current operating range, the voltage drop across R_{ldd} is of few volts. Indeed, considering the typical measured value of 600Ω, and assuming a drain current of 1mA for V_{bg}=5V, the voltage drop V_{bd}-V_{bd1} is 0.6V. This means that the voltage at the channel end of the low-voltage part ranges almost between the high-voltage power supply and ground, which is obviously not acceptable. Actually, the macro-model formed by a simple linear resistor in series with a low-voltage transistor, is valid as far as the lightly doped drain is not fully depleted. But, in Figure 4.29 for instance, it appears clearly that the depletion layers created by the field-plate and the junction merge along a portion of the lightly doped drain. This is due to the thinning of the junction depth and the field oxide thickness at the "bird's beak". Physically, as soon as the merging occurs, most of the potential and electric field lines are confined in this region supporting, thus, the highest voltage drop. Nevertheless, the resistor model remains valid in the undepleted part of the lightly doped drain. Note that the "bird's beak" region is not modelled in the HVNMOS transistor as well, due to the lack of an accurate modelling of the electric field distribution. However, the resulting macro-models discussed in this section allow an acceptable modelling of the main characteristics of the devices, as pointed out in the simulations versus measurements section.

5.2.2 Simulations versus measurements

The implementation of the theoretical considerations developed in the previous section were performed in a SPICE like simulator [8], allowing a C based analog behavioral modelling. The reader interested in the code of both macro-models can refer to *appendix A* for practical details. In this section, the simulations issued from the macro-models are compared to the measured characteristics. The parameters used to fit the data are discussed as well. It must be recalled that since the book is not only a modelling oriented one, the fitting is performed manually without using any optimisation program for parameter extraction. The scope is to point out the influence of the different parameters on the behaviour of the characteristics. Therefore, the macro-model parameters are split into two different sets. The low-voltage one, which is basically the set of the chosen SPICE level, and the high-voltage set, which is given by Equations (5.19), (5.20) and (5.27).

The parameters of the low-voltage part are those of the SPICE level 3, which was chosen for some practical considerations. A better choice would have been a continuous model such as the EKV [9] or the late BSIM version [10], but unfortunately the extracted parameters are not available yet for the used technology. Moreover, some numerical computing problems arose in the EKV model due to the high drain voltage range. Level 3 is, however, a good starting point. The critical parameters that must be tuned to fit the static characteristics are *THETA*, *KAPPA* and *VTO*. The remaining parameters are kept unchanged. *KAPPA* basically controls the output conductance of the high-voltage device, since the channel length modulation is different than the low-voltage case. In the HVNMOS transistor, it is used in conjunction with the high-voltage parameter λ_{hv} of Equation (5.20), which controls the output conductance for high drain voltages. *THETA* is the mobility degradation factor. It controls the absolute value of the current in the I_d-V_d characteristics and the transconductance behaviour. Actually, *THETA*, *KAPPA* and the channel length of the low-voltage transistor, help to cope with the lack of an accurate model taking into account the channel length modulation, the effective channel length due to the lateral diffusion of the lightly doped drain, and the non-modelled section of the "birds beak". Finally, *VTO* is the last parameter of the low voltage set that adjusts the threshold voltage of the HVNMOS transistor only.

In the case of the HVNMOS transistor, the parameters of the high-voltage part are the critical field for velocity saturation E_{sat}, the doping concentration level of the lightly doped drain layer N_D, the length of the field-plate section L_g, the length of the section between the field-plate and the drain contact L_d' and the finite output conductance parameter λ_{hv}. In the case of the HVPMOS transistor, R_{ldd} is the unique fitting parameter. E_{sat}, N_D and R_{ldd} control the I_d-V_d characteristics in the linear and conduction modes, while L_g controls the conduction mode only and L_d' the linear mode. It should be mentioned that the fitting of all the characteristics of the HVNMOS transistor is, generally, more complex to perform due to its higher number of parameters. Starting with the I_d-V_d characteristics, the best manual fitting obtained is shown for both devices in Figure 5.39 and Figure 5.40.

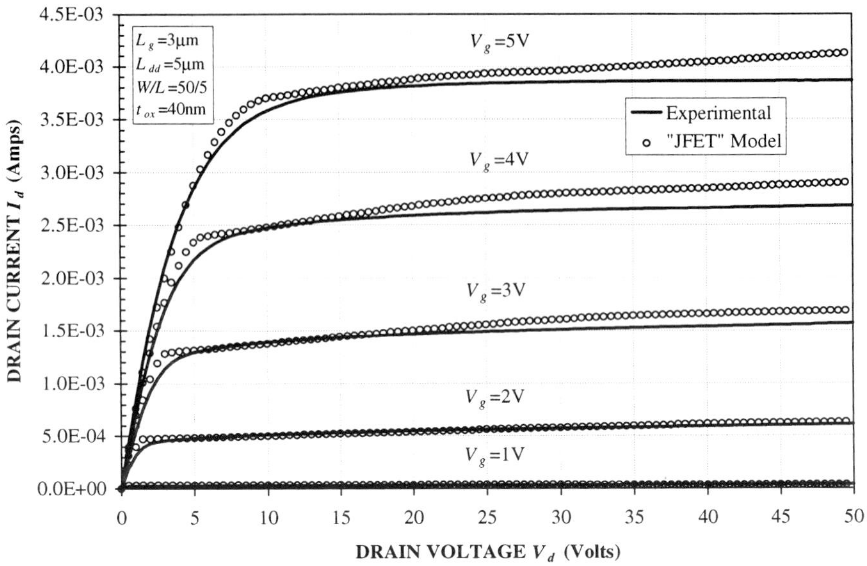

Figure 5.39: Comparison of the experimental I_d-V_d characteristics of the HVNMOS transistor and the simulation data

In the case of the HVNMOS transistor, the discrepancies between the measured and simulated characteristics in the saturation mode are due to self heating effects. Moreover, the slight change of the output conductance between low and high drain voltage values is due to the use of two different output conductance parameters, *KAPPA* and λ_{hv}. This is one of the limiting effects of the macro-modelling approach that should be eliminated in an enhanced model by reducing both parameters to a unique one. In the linear and conduction modes, the discontinuities at the saturation voltage V_{dsat} are related to the SPICE level used. The slight differences, however, are caused by the non-modelled "bird's beak" section of the field-plate.

In the case of the HVPMOS transistor, the comparison shown in Figure 5.40 confirms the theoretical assumption considering that a simple linear resistor in series with a low-voltage transistor models correctly the high-voltage device. As in the case of the HVNMOS transistor the discontinuity at the saturation voltage and the slight differences are related to the SPICE level used and the "bird's beak section. Note that the output conductance fits correctly the measured one in the absence of self heating effects. Moreover, the output conductance is continues between low and high drain voltage values, since *KAPPA* is the only fitting parameter used in the case of the HVPMOS transistor.

As was mentioned previously, only the main characteristics are compared in this section, i.e. those who are considered correctly in the macro-models and SPICE level 3. The substrate current and output conductance are typical opposite examples. It must be recalled that SPICE level 3 is generally used for digital simulations, since its accuracy for analog simulations is very poor. Therefore, in the following paragraphs,

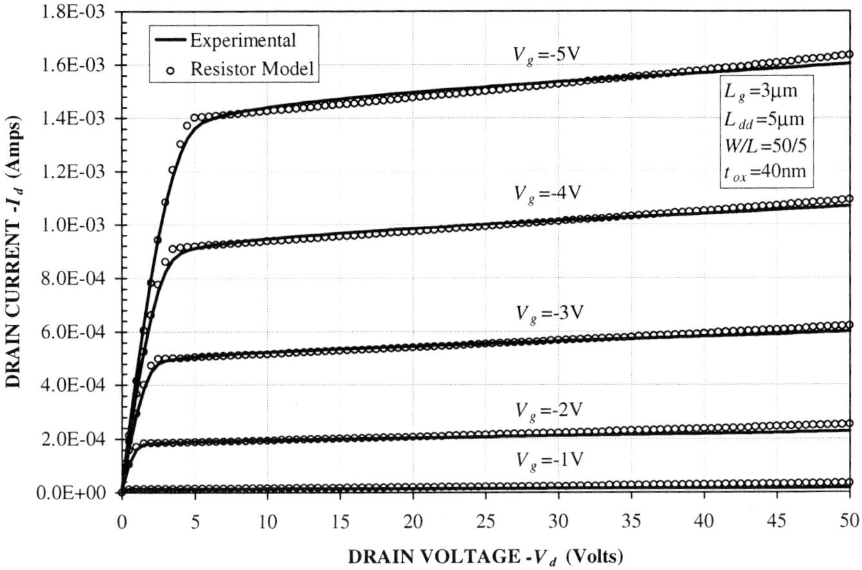

Figure 5.40: Comparison of the experimental I_d-V_d characteristics of the HVPMOS transistor and the simulation data

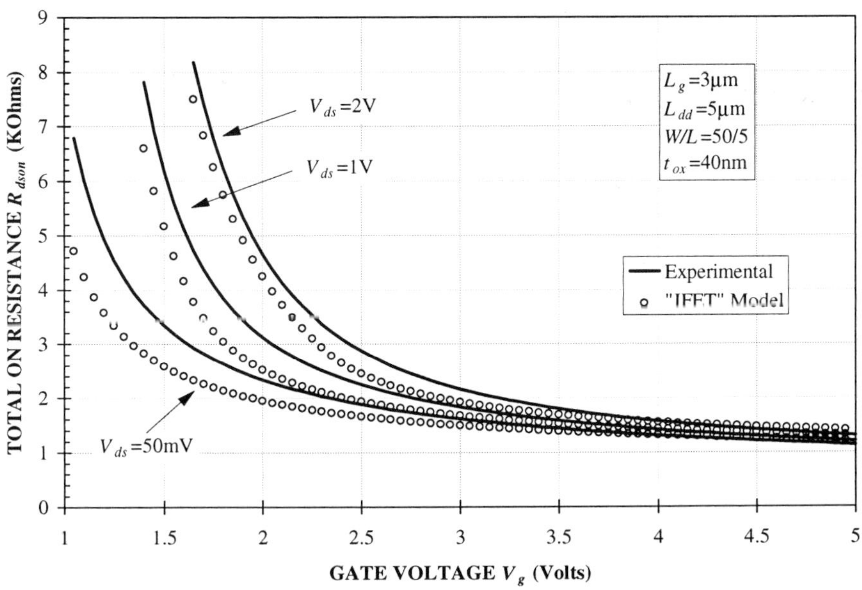

Figure 5.41: Comparison of the measured total on state resistance and the simulation data in the case of the HVNMOS transistor

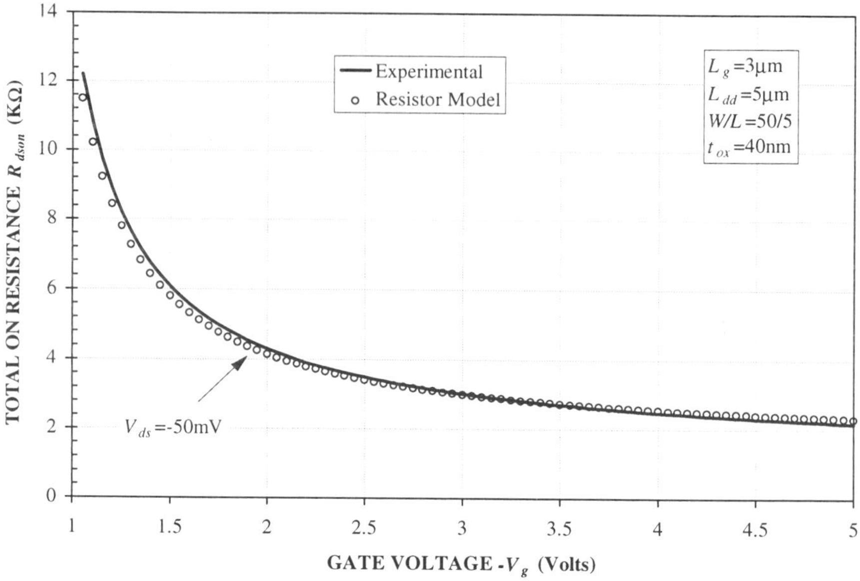

Figure 5.42: Comparison of the measured total on state resistance and the simulation data in the case of the HVPMOS transistor

the total on state resistance and the forward current of the drain junction are the only characteristics presented for the static part. Thus, in Figure 5.41 and Figure 5.42, R_{dson} is plotted as a function of the gate voltage for both devices. In the HVNMOS case, and for V_{ds}=1V or V_{ds}=2V, the model is in general agreement with the measured data in the linear mode. In the conduction mode, the simulated resistance is lower than the measured one, as already observed in the I_d-V_d characteristics. For V_{ds}=50mV, although the transistor is in the linear mode for V_g higher than V_t, the difference is caused by fitting constraints. In fact, in order to cope with the lack of an accurate model in the "bird's beak" section, influencing the characteristics particularly at high V_g values, the total on state resistance was under estimated in the linear mode and for low V_g values.

In the case of the HVPMOS transistor, the agreement between simulation and measurements is better, at $|V_{gs}|$=50mV, than in the HVNMOS case. However, the discrepancy in the conduction mode at higher V_{ds} values remains true for the same reasons.

The last interesting comparison in the static characteristics, is the forward current of the drain-to-bulk junction. It should be noted that the latter is modelled by a discrete diode and not by the extrinsic SPICE parameters, in order to avoid the change of the characteristics of the source junction. The comparison is shown in Figure 5.43 for the HVNMOS case. Due to the high series resistance, the simulated current is higher than the measured one, when the diode is in a strong forward conduction. The main reason is that the series resistance is under-estimated. In fact, since the voltage drop across the series resistance is not negligible for low current

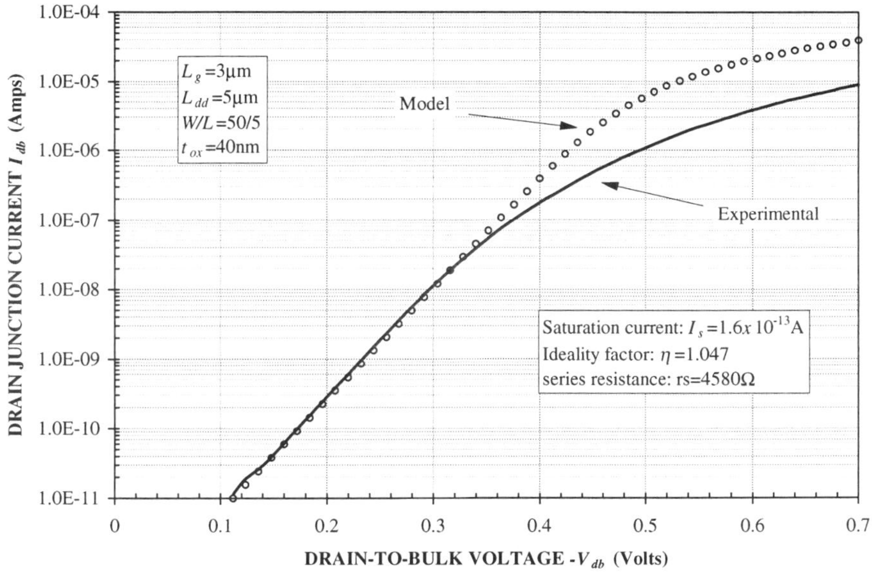

Figure 5.43: Comparison of the measured forward current of the drain junction diode and the simulation data in the case of the HVNMOS transistor

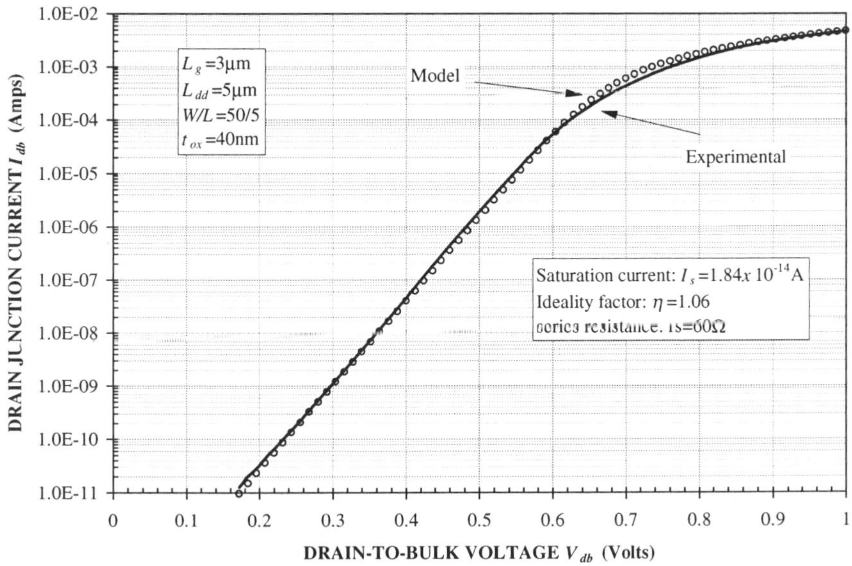

Figure 5.44: Comparison of the measured forward current of the drain junction diode and the simulation data in the case of the HVPMOS transistor

values, the linear regression used to extract r_s should have been performed down to small forward currents (about 0.5μA instead of 10μA). But if such extraction method was applied, the slope of the linear regression will not match the data anymore at high forward currents. The fact is that the linear regression method is inaccurate for such high resistance values, as it fits the parameter over a portion of the curve. The use of a non-linear least-square optimisation method would be more appropriate, as it fits the parameters over all portion of the curve in the current range of interest. However, the fitting shown in Figure 5.43 is a good trade off, since the slopes of the model are equal to those of the measured data at low and high forward currents. In the case of the HVPMOS transistor, the linear regression method yields a better fitting of the model parameter, given the lower series resistance value (60Ω against 4580Ω). The comparison of the measured forward current of the junction and the simulation data is shown in Figure 5.44.

In the following paragraphs some of the measured and simulated dynamic characteristics are discussed. In fact, due to the lack of an accurate estimation of the charges in the high-voltage devices, the capacitive behaviour of the model is discussed only for C_{gd} and C_{db}. A very simple first order model was developed for these capacitors because they play a key role in the design of high-voltage circuits. But first, the transconductance g_m is shown in Figure 5.45 and Figure 5.46 for both devices. In the case of the HVNMOS transistor, the discrepancies are caused by practical fitting considerations, given the number of parameters that must be tuned. It is obvious at this level that an optimisation program for parameters extraction should be used to fit more precisely the static and dynamic characteristics. In the case of the HVPMOS transistor, the fitting is usually more accurate resulting in a better agreement of the simulation data and the measured characteristics.

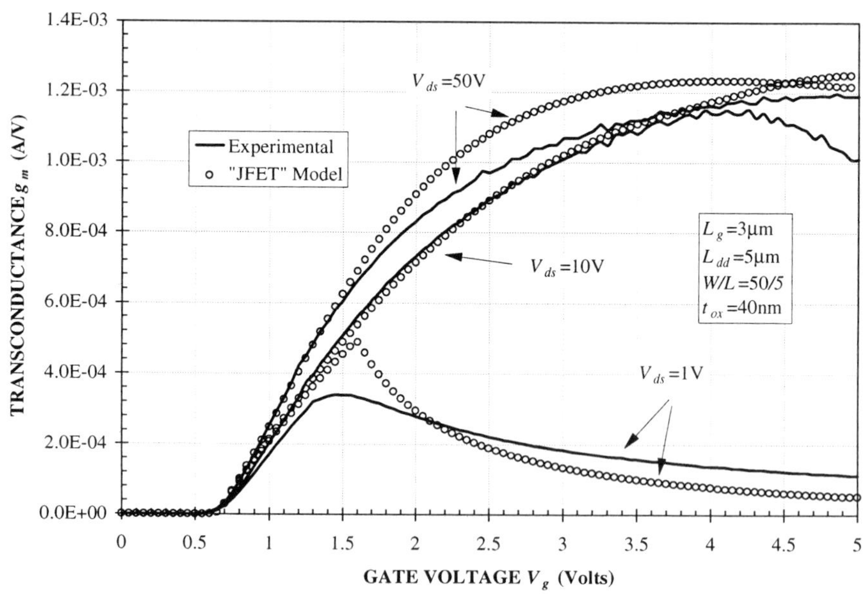

Figure 5.45: Comparison of the measured transconductance g_m and the simulation data in the case of the HVNMOS transistor

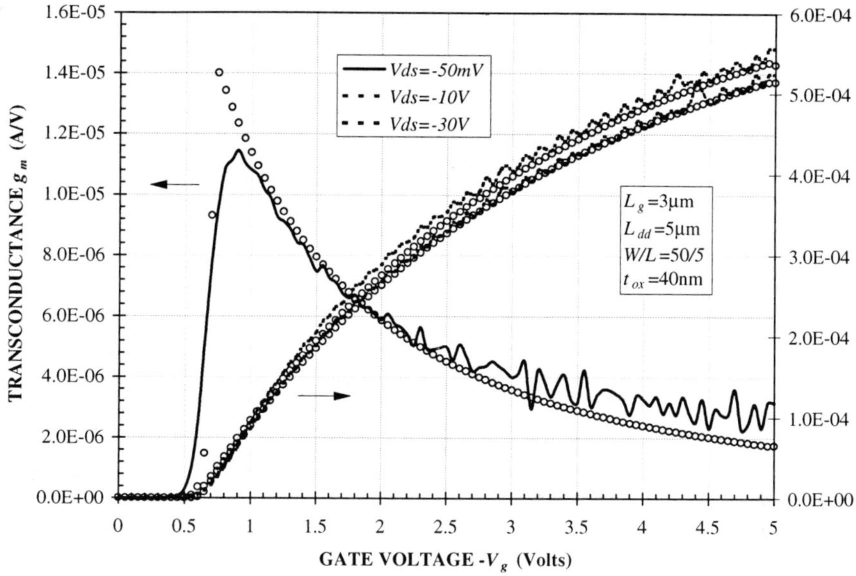

Figure 5.46: Comparison of the measured transconductance g_m and the simulation data in the case of the HVPMOS transistor

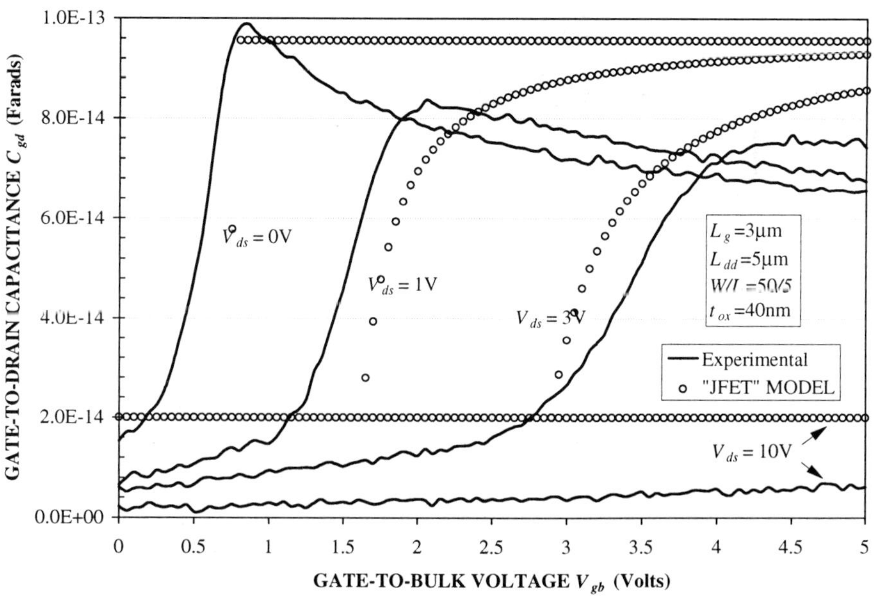

Figure 5.47: Comparison of the measured gate-to-drain capacitance and the simulation data in the case of the HVNMOS transistor

Since the switching range of the drain voltage is given by the high voltage power supply, the Miller capacitance C_{gd} located between the drain and gate terminals cannot be neglected. Because no charge-based capacitance model is available, the intrinsic and extrinsic capacitances of the low-voltage transistor are used to guarantee worst case conditions. Thus, on the one hand, the variations of the capacitance as a function of the gate voltage are given by the intrinsic low-voltage capacitance C_{gd} of SPICE level 3. On the other hand, the field-plate capacitance C_{fp} is approximated by the extrinsic gate overlap parameter *CGDO*. The decrease of C_{fp} as a function of the drain voltage cannot be modelled, since *CGDO* is a constant parameter. Therefore, all the capacitive characteristics start from a worst case constant value of about 20fF for the HVNMOS transistor, and 40fF for the HVPMOS transistor, as shown in Figure 5.47 and Figure 5.48 respectively. Finally, the decrease of the capacitance in the linear mode as a function of the gate voltage is not modelled neither, since the accumulation layer under the field-plate is not considered by the intrinsic capacitance of the low-voltage transistor.

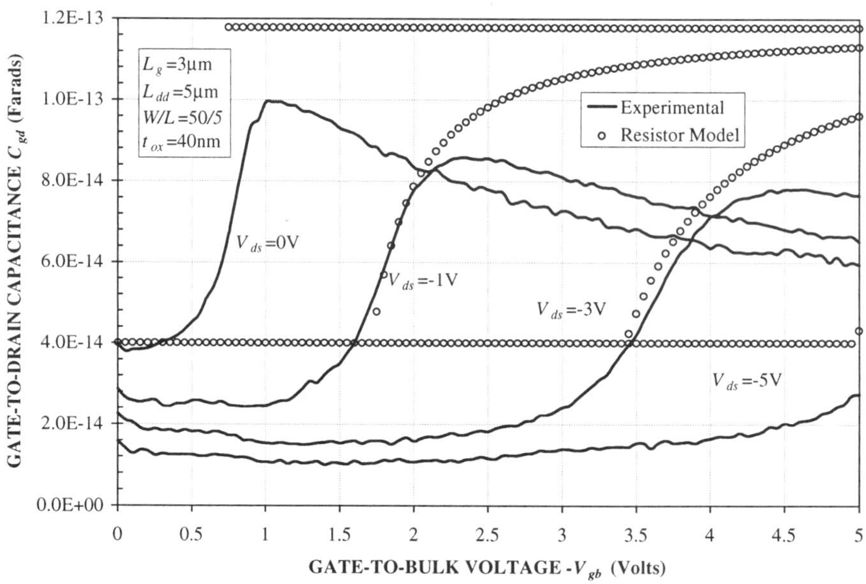

Figure 5.48: Comparison of the measured gate-to-drain capacitance and the simulation data in the case of the HVPMOS transistor

The last dynamic characteristic considered in this section is the drain-to-bulk capacitance C_{db}. In order to avoid the altering of the source-to-bulk capacitance model, C_{db} is modelled by the junction capacitance of diode D_{ldd}. The effect of the drain capacitance of the low-voltage transistor is cancelled by setting A_D and P_D to zero, which are the area and perimeter of the drain junction. The resulting models are shown in Figure 5.49 and Figure 5.50, where the SPICE parameters used to fit the

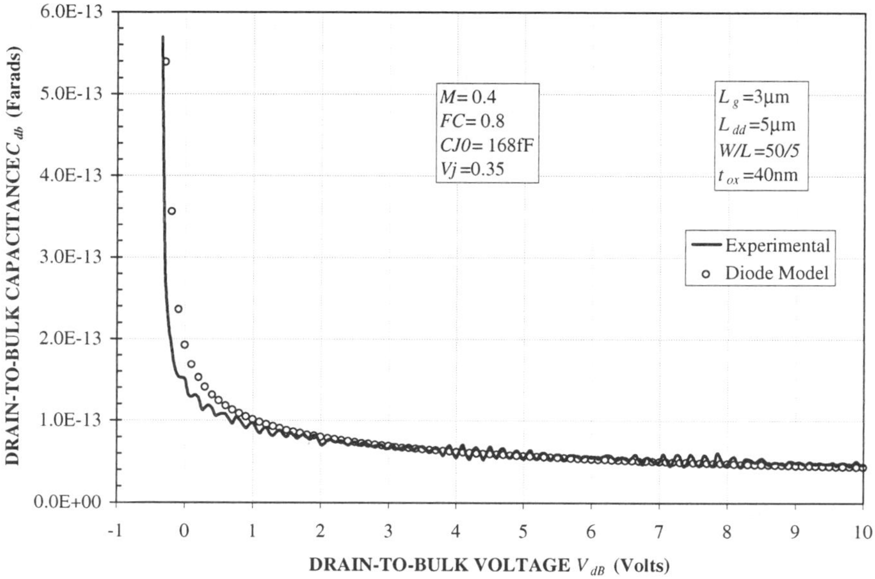

Figure 5.49: Comparison of the measured drain-to-bulk capacitance and the simulation data in the case of the HVNMOS transistor

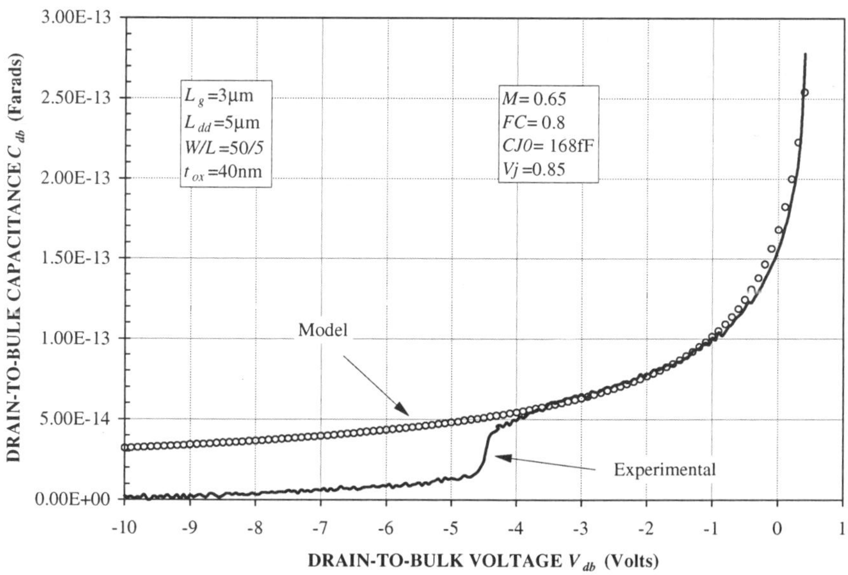

Figure 5.50: Comparison of the measured drain-to-bulk capacitance and the simulation data in the case of the HVPMOS transistor

simulation data and the measured characteristic are summarised. Note that the step of the capacitance in the case of the HVPMOS transistor cannot be simulated when using such first order modelling approach. At this point, it is interesting to summarise the final values of the main parameters of the macro-models. The reader interested in the complete set of low and high voltage parameters can refer to *appendix* ***A***. In Table 5.2 the parameters used to fit the simulation data and the measured characteristics are presented for both high-voltage devices.

GROUP	PARAM.	HVNMOS	HVPMOS
High-voltage	E_{sat}	$3.0x10^4$ V/cm	-
	N_D	$2.5x10^{15}$ cm^{-3}	-
	$L_{g,}$	4.25 μm	-
	L_d	0.25 μm	-
	λ_{hv}	0.01	-
	R_{ldd}	-	1000 Ω
Low-voltage	*KAPPA*	1.0	10
	THETA	0.06 V^{-1}	0.15 V^{-1}
	VTO	0.7 V	-0.7 V
	CGDO	400 pF/m	800 pF/m
Diode D_{ldd}	I_s	$1.6x10^{-13}$ A	$1.84x10^{-14}$ A
	r_s	4580 Ω	60 Ω
	η	1.047	1.06
	M	0.4	0.65
	FC	0.8	0.8
	CJ0	168 fF	168 fF
	Vj	0.35 V	0.85 V

Table 5.2: Summary of the parameters used to fit the simulation data and the measured characteristics of the high-voltage devices

5.3 Layout precautions against parasitic effects

Since the technology used to implement the high-voltage devices is basically a low-voltage n-WELL CMOS process, particular care must be taken when designing a high-voltage integrated circuit. In fact, different problems may result from handling high-voltage power supplies on the chip. On the one hand, the higher substrate current generated by the high-voltage devices may cause latchup under particular bias conditions. On the other hand, the wide depletion regions, on each side of the lightly doped drains, can lead to punchthrough breakdown between junctions. Additionally, parasitic field MOSFETs, resulting from metal wires carrying the high-voltage signals, may cause short circuits between isolated junctions. Such parasitic effects can be avoided by using guard rings at a device and circuit levels, shielding the high-voltage wires and, finally, redefining the layout rules related to the high-voltage devices. These solutions are discussed in further detail in the following paragraphs.

5.3.1 Layout rules and precautions at a device level

The scaling of the high-voltage transistors is similar to the one discussed in chapter ***3*** for lateral DMOSFETs. Thus, except for the high-voltage parameters L_c, L_g and L_{dd} simulated in chapter ***4***, the remaining layout rules, such as the distance L_d from the drain contact to the lightly doped drain edge and the one from the lightly doped drain edge to the guard ring L_{gr}, are given by punchthrough conditions. It corresponds to the merging of the depletion layers on each side of the lightly doped drain with the diffusion contact and the guard ring contact. These rules are shown in Figure 5.51 and Figure 5.52 for both high-voltage devices. Assuming a step approximation of the doping profile of the lightly doped drains, L_d in the case of the HVNMOS transistor can be expressed as

$$L_{d(n)}, L_{gr(p)} = \sqrt{\frac{2\varepsilon_s N_A}{qN_D(N_D + N_A)}|V_{d\max}|} \tag{5.29}$$

which is also the expression of L_{gr} in the case of the HVPMOS transistor. L_{gr} is given by

$$L_{gr(n)}, L_{d(p)} = \sqrt{\frac{2\varepsilon_s N_D}{qN_A\left(N_A + N_D\right)}|V_{d\max}|} \tag{5.30}$$

which is also the expression of L_d in the case of the HVPMOS transistor. It is obvious that these layout rules reduce to those of the low-voltage process, when the high-voltage power supply equals the low-voltage one. Therefore, it is more convenient to define them as a function of the ratio of the power supplies. Thus, assuming the same step approximation of the doping profiles, and considering $|V_{dmax}|=V_{DD}$, where V_{DD} is the low-voltage power supply, Equations (5.29) and (5.30) becomes in the case of the HVNMOS transistor for instance

$$L_{dlv} = \sqrt{\frac{2\varepsilon_s N_A}{qN_D\left(N_D + N_A\right)}V_{DD}} \tag{5.31}$$

and

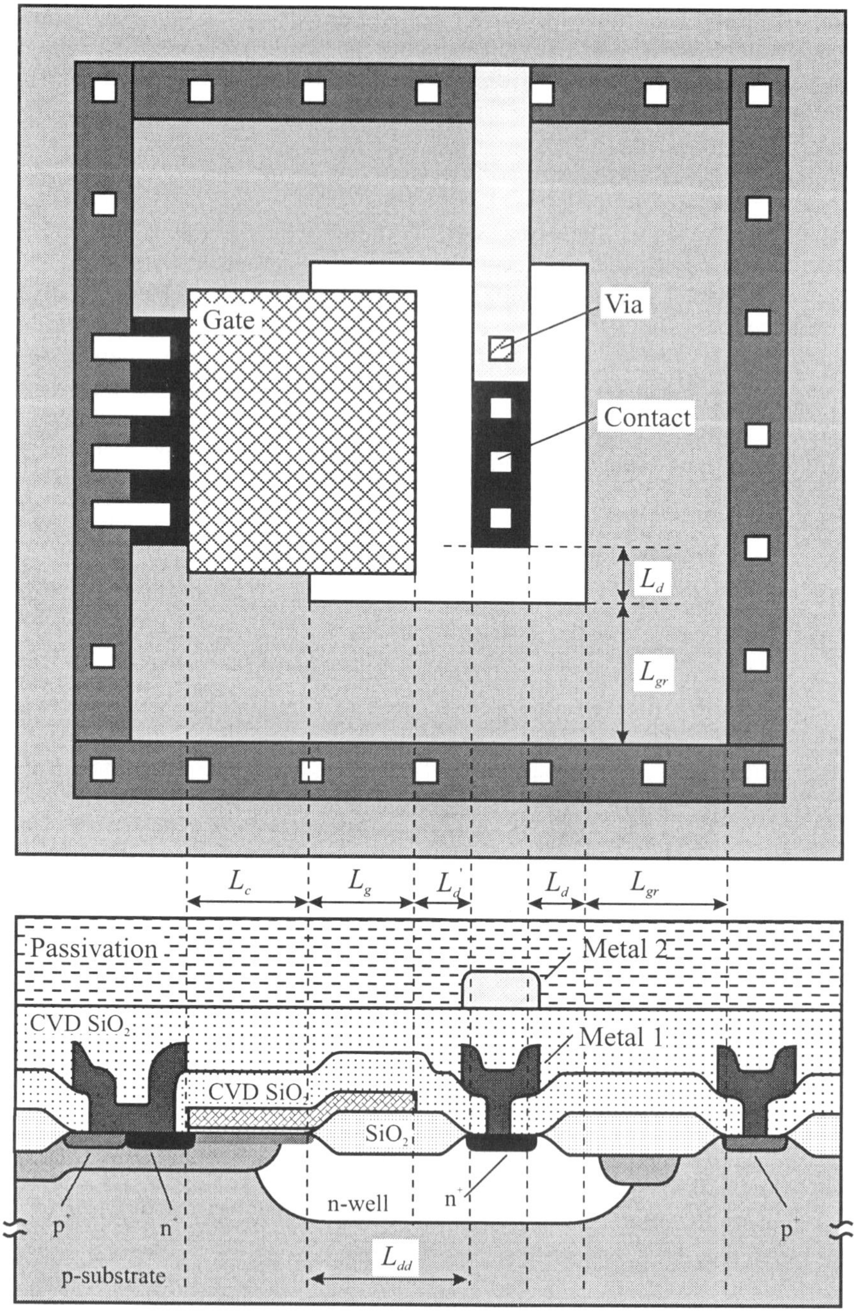

Figure 5.51: Layout rules at a device level with L_c=5µm, L_g=3µm, L_{dd}=5µm, L_d=5µm, L_{gr}=10µm for V_{dmax}=30V and V_{DD}=5V (HVNMOS)

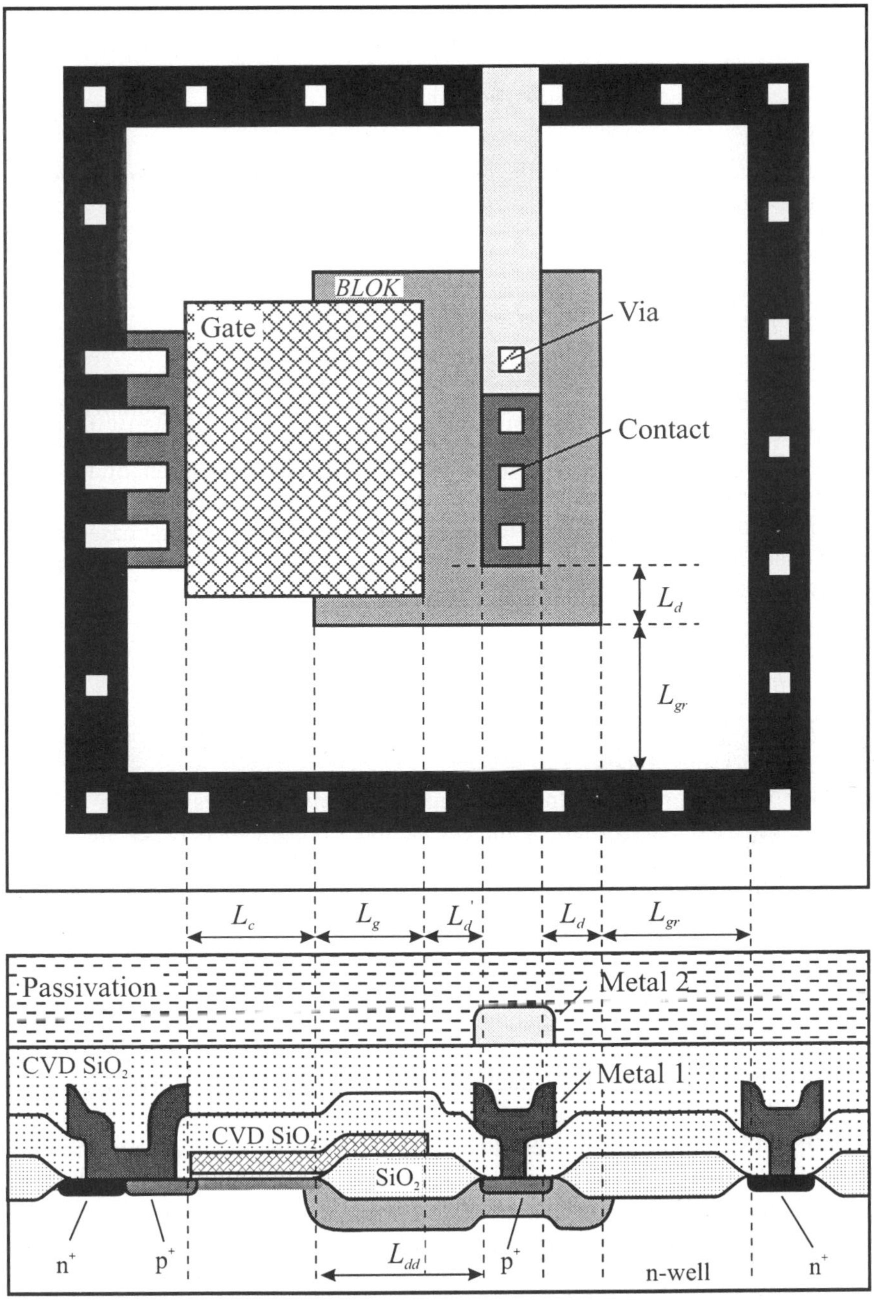

Figure 5.52: Layout rules at a device level with L_c=5μm, L_g=3μm, L_{dd}=5μm, L_d=3μm, L_{gr}=10μm for V_{dmax}=30V and V_{DD}=5V (HVPMOS)

$$L_{grlv} = \sqrt{\frac{2\varepsilon_s N_D}{qN_A\left(N_A + N_D\right)} V_{DD}} \tag{5.32}$$

Consequently, the ratio between the high-voltage and low-voltage layout rules reduce to a single relationship given by

$$\frac{L_{hv}}{L_{lv}} = \sqrt{\frac{\left|V_{d\max}\right|}{V_{DD}}} \qquad \text{Abrupt junctions} \tag{5.33}$$

which is independent of the considered high-voltage rule assuming that the corresponding low-voltage one exists. It must be recalled that the square-root law of the power supplies ratio results from a crude step approximation of the doping profile of the considered layer. This is valid in the case of shallow junctions with high surface doping concentration levels only. It is, however, a worst case first order approximation, since power laws resulting from other doping profiles are smaller. A linearly graded junction for instance would lead to

$$\frac{L_{hv}}{L_{lv}} = \left(\frac{\left|V_{d\max}\right|}{V_{DD}}\right)^{1/3} \qquad \text{Linearly graded junctions} \tag{5.34}$$

Since the doping profile of the junctions in a CMOS process usually follows a Gaussian law, the actual ratio of the layout rules is located between the previous theoretical boundaries. Therefore, when a higher accuracy of the layout rules is required, due to surface constraints for instance, a two dimensional simulation should be performed for each case. In this book the square root law resulting from the step approximation, is used for all the layout rules. Thus, assuming a low-voltage power supply V_{DD}=5V and a maximum drain voltage $|V_{dmax}|$=30V, the ratio of the layout rules is about 2.5. Considering that the low-voltage layout rule corresponding to L_d is given by the minimum distance of the edge of an n^+ diffusion to the edge of the n-well layer in which it is implanted (p^+ edge to p-well edge), the high-voltage parameter L_d is 7.5μm in the case of the HVNMOS transistor and 5μm in the case of the HVPMOS transistor. However, it is useless to set these lateral spacings to a value higher than the depth of their corresponding layers. Therefore, they are limited to 5μm and 3μm respectively. Parameter L_{gr} is given by the minimum spacing between the n^+ and p^+ diffusions, leading to L_{gr}=10μm in both cases. The reader interested in the complete set of the layout rules can refer to *appendix* ***A***, where the high-voltage rules are specified for $|V_{dmax}|$=30V.

Finally, the use of guard rings at a device level is necessary for two main reasons. On the one hand, since the threshold voltage of parasitic field MOSFETs ranges between 15V and 30V, an inversion layer can form in the substrate (bulk) under the metal wire connecting the drain electrode. Therefore, the guard ring is used to stop such parasitic channels, avoiding thus short circuits between isolated junctions. As will be discussed in section 5.3.2, guard rings are also used as routing channels for high-voltage metal wires at a circuit level. On the other hand, the guard ring at a device level is a majority carrier one, i.e. it consists of an n^+ diffusion in an n-well or a p^+ diffusion in a p-type substrate. It is particularly efficient in avoiding latchup problems by minimising voltage drops created by majority carrier currents. This is illustrated schematically in Figure 5.54 for both guard ring types, but first, the simplest structure that can exhibit latchup is recalled in Figure 5.53. It is a 4-terminal PNPN structure and it represents the most sensitive part of any CMOS circuit. The corresponding lumped element circuit model is also superposed in Figure 5.53.

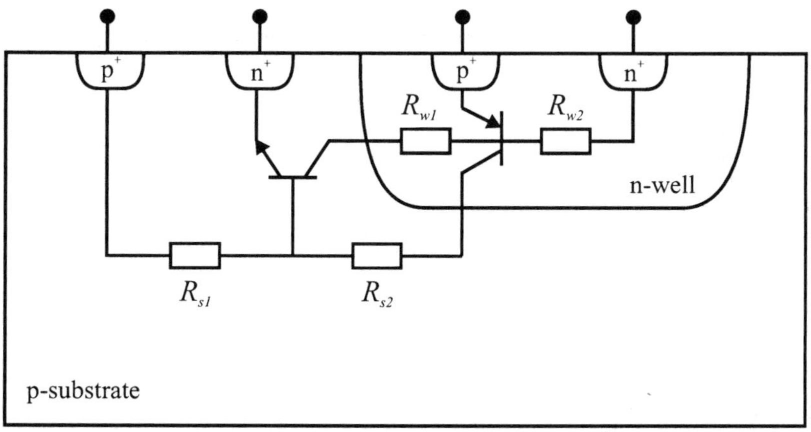

Figure 5.53: Simplest 4-terminal PNPN structure that can exhibit latchup, and most common version of the lumped element circuit model

The causes responsible of the switching from the blocking state to the latched state of the PNPN structure must satisfy the following conditions necessary for latchup

- The loop gain of the PNPN structure must exceed unity
- Both bipolar transistors must be turned on
- The current through the bipolar transistors must exceed the switching condition
- The power supplies must be capable of supplying the holding current in the latched state

Any precaution allowing the elimination of one of these conditions will prevent latchup in CMOS circuits. Thus in Figure 5.54 the majority carrier guard rings help to prevent latchup by prohibiting one of the parasitic bipolars to turn on. In fact, in Figure 5.54a the n^+ guard in the n-well surrounding the p^+ emitter reduces the base resistance of the vertical PNP transistor and steers the current away from the p^+ emitter. Consequently, the voltage drop across R_{w1} and R_{w2} resulting from a majority

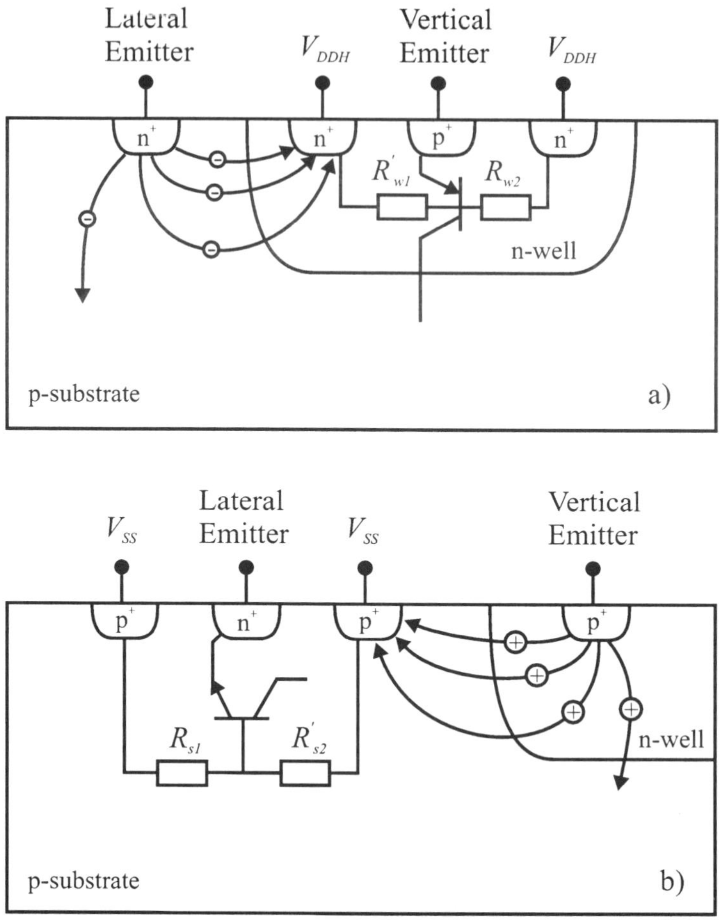

Figure 5.54: Majority carrier guards preventing latchup by reducing the base resistance for all parasitic emitters and steering the current away from these emitters

carrier current is reduced, and the turn on of the vertical bipolar transistor is delayed. Such majority carrier current can be generated by impact ionisation of holes in a HVPMOS transistor (bulk current), or can result from electron injection of the lateral NPN emitter (source/drain of an nMOS transistor). Symmetrically, the p^+ guard in p-type substrate reduces the base resistance of the lateral NPN transistor and steers the current away from the n^+ emitter. The majority carrier current results from impact ionisation of electrons in a HVNMOS transistor (substrate current), or from holes injection of the vertical PNP emitter (source/drain of a pMOS transistor).

5.3.2 Layout rules and precautions at a circuit level

As was mentioned in the previous section all the high-voltage layout rules must satisfy the power law given by Equation (5.33). Hence, the rules given in *appendix* ***A***

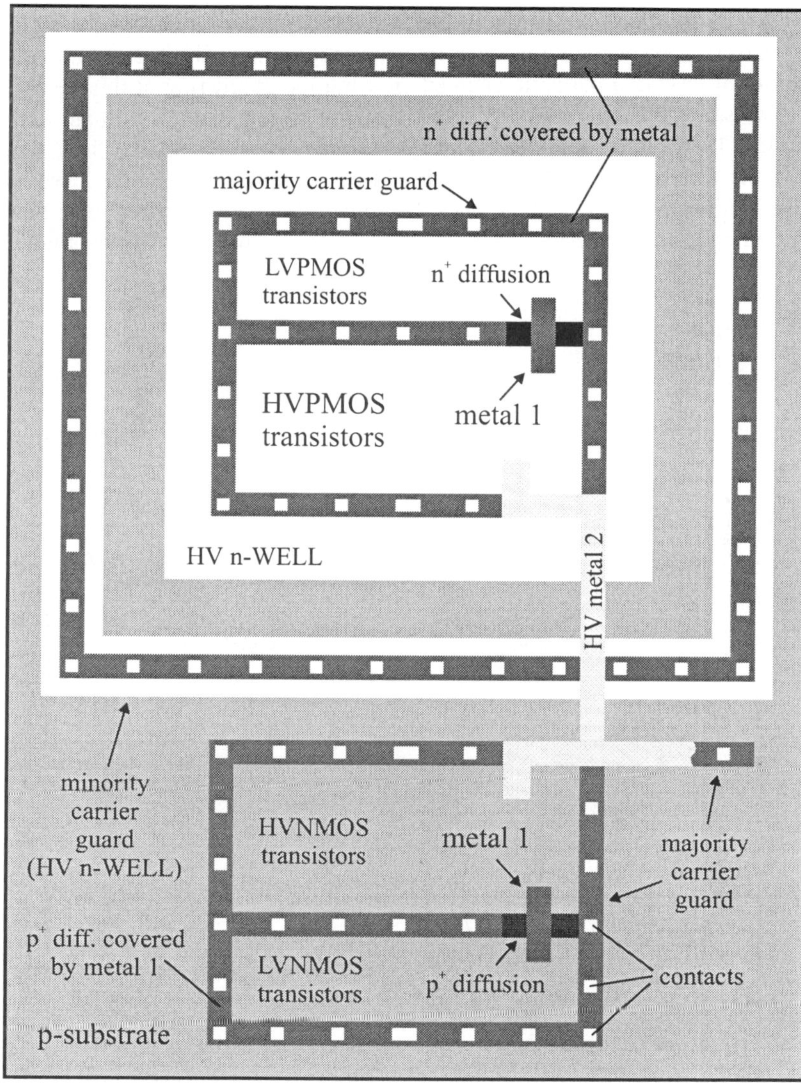

Figure 5.55: Typical topology to be used for a single high-voltage cell

remain valid at a circuit level. Additional precautions, however, must be taken to avoid parasitic effects. They can be considered at two different levels. The first level consists of a simple high-voltage cell, where the topology shown in Figure 5.55 should be used. The low-voltage nMOS section of the cell must be separated from the HVNMOS section by the use of majority carrier guard rings, in order to avoid short circuits resulting from parasitic field MOSFETs. The connections between the low-voltage nMOS section and the high-voltage one can be performed with metal 1 wires since such connections are simple low-voltage ones. Care must be taken, however, to cross the guard rings. In fact, it is preferable to interrupt the metal wire of the guard ring contacts, rather than interrupting the majority diffusion, even if the probability of having a parasitic channel crossing the guard ring at this level is very small. The same precautions considered so far between the low-voltage and high-voltage nMOS sections, apply to the pMOS sections as well. Finally, the connection between the nMOS and pMOS parts should be achieved with metal 2 wires since such connections are high-voltage ones. Moreover, the metal 2 wires must be routed over guard rings as far as possible in order to avoid inversion layers under these wires.

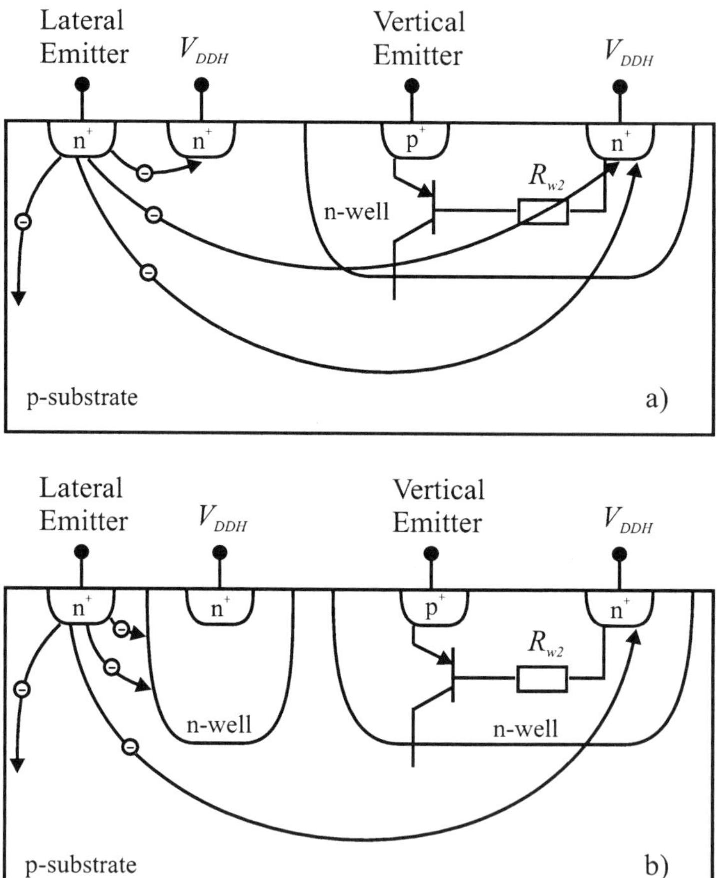

Figure 5.56: Minority carrier guards in substrate using a) shallow n^+ diffusion and b) deep n-WELL layer

An additional precaution suitable at a single cell level, appearing in Figure 5.55, is the use of a minority carrier guard in substrate rounding the HVPMOS section. Minority carrier guards consist of n^+ in a p-type substrate or p^+ in an n-WELL. They are usually used to pre-collect injected minority carriers before they can cause a problem, as shown in Figure 5.56a. In fact, minority carriers injected into the substrate could be collected by a the n-WELL/substrate junction where they become majority carriers. The voltage drop caused by such carriers, could turn on the vertical PNP transistor starting thus a latchup condition. Since n^+ and p^+ diffusions are very shallow their efficiency is limited in pre-collecting the minority carriers which can flow under the diffusion. Therefore, n-WELL guards are more efficient for this task because they can reach deeper into the substrate, as shown in Figure 5.56b. It is obvious that such guards must not contain p^+ diffusions in order to avoid vertical parasitic bipolar transistors. Since no equivalent deep layer exists in the n-WELL and since the efficiency of a surface p^+ guard is limited, only minority carrier guards in the substrate are used. Note that the surface loss due to the use of minority carrier guards is not negligible.

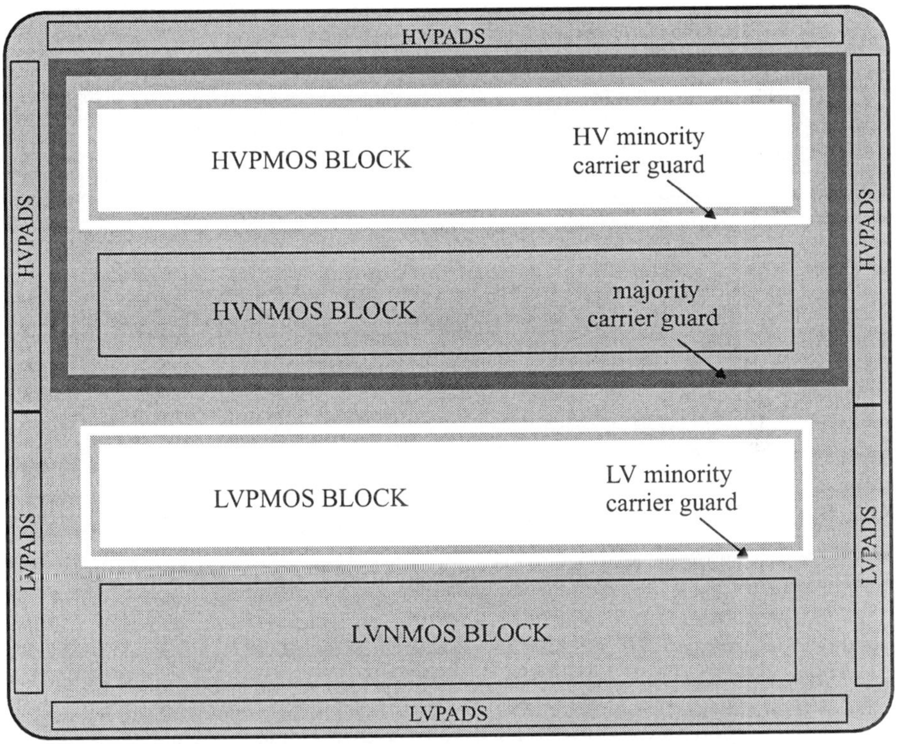

Figure 5.57: Layout precautions when separating low and high voltage blocks at a chip level

The second level is the chip level where high-voltage and low-voltage circuits are combined on the same substrate. Two critical points must be considered at this level. The partitioning of the chip between high and low voltage blocks, and the routing of the high-voltage wires. It is obvious at this level that it is not possible to provide

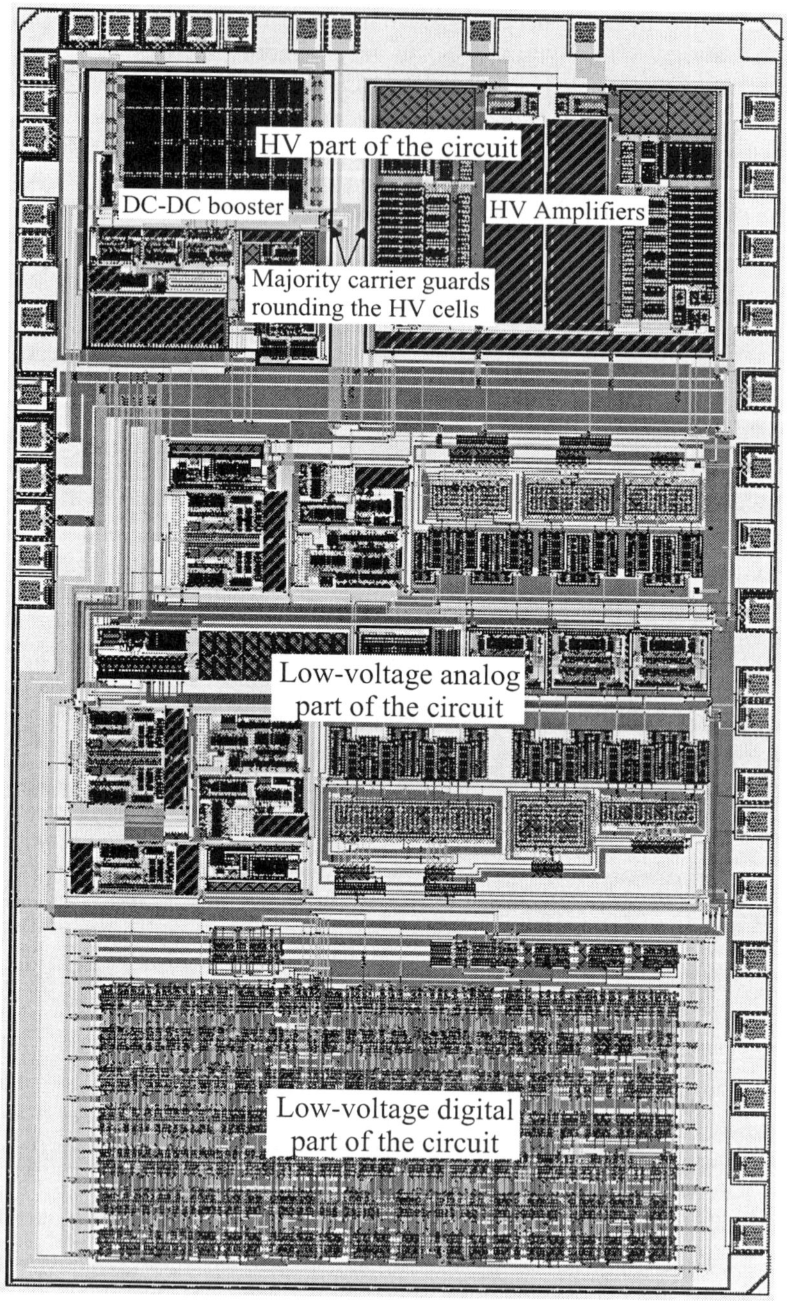

Figure 5.58: Typical example of a high-voltage integrated circuit using different layout precautions (printed with permission of MEAD S.A [11])

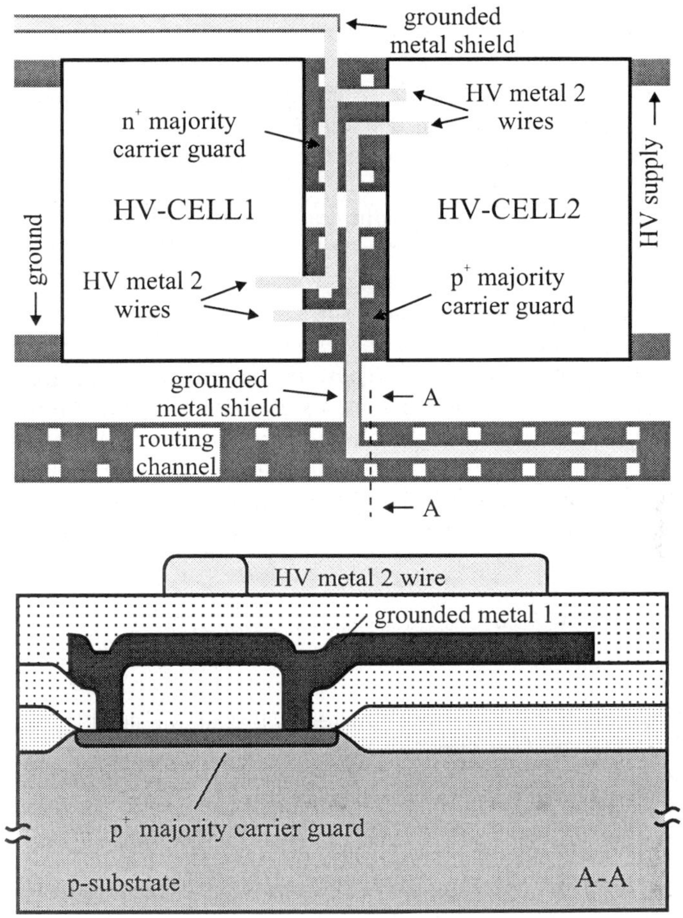

Figure 5.59: layout precautions avoiding parasitic field MOSFETs arising from high-voltage wires

detailed design rules valid for all designs, but it is however possible to state some simple layout precautions that should be observed to avoid latchup problems. These precautions are basically similar to those considered at a single cell level. Starting with the separation between high and low voltage blocks, it should be performed using a majority carrier guard which surrounds the high-voltage block, as shown in Figure 5.57. Such precaution is useful to prevent any excessive substrate current (holes) from reaching the low-voltage block, minimising thus interactions between high and low voltage parts of the circuit. Note that minority carriers (electrons) injected into the substrate should be collected by the minority guards rounding the pMOS sections of each block, as shown in Figure 5.57 as well. Any partitioning between the low and high voltage blocks of a high-voltage integrated circuit should observe the previous layout precautions. A concrete example of a high voltage integrated circuit using such layout precautions is shown in Figure 5.58. The chip consists of a piezzo-electric transducer driver where the high-voltage part is formed by a high-voltage amplifier and a DC to DC voltage booster. It is surrounded with a majority carrier guard connected to ground. The low-voltage part of the circuit implements some complex analog and digital signal processing features such as filters synchronising logic and other functions. No minority carrier guards were used yet in this circuit.

The last precaution consists of avoiding parasitic field MOSFETs arising from the high-voltage wires. Therefore, the highest metal level should be used to perform the interconnections between the different high-voltage cells or to route the high-voltage signals to the output pads. Moreover, since the high-voltage power supply can exceed the threshold voltage of the field MOSFET formed by the highest metal level, the shielding of the wires is necessary. Thus, the majority carrier guards are the most frequently used for this purpose. In some cases routing channels formed by large majority carrier guards between high-voltage cells are used to achieve more complex interconnections. This is shown schematically in Figure 5.59. Finally, when the majority carrier guards cannot be used to route the high voltage wires, the shielding can be performed with a grounded wire at a lower level, as shown in the cross section of Figure 5.59.

5.4 References

[1] M. J. Declercq, J. D. Plummer, "Avalanche Breakdown in High-Voltage DMOS Devices" *IEEE Trans. Elec. Dev.*, Vol. ED-23, January 1976, pp. 1-4

[2] B. J. Baliga, "Modern Power Devices", Krieger Publishing Company, Folrida, 1987

[3] C. Hu, S. C. Tam, F. C. Hsu, P. K. Ko, T. Y. Chan, K. W. Terrill, "Hot-Electron-Induced MOSFET Degradation—Model, Monitor, and Improvement" *IEEE J. Solid-State Circuits*, Vol. SC-20, No. 1 February 1985

[4] N. Arora, "MOSFET Models For VLSI Circuit Simulation Theory and Practice", Springer-Verlag, Wien New York, 1993

[5] S. M. Sze, "Physics of Semiconductor Devices", Wiley-Interscience, New York, 1981

[6] H. R. Claessen, P. V. D. Zee, "An Accurate DC Model for High-Voltage Lateral DMOS Transistors Suited for CACD" *IEEE Trans. Elec. Dev.*, Vol. ED-33, December 1976, pp. 1964-1970

[7] Y. S. Kim, J. G. Fossum, R. K. Williams, "New Physical Insights and Models for High-Voltage LDMOST IC CAD" *IEEE Trans. Elec. Dev.*, Vol. ED-38, July 1991, pp. 1641-1649

[8] Dolphin integration, Meylan Cedex, France

[9] Enz-Krummenacher-Vittoz Model, release 2.3, *Electronics Laboratories*, Swiss Federal Institute of technology, Lausanne, Switzerland

[10] Berkeley Short-Channel IGFET Model, *Electronics Research Laboratory*, University of California/Berkeley

[11] MEAD Microelectronics S.A, St-Sulpice Switzerland

6 HIGH-VOLTAGE ANALOG AND DIGITAL OUTPUT INTERFACES

Abstract:
The second part of this book describes the design of high-voltage integrated circuits. One of the most common solutions for high-voltage applications using the SVX devices, involves complex low-voltage signal processing communicating with an external high-voltage environment. This communication is usually performed by high-voltage digital or analog output interfaces. These are discussed in this chapter, where circuit design techniques overcoming the voltage limitations of the devices are presented as well.

Although the utilisation of a thin gate oxide for the high-voltage devices allows high current gains and standard threshold voltage values, the cost is a severe limitation of the gate voltage swing. In fact, since the SVX devices can handle high-voltage power supplies, the gate voltage range must be limited to the maximum value set by the technology to avoid hot-carrier effects and oxide breakdown. Considering a standard 2µm n-WELL CMOS process, the maximum gate voltage swing with respect to the bulk is 5V. The use of classic low-voltage CMOS circuits to drive the gate of the HVNMOS transistor overcomes this limitation without major difficulties. In the case of the HVPMOS transistor, however, the circuit driving the gate of the device is not straightforward, since the voltage swing of 5V is referenced to the high-voltage power supply V_{DDH}. Therefore, level shifting techniques must be used to cope with this limitation. They are discussed in sections 6.1 and 6.2 for high-voltage digital and analog output interfaces. Their static and dynamic limits are also considered.

6.1 High-voltage digital output interfaces

The high-voltage CMOS (HVCMOS) digital output interface [12] shown in Figure 6.1 is similar to a simple CMOS inverter, except for the gate connections. These are driven by two signals issued or created from low-voltage logic. The gate of the HVNMOS transistor can be readily controlled by the low-voltage logic levels 0 and V_{DD}, where V_{DD} is the low-voltage power supply. The gate of the HVPMOS transistor needs a level shift to operate between V_{DDH} and V_{DDH} - V_{DD}. This is achieved by one of the level shifter cells discussed in sections 6.1.1 and 6.1.2, where alternative solutions are presented as a function of the required power consumption. Transients considerations of the complete digital output interface are discussed in section 6.1.4.

The measurements shown in Figure 6.2 were performed on a digital output interface using the basic static level shifter cell presented in section 6.1.1. The maximum output current capability of the HVCMOS output stage is I_{omax}=20mA assuming $(W/L)_n$=400/10 and $(W/L)_p$=1000/10. Considering a capacitive output load of 30pF, the rise and fall times are about 80ns with 50V high-voltage power supply. Note that in the upper part of the graph the limited voltage swing (5V) of the level shifted signal S_H driving the gate of the pMOS device. The static power consumption of such an interface is limited to the level shifter only. A typical value is about 1 to 5% of the maximum output drive capability, depending on the speed requirements. This value can be reduced significantly by the use of the low-standby current level shifter presented in section 6.1.2. Thus, the most critical part of the high-voltage digital output interface is the level shifter cell, while the output stage formed by M_1 and M_2 is designed similarly to a low-voltage CMOS inverter.

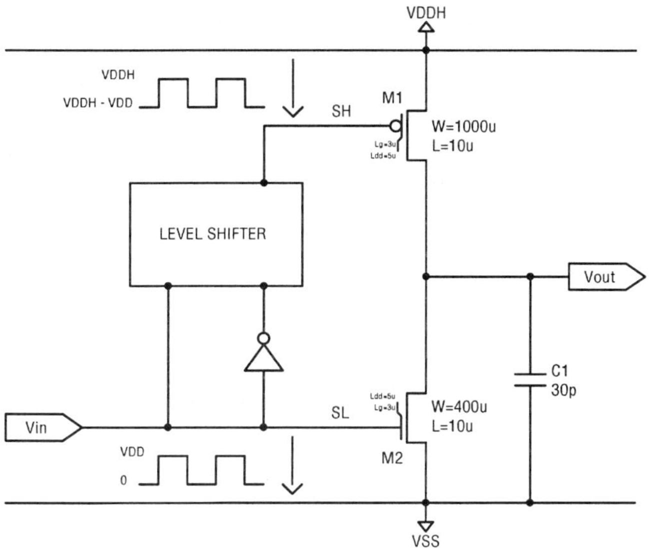

Figure 6.1: Schematic of the high-voltage digital output interface principle

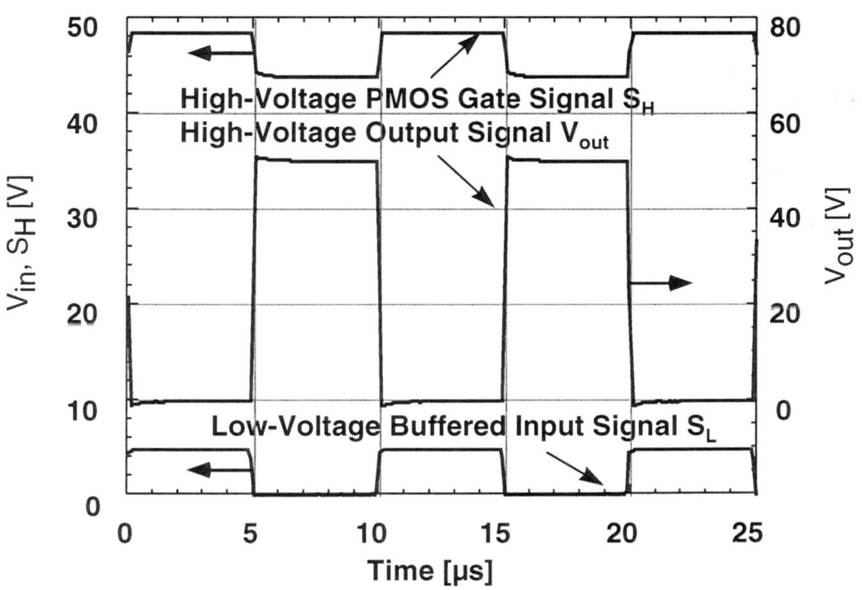

Figure 6.2: Measured characteristics of the high-voltage digital output interface

6.1.1 Basic static level shifter cell

The schematic of the simplest static level shifter cell [13] is shown in Figure 6.3a. It is based on cross coupled "voltage mirrors". The voltage mirror is defined as a basic circuit block, connected between V_{DDH} and ground V_{SS}, delivering an output voltage V_x such that (V_{DDH} - V_x) is equal to the input voltage V_{in}, as shown in Figure 6.3b. It consists of a HVNMOS driver M_9 loaded by a low-voltage pMOS transistor M_8

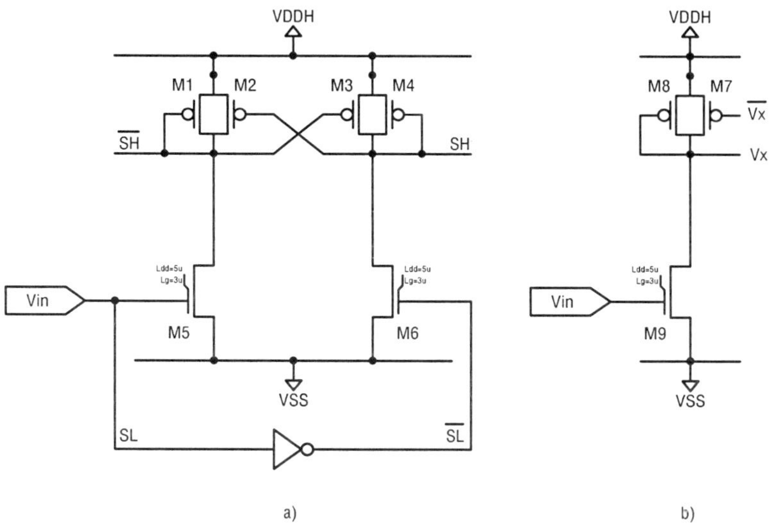

Figure 6.3: Schematic of a) Basic level shifter cell b) Voltage mirror circuit block

connected in a diode configuration. In parallel to M_8, a pull-up transistor M_9 is necessary to drive V_x to V_{DDH} in the high-state, otherwise it could never reach the high-voltage power supply. The driving signal of the gate of M_9 must be complementary to V_x. This is achieved by cross coupling two voltage mirrors, thus forming a high-voltage flip-flop, where each voltage mirror generates the driving voltage for the opposite pull-up transistor, as shown in Figure 6.3a. Consequently, the resulting level shifted signal S_H at the output of the level shifter, exhibits a full swing of V_{DD} between (V_{DDH} - V_{DD}) and V_{DDH} when the input signal V_{in} switches between V_{SS} and V_{DD}, as it appears in Figure 6.2. This can be demonstrated through simple theoretical considerations as well. In fact, the current expressions through M_8 and M_9 yield

$$\frac{\beta_n}{2}\left(V_{in}-V_{tn}\right)^2=\frac{\beta_p}{2}\left[\left(V_{DDH}-V_x\right)-V_{tp}\right]^2 \tag{6.1}$$

as far as M_9 is in the saturation mode, i.e. $V_x > V_{dsat}$. Assuming that $V_{in} = V_{DD}$, $\beta_n = \beta_p$ and $V_{tn} = |V_{tp}|$, then the voltage at the output of the level shifter is given by

$$V_x = V_{DDH} - V_{DD} \quad for \quad \begin{cases} V_x > V_{dsat} \\ V_{in} = V_{DD} \end{cases} \tag{6.2}$$

If V_{DDH} is reduced such that $V_x < V_{dsat}$, Equation (6.1) becomes

$$\beta_n \left[(V_{in} - V_{tn}) V_x - \frac{V_x^2}{2} \right] = \frac{\beta_p}{2} \left[(V_{DDH} - V_x) - V_{tp} \right]^2 \tag{6.3}$$

If V_{DDH} is reduced further down to V_{DD}, and assuming that $V_{in} = V_{DD}$, $\beta_n = \beta_p$ and $V_{tn} = |V_{tp}| = V_t$, then the voltage at the output of the level shifter is given by

$$V_x = \left(1 - \frac{\sqrt{2}}{2}\right)(V_{DD} - V_t) \quad for \quad \begin{cases} V_x < V_{dsat} \\ V_{DDH} = V_{DD} = V_{in} \end{cases} \tag{6.4}$$

which is about 3.8V for $V_{DD} = 5$V and $V_{tn} = |V_{tp}| = V_t = 0.8$V. Consequently, the basic static level shifter cell, shown in Figure 6.3a, continus to drive the gate of the HVPMOS transistor for high-voltage power supplies as low as V_{DD}. The resulting current drive capability, however, is lower than the HVNMOS transistor.

Considering normal high-voltage operating conditions where $V_{DDH} >> V_{DD}$, the voltage swing resulting from Equation (6.2) assumes a perfect matching of the electrical parameters of the devices. Actually, the relative deviation of the full swing of S_H around a typical value V_{typ} depends on the matching of the threshold voltage values V_{tn} and V_{tp}, and the matching of β_n and β_p. In fact, assuming $\beta_n \neq \beta_p$ and $V_{tn} \neq |V_{tp}|$, Equation (6.1) yields

$$V_x = V_{DDH} - \left[\sqrt{\frac{\beta_n}{\beta_p}} (V_{in} - V_{tn}) + |V_{tp}| \right] \tag{6.5}$$

As far as the relative deviation of the full swing remains within the limits of the low-voltage power supply tolerances, typically ±10% of V_{DD} for standard 5V CMOS technologies, the output of the level shifter safely drives the gate of the HVPMOS transistors. The expression of the relative deviation can be deduced easily from Equation (6.5) as

$$\frac{\Delta V}{V_{typ}} = 1 - \sqrt{\frac{\beta_{pt}}{\beta_p} \cdot \frac{\beta_n}{\beta_{nt}}} \cdot \left(1 - \frac{|V_{tpt}|}{V_{typ}}\right) - \sqrt{\frac{\beta_n}{\beta_p}} \cdot \frac{V_{tnt} - V_{tn}}{V_{typ}} - \frac{|V_{tp}|}{V_{typ}} \tag{6.6}$$

where the t mark stands for the typical set of parameters. Note that the maximum relative deviation resulting from Equation (6.6) occurs when the deviation of the electrical parameters of the nMOS and pMOS transistors is

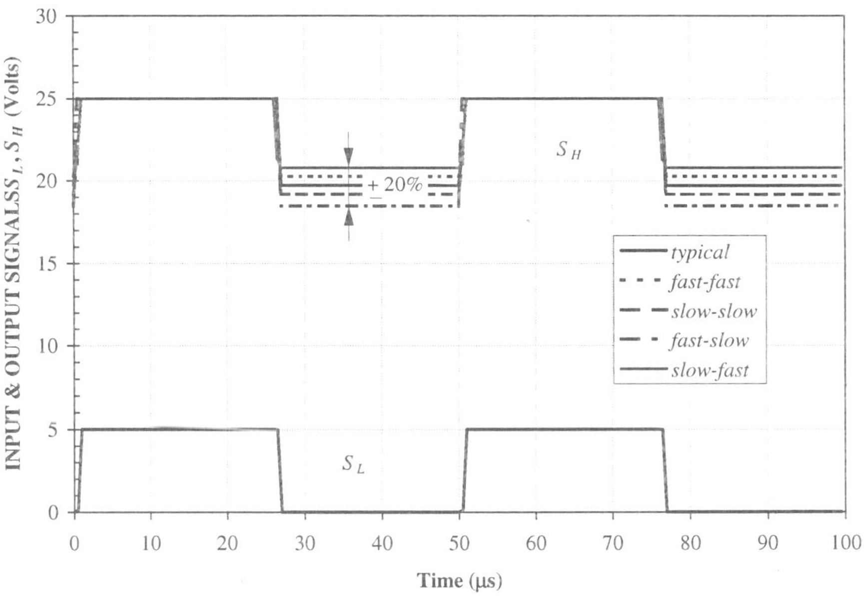

Figure 6.4: Input and output signals of the basic level shifter cell for all possible combinations of the electrical parameters deviations

opposite with respect to the typical value. Thus, considering the electrical parameters given in *appendix A*, the maximum relative deviation of the full swing is about ±12%. It should be mentioned at this point that the finite output conductance of the devices was not considered in Equation (6.1). Consequently, the actual relative deviation should be higher than the one computed with Equation (6.6). In fact, the simulation data of Figure 6.4 shows that the maximum relative deviation of the full swing is ±20% around the typical value, considering the worst case deviations of the electrical parameters. According to Equation (6.5), the *typical* values yield a full swing at the output of the level shifter of about V_{DD} - V_{tnt} + $|V_{tpt}|$, since it is reasonable to assume that $(\beta_{nt}/\beta_{pt}) \approx 1$. The relative deviation of the full swing around this typical value remains within the limits of the low-voltage power supply tolerances (±10%) in the *fast-fast* and *slow-slow* cases. The most constraining combinations are the *slow-fast* and *fast-slow* sets. Although the corresponding voltage levels V_x are out of typical industrial specifications, no reliability aspects should be considered here. In fact, assuming that V_{DD}=5V, the *fast-slow* case (nMOS-*fast*, pMOS-*slow*) would lead to a full voltage swing at the output of the level shifter of about 6V. Since S_H drives the gate of the HVPMOS transistor, no hot-carrier effects arise from the corresponding voltage level, particularly for a pMOS device with 400Å gate oxide thickness. Moreover, the typical design can be centred such that a worst case condition of the electrical parameters does not lead to a full voltage swing 10% higher than the low-voltage power supply value.

The relative deviation of ±20%, however, is theoretically cancelled by the use of the improved static level shifter cell discussed in section 6.1.2, so that the full voltage swing of S_H meets the industrial specifications. It should be mentioned at this point, that the previous considerations apply for steady state conditions only. The transient constraints are discussed further in section 6.1.4, where the upper limits of the static level shifter are pointed out.

6.1.2 Improved static level shifter cell

In order to reduce the relative deviation of the full swing around its typical value, the dependence of the voltage level V_x on the matching of the electrical parameters between n-type and p-type transistors should be eliminated. In fact, the maximum relative deviation in the basic static level shifter cell occurs when the deviation of the electrical parameters of both devices is opposite (*slow-fast* and *fast-slow*). If the voltage level V_x was dependent on the deviation of the electrical parameters of a single transistor type, the relative deviation of the full swing around

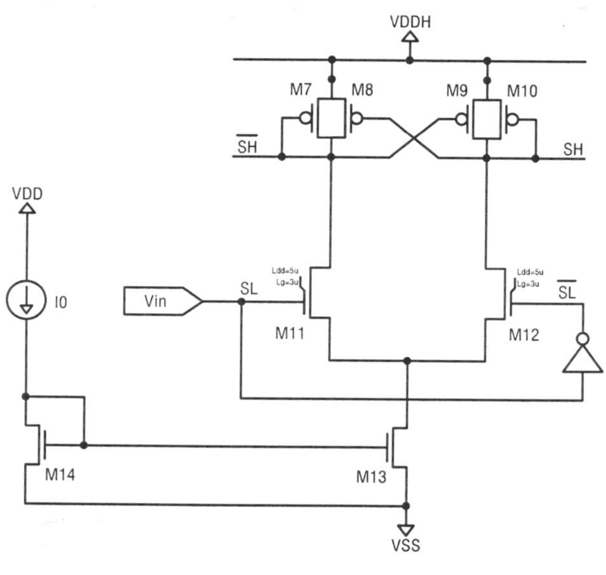

Figure 6.5: Improved static level shifter cell

its typical value should be reduced. Indeed, this is achieved in the improved static level shifter cell shown in Figure 6.5, where current source I_0 is added for this purpose. In this case, the voltage level V_x at the outputs of the level shifter is fixed by the current flowing through transistors M_{10} and M_7 according to

$$V_x = V_{DDH} - \left(\sqrt{\frac{2nI_0}{\beta_p}} + \left| V_{tp} \right| \right) \tag{6.7}$$

Current I_0 is switched between M_{10} and M_7 by the high-voltage differential pair M_{11}-M_{12}. Assuming that I_0 is constant for all possible deviations of the electrical parameters, the maximum relative deviation of V_x around the typical value is given by

$$\frac{\Delta V}{V_{typ}} = 1 - \sqrt{\frac{K_{pt}}{K_p}} \left(1 - \frac{\left| V_{tpt} \right|}{V_{typ}} \right) - \frac{\left| V_{tp} \right|}{V_{typ}} \tag{6.8}$$

where the t mark stands for the typical set of parameters and $\beta_p = K_p{*}(W/L)_p$. Considering the electrical parameters given in *appendix* **A**, and assuming a typical voltage swing of V_{DD}, the maximum relative deviation resulting from Equation (6.8) is about ±7%. If current I_0 is created such that the deviations of the electrical parameters of the pMOS device are cancelled, the relative deviation of the full swing around the typical value is eliminated. This achieved by generating I_0 with a pMOS transistor connected to V_{DD} and which is perfectly matched to the pMOS load of the level shifter, as shown in Figure 6.6. In this case, current I_0 generated by M_{15} is given by

$$I_0 = \frac{\beta_p}{2n} \left(V_{DD} - \left| V_{tp} \right| \right)^2 \tag{6.9}$$

Replacing I_0 in Equation (6.7) by (6.9), the voltage level at the output of the level shifter reduces to

$$V_x = V_{DDH} - V_{DD} \tag{6.10}$$

which meets the initial objective of the basic static level shifter cell. Moreover, the relative deviation of this voltage level around V_{DD} is directly given by the tolerance of the low-voltage power supply. This is confirmed by the simulation data shown Figure 6.7, where the full swing of S_H is identical for all possible combinations of the electrical parameters deviations. The residual relative deviation of the full swing can

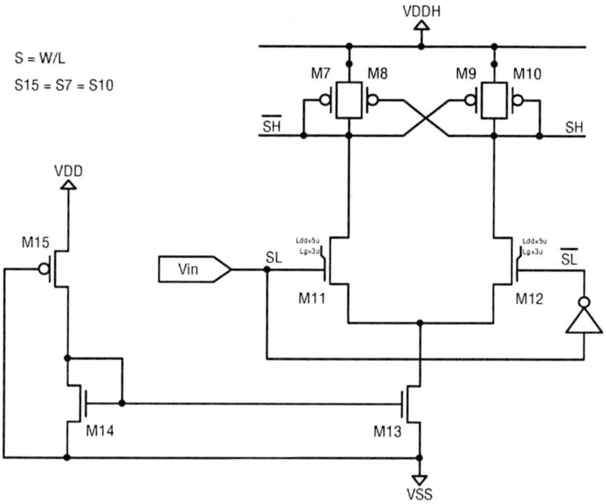

Figure 6.6: Generation of I_0 such that the improved static level shifter cell becomes insensitive to the deviations of the electrical parameters

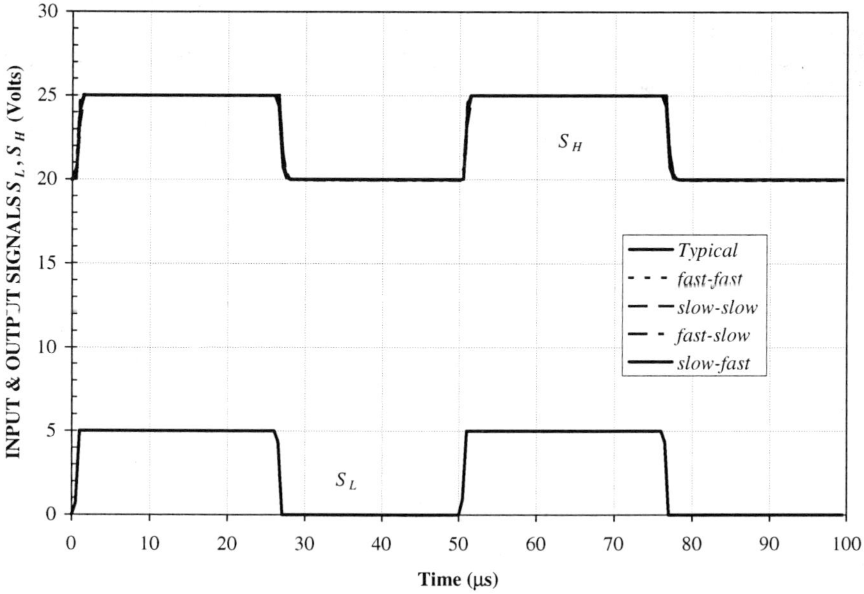

Figure 6.7: Input and output signals of the improved static level shifter cell for all possible combinations of the electrical parameters deviations

result from the mismatch of the current flowing through M_{15} and M_{10}. In fact, since the drain voltage values of both devices are not perfectly matched, the output conductance leads to slight current differences. Moreover, the physical mismatch of current mirror M_{14}-M_{13} increases such differences. The resulting relative deviation, however, is negligible with respect to the full voltage swing of S_H.

6.1.3 low standby current static level shifter cell

One of the most important aspects of high-voltage integrated circuits is the static current consumption. In fact, given the high-voltage power supply values involved in these circuits, the resulting power dissipation is not negligible. Therefore, as far as static currents are negligible, the basic or improved cells can be used. When speed requirements are such that static currents are not negligible anymore, the low standby current level shifter cell shown in Figure 6.8 must be used. Consider for instance a power output stage in which the gate capacitance of the pMOS device is about 420pF. Such a device is used in the full bridge of the class-D audio amplifier discussed in chapter 7. Assuming a switching frequency of 500kHz, the current driving capability of the level shifter cell is about 50mA. If either basic or improved cells are used, the static power dissipation on the chip would be as high as 550mW, assuming a 12V high-voltage power supply value. This is obviously not acceptable. The use of the level shifter shown in Figure 6.8 reduces the static consumption by a factor of 40, leading thus to a static power dissipation of 15mW. This is achieved by inserting a high-voltage buffer stage between the HVPMOS transistor of the output stage and the static level shifter, which is not straightforward since the voltage swing must be limited to V_{DD}. Therefore, the high-voltage buffer stage is formed by the high-voltage source follower transistor M_{15}, the pull-up transistor M_8 and the current source I_1. It dynamically drives the transistor of the output stage as follows. M_{15} and M_8 are controlled by S_{H2} and $\overline{S_{H1}}$ which are in phase opposition. Ideally, M_8 should be driven by the complementary signal $\overline{S_{H2}}$, which is actually not the case since the full swing of $\overline{S_{H2}}$ is higher than V_{DD}, as will be pointed out further. When $\overline{S_{H1}}$ is active low, the output is pulled-up to V_{DDH} by discharging the gate capacitance of the output pMOS device. When S_{H2} is active low, S_H is pulled down until M_{15} enters in weak inversion, thus charging the gate capacitance to a voltage level V_x. While the maximum voltage swing of S_H is limited to V_{DD} by the static level shifter cell, its voltage level is maintained at (V_{DDH} - V_x) by the current source I_1. The static value of I_1 is given by the minimum current required to keep M_{15} in weak inversion, depending on the scaling of the device and the chosen inversion factor. In the considered example, less than 25µA are necessary, assuming $(W/L)_{M15}$ = 4000/5 and an inversion factor of 0.8. The complementary output $\overline{S_H}$ is controlled symmetrically.

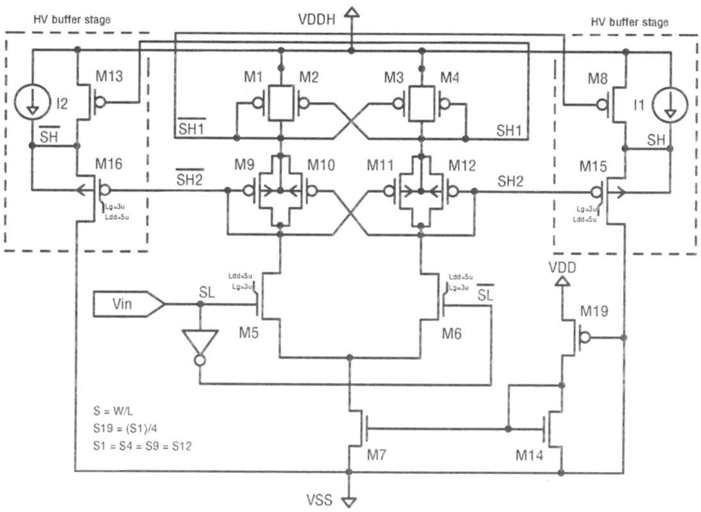

Figure 6.8: Low-standby current static level shifter cell

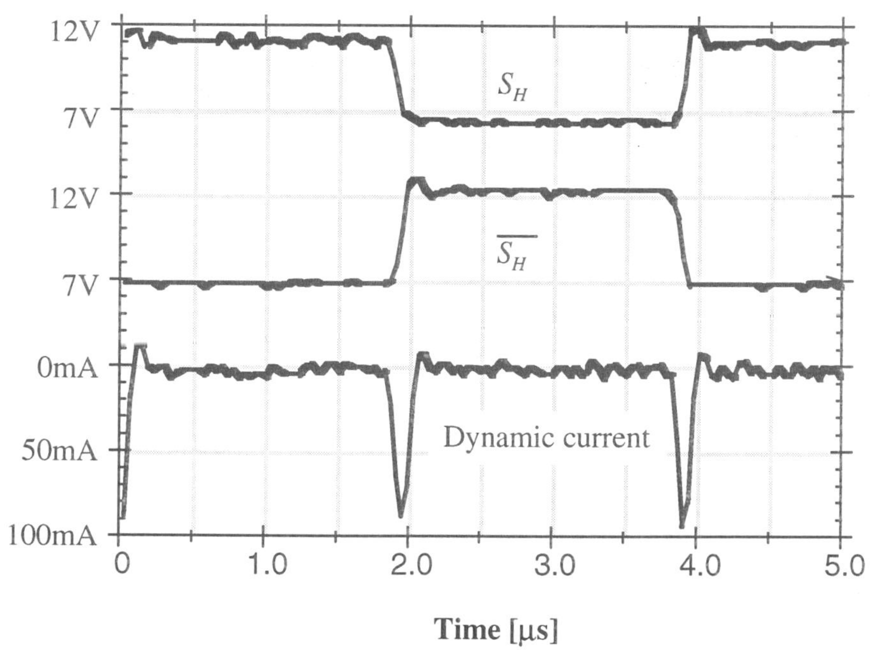

Figure 6.9: Measured dynamic current of the low-standby current static level shifter cell

The splitting of the cross coupled pair in the static level shifter cell is necessary for voltage levels compatibility. In fact, assuming that the full swing of S_H is V_{DD}, the corresponding full swing of S_{H2} should be $(V_{DD} + |V_{tp}|)$. In order to avoid stressing conditions on the gate oxide of M_{15}, the cross coupled pair is divided in two parts to share the full voltage swing. Note that the bulk of the pMOS transistors, handling gate voltage levels higher than $(V_{DDH} - V_{DD})$, is connected to the source of the device instead of the high-voltage power supply. Considering a high-voltage source follower transistor with $(W/L)_{M15} = 4000/5$ and a pull-up transistor with $(W/L)_{M8} = 1800/2$, the current driving capability of a single high-voltage buffer stage is 45mA. In Figure 6.9 the measured dynamic current stands for both buffer stages, where a peak of 90mA appears at the switching tops. The static current of the static level shifter is about 1.2mA, and the switching frequency is 500kHz.

As in the case of the improved static level shifter cell, the voltage level at the output S_{H2} is fixed by the current flowing through transistors M_4 and M_{12}. Assuming that transistor M_{15} is kept in weak inversion mode, the voltage level V_x at the output S_H is given by

$$V_x = V_{DDH} - \left[\sqrt{\frac{2nI_0}{K_p}} \left(\frac{1}{\sqrt{S_4}} + \frac{1}{\sqrt{S_{12}}} \right) + |V_{tp4}| + |V_{tp12}| - |V_{tp15}| \right] \tag{6.11}$$

where $S = (W/L)$. A reasonable assumption is that the threshold voltage values of transistors M_{12} and M_{15} are perfectly matched, it can be shown that the relative deviation of V_x around the typical value is given by

$$\frac{\Delta V}{V_{typ}} = 1 - \sqrt{\frac{K_{pt}}{K_p}} \left(1 - \frac{|V_{tpt}|}{V_{typ}} \right) - \frac{|V_{tp}|}{V_{typ}} \tag{6.12}$$

which is identical to the expression of the improved static level shifter cell. Thus, considering that I_0 is constant for all possible deviations of the electrical parameters and assuming a typical voltage swing of V_{DD}, the maximum relative deviation is about ±7% as well. Therefore, I_0 should be generated by a pMOS transistor similar to the one in the improved static level shifter cell, except that the scaling is different. In fact, since M_{19} is connected to V_{DD} and the full swing of S_{H2} is higher than V_{DD}, actually $V_{DD} + |V_{tp}|$, the transistor should be smaller than the pMOS load transistors of the static level shifter. Thus, the ratio between S_{19}, S_4 and S_{12} is defined as follows. Assuming that $S_1 = S_4 = S_9 = S_{12} = S$, $|V_{tp4}| = |V_{tp12}| = |V_{tp15}| = |V_{tp}|$ and that the full voltage swing at the output S_H is V_{DD}, the biasing current I_0 resulting from Equation (6.11) can be expressed as

$$I_0 = S_{19} \frac{K_p}{2n}\left(V_{DD} - \left|V_{tp}\right|\right)^2 = \frac{S}{4} \cdot \frac{K_p}{2n}\left(V_{DD} - \left|V_{tp}\right|\right)^2 \tag{6.13}$$

Thus, if the biasing transistor M_{19} is scaled such that $S_{19} = S/4$ the voltage level at the output of the level shifter reduces to

$$V_x = V_{DDH} - V_{DD} \tag{6.14}$$

as in the case of the improved static level shifter cell. The simulation data for all possible combinations of the electrical parameters deviations is shown in Figure 6.10, where the full swing of S_H is V_{DD} in all cases. It should be mentioned at this point that since the full voltage swing of S_{H2} is created by a proper scaling of the widths ratio between M_{19}, M_4 and M_{12}, the absolute value of the swing can differ slightly from V_{DD} if the threshold voltages of M_{12} and M_{15} are not perfectly matched. Moreover, the difference between the drain voltage operating conditions of M_{19}, M_4, and M_{12} as well as the physical mismatch of current mirror M_7-M_{14} result in a negligible relative deviation of the full voltage swing of S_H.

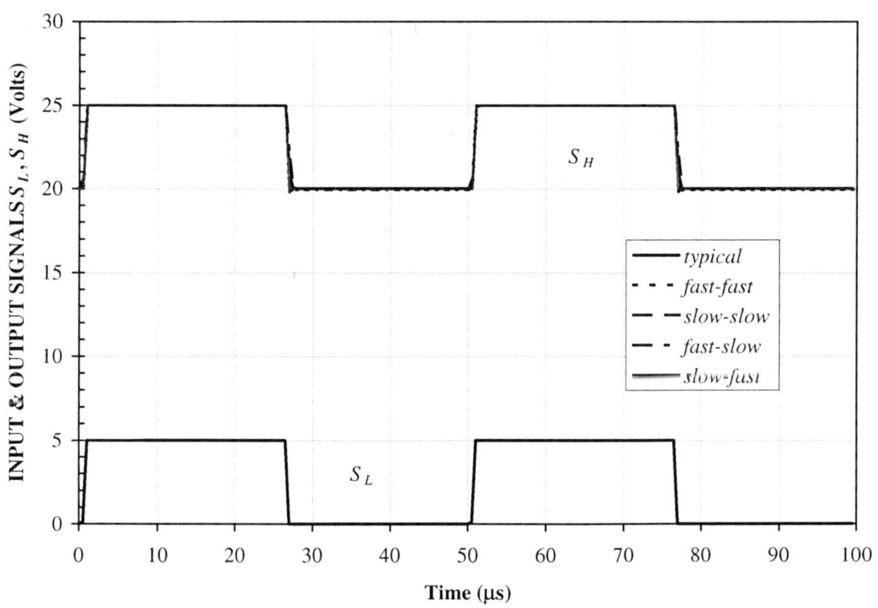

Figure 6.10: Input and output signals of the standby static level shifter cell for all possible combinations of the electrical parameters deviations

6.1.4 Transients considerations

In high-voltage integrated circuits, the power supply transients and the wide switching voltage range of the output stages can lead to reliability problems. In fact, due to capacitive coupling, such transients can induce voltage spikes at the gate nodes of the transistors. These nodes will be called *hot-nodes*, where undesirable effects such as hot-carriers generation or gate oxide breakdown must be considered and avoided. The scope of this section consists in analysing the amplitude and slew-rate of these spikes in order to characterise the robustness of the high-voltage digital output interface. Since a thin gate oxide solution is chosen for the high-voltage devices, they must satisfy the same safe operating conditions of gate transients as for low-voltage devices.

Considering the digital output interface shown in Figure 6.11 and using the improved static level shifter cell, the *hot-nodes* are the gates of the pMOS devices connected to V_{DDH} (nodes S_H and $\overline{S_H}$) and the gates of the nMOS devices connected to the HVNMOS transistor of the output stage (node S_L). While S_H and $\overline{S_H}$ are sensitive to the power supply transients through the gate capacitances C_{gp} and $\overline{C_{gp}}$, S_L and S_H are sensitive to the switching of the high-voltage output stage through the Miller capacitances C_{mp} and C_{mn}. The induced voltage spikes are analysed for both cases in the following sections.

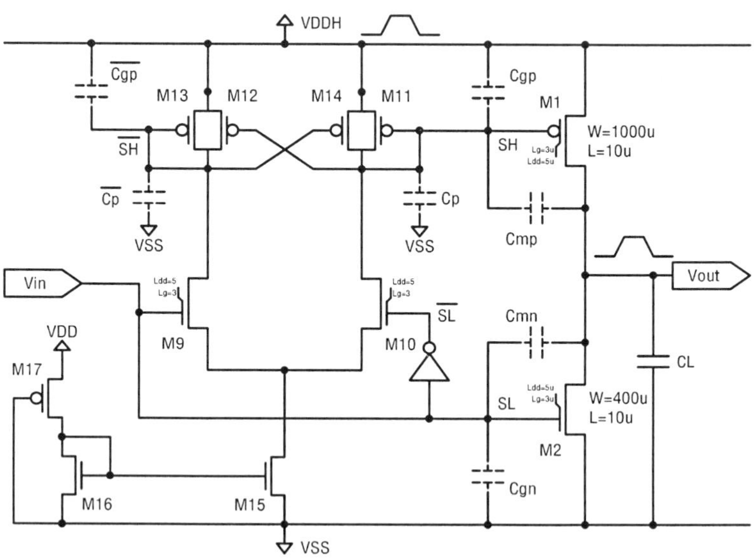

Figure 6.11: Schematic of a complete digital output interface using the improved static level shifter. All parasitic capacitances involved in power supply and output stage transients are included

1) Power supply transients: Starting with the power supply transients, Figure 6.12 shows the behaviour of the circuit to a typical switch-on condition of a stabilised high-voltage power supply. It starts from a voltage level of 24V and reaches 40V in less than 300μs. The shifted signal S_H adapts its level to the transients of the power supply, thus keeping the full voltage swing of S_H limited to V_{DD}. It appears clearly in the curve where the difference between V_{DDH} and S_H is computed continuously. The full swing (V_{DDH} - V_{SH}) is kept within the limit of 5V during the rising and falling edges of the power supply. The lower limit of the rise and fall times, beyond which the full swing exceeds 5V, depends on the static current of the level shifter and the parasitic capacitances at node S_H.

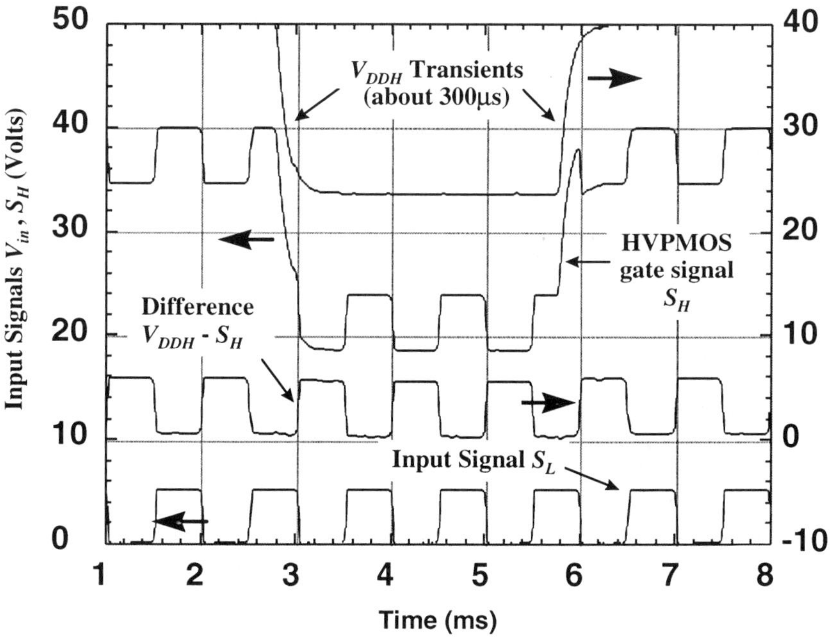

Figure 6.12: Behaviour of the level shifter during a typical switch-on condition of a stabilised high-voltage power supply

A first order approximation considers that the voltage spike is induced through the simple capacitive divider formed by C_{gp} and C_p. Such an approximation is based on the assumption that the impedance at node S_H is infinite. Thus, during the power supply transient the voltage at node S_H tracks V_{DDH} according to

$$v_{SH}(t) = v_{DDH}(t)\left(\frac{C_{gp}}{C_p + C_{gp}}\right) \tag{6.15}$$

The difference between the power supply voltage and the voltage at node S_H yields the magnitude of the spike $\Delta v_{sp}(t)$. Considering the full voltage swing of the power supply transient ΔV_{DDH}, the corresponding relative magnitude of the spike can be expressed as

$$\frac{\Delta V_{sp}}{\Delta V_{DDH}} = \frac{C_p}{C_p + C_{gp}} \tag{6.16}$$

which applies to the complementary output $\overline{S_H}$ as well. Consequently, in order to minimise the magnitude of the resulting spikes, the parasitic capacitance C_p must be minimised with respect to the equivalent gate capacitance of node S_H. However, the safe operating rules that apply to low-voltage gate transients, also fix the upper limit of ΔV_{sp}. Thus, considering a standard input protection formed by two diodes and a series resistor, the maximum spikes amplitude should not exceed a junction built-in potential V_{bi}. In the example of the digital output interface shown in Figure 6.11, the

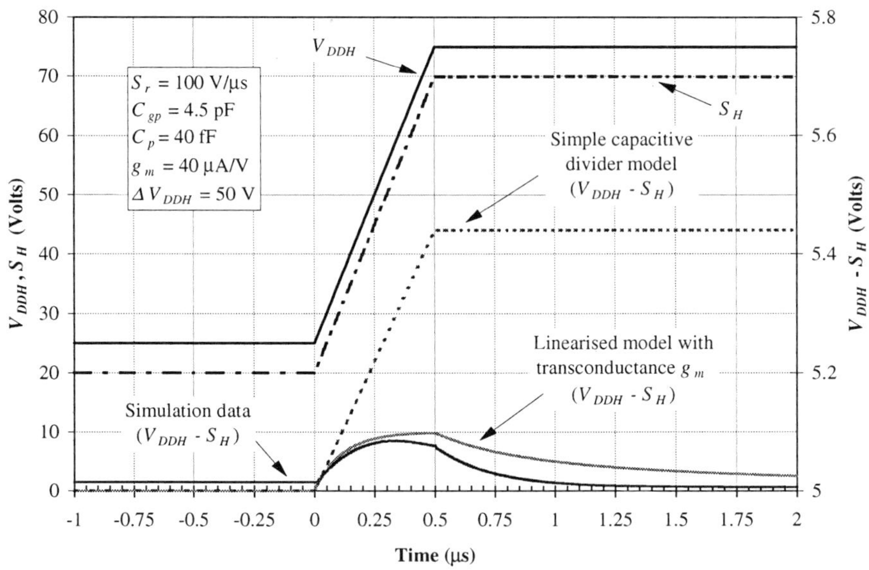

Figure 6.13: Induced voltage spikes models resulting from a high-voltage power supply transient in comparison to the simulation data of the improved static level shifter ($C_L << C_{mp}$)

equivalent gate capacitance at node S_H is about 4.5pF and the parasitic capacitance C_p is 50fF. Considering a maximum acceptable spike amplitude value of 0.55V, the high-voltage power supply transients should not exceed 50V, as shown in Figure 6.13. Note that the simple capacitive model is neither dependent on the nature of the power supply transient (step, exponential, linear) nor on its slew-rate S_r. Moreover, since the complementary output is not loaded with an output stage, the parasitic capacitance $\overline{C_p}$ is almost as high as $\overline{C_{gp}}$. Consequently, according to Equation (6.16) and assuming the same previous considerations, the power supply transient should be limited to 1.1V! This is obviously not correct due to the lack of the resistive component in the equivalent impedance of nodes S_H and $\overline{S_H}$. Actually, such an asymptotic model is valid only when $1/g_m$ (R_{on}) is negligible with respect to the impedance of C_{gp} ($\overline{C_{gp}}$). Thus, a more accurate modelling approach during the

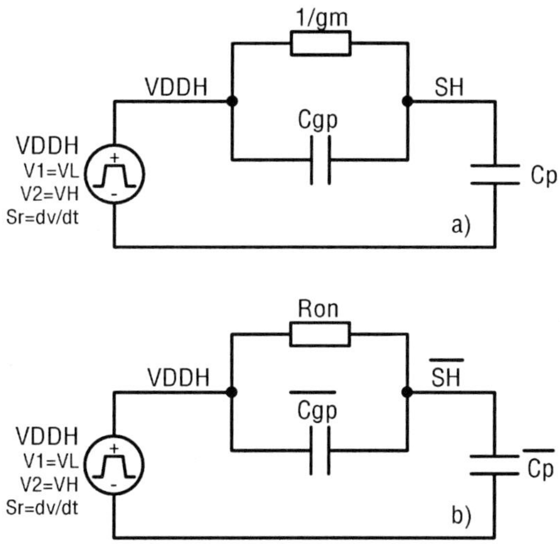

Figure 6.14: Equivalent small signals circuits of nodes a) S_H and b) $\overline{S_H}$ in the case of the improved static level shifter cell ($C_L << C_{mp}$)

power supply transient, requires considering the equivalent small signals circuits of nodes S_H and $\overline{S_H}$, as shown in Figure 6.14 where C_L is considered negligible with respect to C_{mp}. Writing the Kirchhoff's law at node S_H for instance, results in the following differential equation

$$\frac{dv_{SH}}{dt} + \frac{g_m}{C_{pt} + C_{gp}} v_{SH} = \frac{dv_{DDH}}{dt} \cdot \frac{C_{gp}}{C_{pt} + C_{gp}} + \frac{g_m}{C_{pt} + C_{gp}} v_{DDH} \qquad (6.17)$$

the solution of which depends on the nature of the power supply transient. If the rise and fall times of the power supply transient are larger than the time constant of the equivalent circuit, it is reasonable to assume that $v_{DDH}(t)$ is linear during these times, otherwise it should be considered as a step function of amplitude ΔV_{DDH}. Thus,

$$\begin{cases} v_{DDH}(t) = S_r \cdot t & (t_{r,f} > \tau) \\ v_{DDH}(t) = \Delta V_{DDH} \cdot \varepsilon(t) & (t_{r,f} < \tau) \end{cases} \tag{6.18}$$

where S_r is the slew-rate of the transient and $\varepsilon(t)$ is the unit step function. Under such conditions, the solutions to the differential equation are given by

$$\begin{cases} v_{SH}(t) = S_r \cdot t - S_r \cdot \tau_p \left(1 - e^{-t/\tau}\right) \\ v_{SH}(t) = \Delta V_{DDH}\left(1 - e^{-t/\tau}\right) \end{cases} \tag{6.19}$$

where

$$\tau_p = \frac{C_p}{g_m} \qquad \textit{and} \qquad \tau = \frac{C_{gp} + C_p}{g_m} \tag{6.20}$$

The difference between the power supply voltage and the voltage at node S_H yield the expression of the voltage spike $v_{sp}(t)$ as a function of time

$$\begin{cases} v_{sp}(t) = S_r \cdot \tau_p \left(1 - e^{-t/\tau}\right) & (t_{r,f} > \tau) \\ v_{sp}(t) = \Delta V_{DDH} \cdot e^{-t/\tau} & (t_{r,f} < \tau) \end{cases} \tag{6.21}$$

Thus, according to Equation (6.21), the induced voltage spikes increase significantly when the rise and fall times of the power supply transient become smaller than the time constant of the circuit itself. Therefore, such a situation should be avoided either by increasing the duration of these times or by minimising the circuit time constant through C_{gp} and g_m. Since τ is generally fixed by the required circuit performance, the minimum acceptable rise and fall times are fixed as well. In the example of the digital output interface considered above, the circuit time constant is about 100ns assuming that the output load is negligible with respect to C_{mp}. In this case, the minimum rise and fall times are set to at least 250ns (~2.3τ). Assuming that this condition is satisfied, the spike amplitude must be limited to the safe operating conditions of low-voltage gate transients. This is achieved by setting the upper limit of the slew-rate S_r, according to Equation (6.21). Thus, considering a maximum acceptable spike amplitude of 0.55V and a parasitic time constant $\tau_p \approx 1.25$ns, the maximum slew-rate is set to 450V/μs. This case is shown Figure 6.13, where the induced voltage spikes resulting from Equation (6.21) and from the simulation data are plotted for comparison. The discrepancies are mainly related to the parasitic capacitance C_p. In fact, the actual parasitic capacitance decreases as a function of the voltage V_{SH}, since it is given by the junction capacitance of the drain-to-bulk diffusion of transistor M_{10}. The higher the reverse voltage, the lower the capacitance value, resulting in a slight decrease of the spike amplitude. The latter is, however, in general agreement with the simulation data.

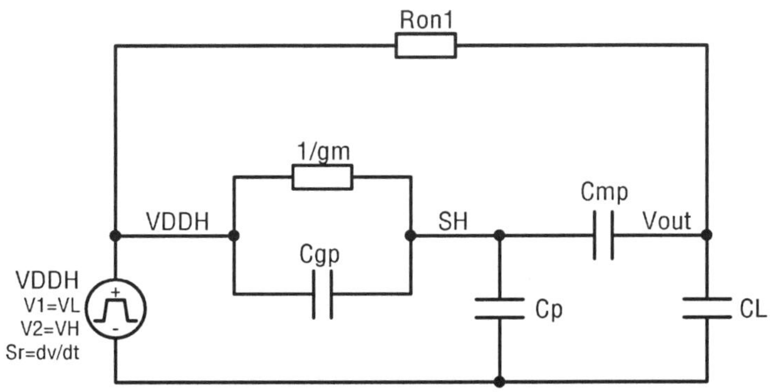

Figure 6.15: Equivalent small signals circuits of node S_H in the case of the improved static level shifter cell ($C_L \geq C_{mp}$)

In the case where the capacitive load is negligible anymore with respect to the Miller capacitance C_{mp}, the equivalent small signals circuit at node S_H is shown in Figure 6.15. In this case, the time constant τ_{on1} associated to the on state resistance R_{on1} of the HVPMOS device and the output load C_L, is important in determining the maximum slew-rate and safe transients conditions. In fact, if the slew-rate of the power supply is such that $t_{r,f} >> \tau_{on1}$, then the output voltage follows the transient without any delay and the transistor remains in the linear mode. Thus, the Miller capacitance is given by

$$C_{mp} = C_{gd} + C_{gdo} \qquad \left(t_{r,f} >> \tau_{on1}\right) \tag{6.22}$$

where C_{gd} is the gate-to-drain capacitance of the device, C_{gdo} the gate overlap of the field-plate, and τ_{on1}

$$\tau_{on1} = R_{on1} \cdot C_L \tag{6.23}$$

If the rise and fall times are faster than the time constant τ_{on1}, then the output follows the transient with a delay forcing the transistor into the saturation mode. In this case, the Miller capacitance reduces to C_{gdo}

$$C_{mp} = C_{gdo} \qquad \left(\tau << t_{r,f} << \tau_{on1}\right) \tag{6.24}$$

Note that Equation (6.24) assumes that the rise and fall times remain higher than the time constant of node S_H. The voltage spikes resulting from different load conditions are shown in Figure 6.17 where the total gate-to-drain capacitance of the device is plotted as well. The corresponding output voltage V_{out} is shown in Figure 6.16 where the delay between the power supply transient and V_{out} increases with increasing C_L. The asymptotic case occurs for $C_L = 1000xC_{mp}$ where the Miller capacitance reduces to C_{gdo} and the spike amplitude reaches a maximum, assuming that the slew-rate and the rise and fall times are kept constant. The maximum acceptable load value beyond which the transient condition is no longer safe is about 100pF ($20xC_{mp}$).

Finally, the induced voltage spike at the complementary output $\overline{S_H}$ is given by the same expressions as in Equation (6.21) except that the time constants are changed

$$\tau_p = R_{on} \cdot \overline{C_p} \qquad and \qquad \tau = R_{on}\left(\overline{C_p} + \overline{C_{gp}}\right) \tag{6.25}$$

Since R_{on} is generally smaller than $1/g_m$, the time constants associated to node $\overline{S_H}$ are negligible with respect to those of node S_H. Consequently, the former is less sensitive to power supply transients. It must be recalled at this point that the impedance of nodes S_H and $\overline{S_H}$ switches alternately between R_{on} and $1/g_m$ at the rate of the input signal. Note that since the output $\overline{S_H}$ does not drive a HVCMOS stage, the voltage spikes are independent of the load condition.

The transients considerations developed above for the improved static level shifter cell can be directly transposed to the case of the basic cell. They lead to the same results as those presented in Figure 6.13, assuming comparable performances of the digital output interface. In the case of the low-standby current cell, things becomes more complicated due to the high-voltage buffer stage, as shown in Figure 6.18. In fact, the equivalent small signals circuit of node S_H is shown in Figure 6.19, where all the components involved during the power supply transients are shown. Note that the output load is not critical in this case since the digital interface is supposed to drive a speaker load through an LC filter, as will be discussed in chapter 7. Thus, the impedance of the inductor during the power supply transient can be neglected to a first order approximation. The crossed components can be

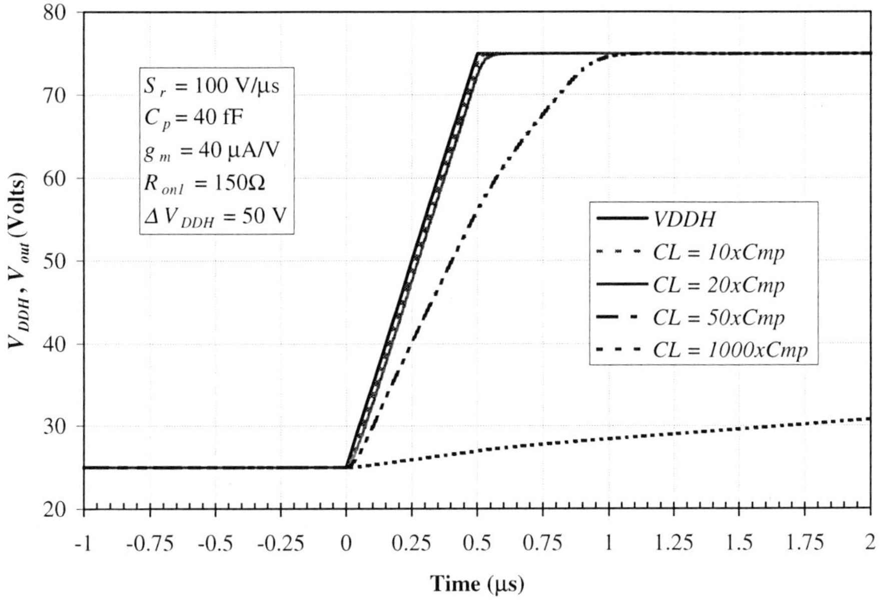

Figure 6.16: Delay between power supply transient and output voltage with increasing capacitor values

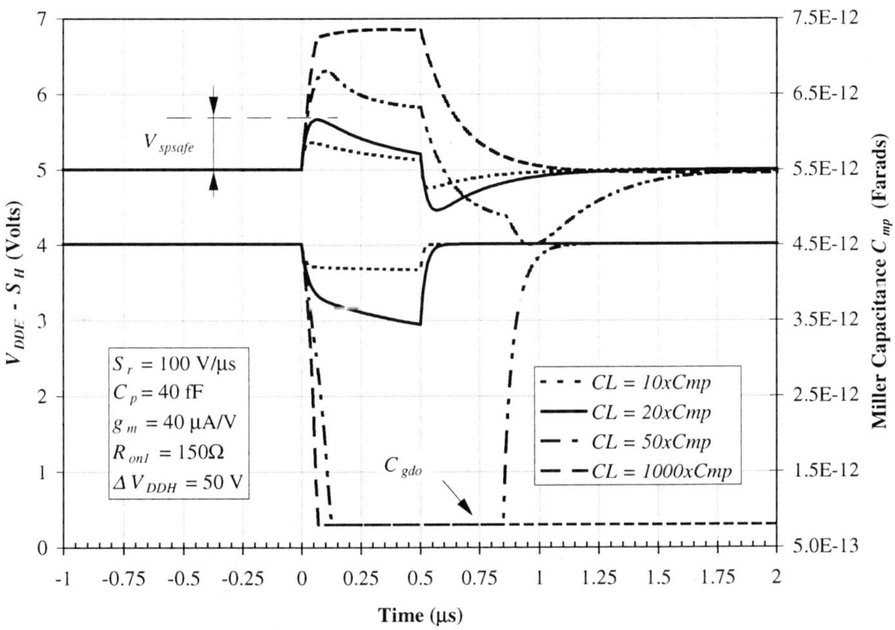

Figure 6.17: Induced voltage spikes and Miller capacitance of transistor M_1 for different capacitive load values

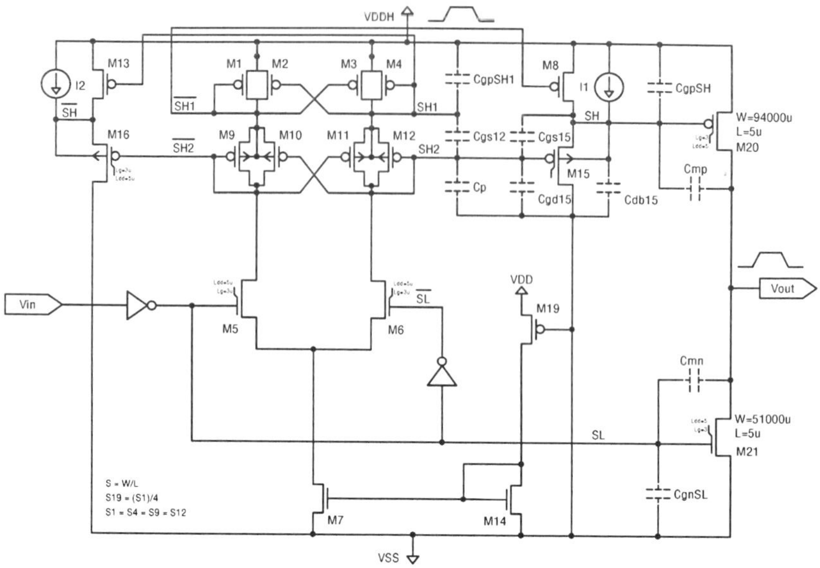

Figure 6.18: Schematic of a complete digital output interface using the low-standby current level shifter. All parasitic capacitances involved in power supply and output stage transients are included

reasonably neglected as well. Indeed, given the size of the transistors of the high-voltage buffer stage ($S_{15} = S_{16} = 4000/5$ and $S_{13} = S_8 = 1800/2$), the parasitic capacitances such as the well junction capacitance C_{b15} and the parasitic capacitance C_p at node S_{H1} are negligible with respect to the huge gate-to-drain capacitance C_{gd} of M_{15}. Even if these parasitic components are neglected, the solution to the differential equation resulting from the Kirchhoff's law at node S_H is not straightforward. However, the qualitative behaviour of the voltage spikes can still be predicted, and the exact value can be simulated. Thus, the worst condition occurs when the transconductance values of M_4 and M_{12} are so small that the corresponding impedance is infinite. In such an asymptotic case, the series connection of C_{gs12} and C_{gpSH1} leads to the equivalent small signals circuit shown in Figure 6.20, where the power supply transients are transmitted to node S_{H2} through the capacitive divider formed by C_{geq} and C_{gd15}. Thus, if C_{gd15} is significantly higher than C_{geq}, the induced voltage spikes can reach several volts at the gate node. Therefore, to avoid such a destructive situation, the transconductance of M_4 and M_{12} must be as high as possible to reduce the impedance of nodes S_{H1} and S_{H2}. If this condition is satisfied, the low-standby current level shifter cell behaves safely under power supply transients. The exact amplitude of the spikes as well as the upper limit of the slew-rate of the power supply can be computed by using a circuit simulator. Note that the increase of the transconductance of transistors M_4 and M_{12} is a trade-off between power consumption and power supply transients. In the example considered in Figure 6.8,

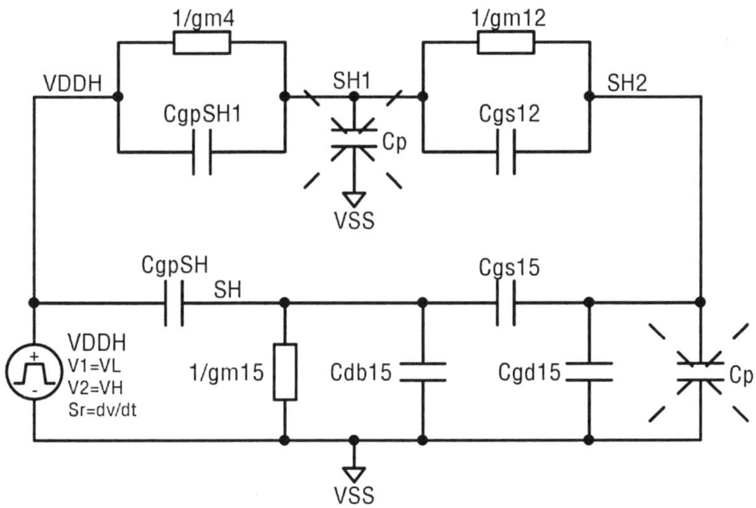

Figure 6.19: Equivalent small signals circuit of node S_H in the case of the low-standby current level shifter cell

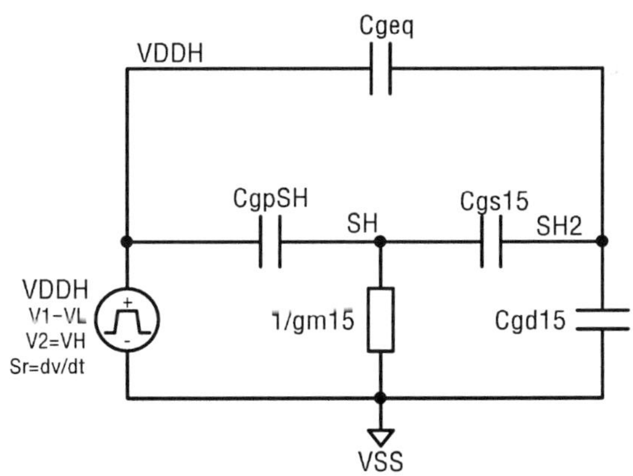

Figure 6.20: Equivalent small signals circuit of node S_H in the asymptotic case of the low-standby current level shifter where $1/g_{m4} = 1/g_{m12} = \infty$

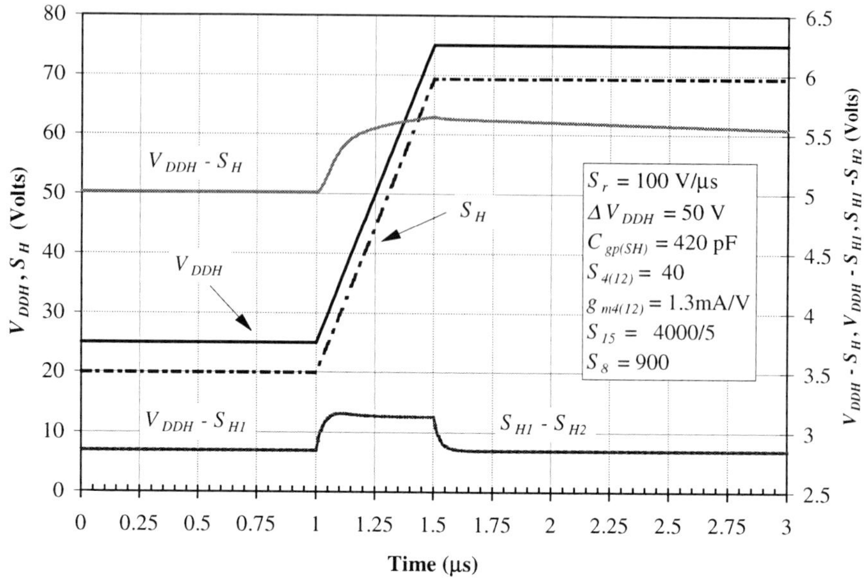

Figure 6.21: Simulation data of induced voltage spikes resulting from a power supply transient in the case of the low-standby current level shifter

the transconductance value of M_4 and M_{12} is about 1.3mA/V assuming $S_4 = S_{12} = 40$. The corresponding standby current is 1.5mA. Considering the same power supply transient condition as in the case of the improved static level shifter cell (S_r = 100V/μs and ΔV_{DDH} = 50V), the induced voltage spike amplitude at node S_H is 0.6V, as shown in Figure 6.21. For more constraining power supply transients the transconductance values can be increased further, as far as the corresponding standby current remains negligible in comparison to the 50mA required initially by the static cell. The amplitude of the spikes at nodes S_{H1} and S_{H2} is lower than the one at node S_H. Moreover, since the full voltage swing of $V_{DD} + |V_{tp}|$ is shared between M_4 and M_{12}, the tolerable spike amplitude at the corresponding nodes is higher than that of node S_H. Therefore, if the circuit is scaled to minimise the spikes at node S_H, those of nodes S_{H1} and S_{H2} are not critical.

The main rule resulting from this section, to be kept in mind to avoid significant voltage spikes at the gate nodes of the level shifters, consists of maximising the transconductance of the pMOS diode loads and minimising the parasitic capacitances to ground. It will be shown in the next section that the same rule applies for output transients as well.

2) Output stage transients: As was mentioned previously, the output stage transients can induce voltage spikes at the gate nodes of the pMOS and nMOS transistors, through the Miller capacitance of the high-voltage devices. The wider the transistors are the higher is the coupling of the transients. If some particular precautions are not considered, the resulting spikes can lead to hot-carriers generation, gate oxide breakdown and latchup triggering. The latter effect is additional in comparison to the case power supply transients. It is related to the conduction of the body diodes of the transistors of the gate drivers, as shown in Figure 6.22. Note that positive or negative voltage spikes at the gate of the HVNMOS transistor are clamped by the body diodes D_{p19} and D_{n18}. Thus, the gate node of this transistor is self protected, but, the layout precautions discussed in chapter ***5*** should be respected at the driver level to avoid latchup problems.

However, while node S_L is self protected, the one of the HVPMOS transistor is not. In fact, during the falling edge of the transient, no clamping of the induced voltage spike is performed, since the body diodes D_{p11} and D_{p14} limits the positive ones only. Therefore, the amplitude of the spike, resulting from the falling edge of the output transients, should be analysed as well to avoid stressing conditions of the gate of the pMOS transistors. During the falling edge of the output transient the HVNMOS transistor is supposed to switch on and the HVPMOS transistor is

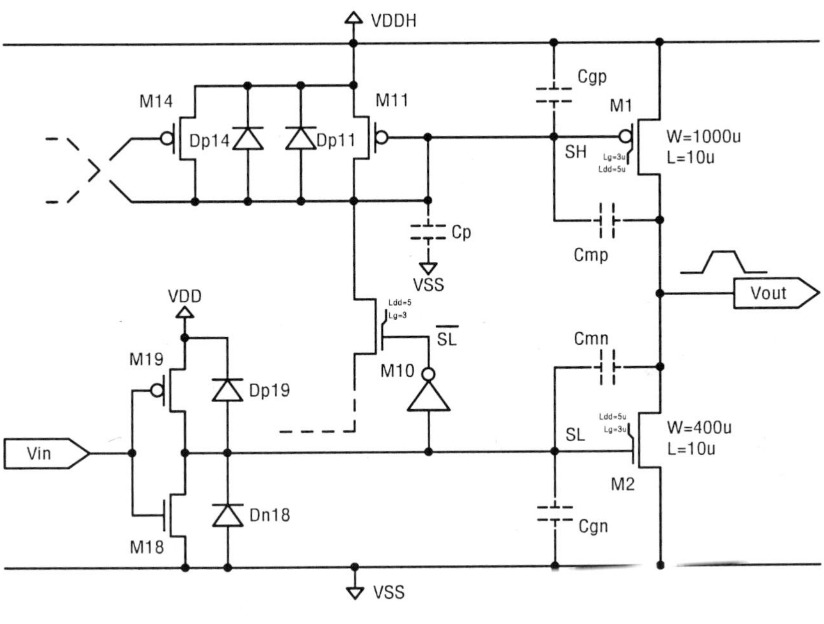

Figure 6.22: Detailed schematic of the digital output interface showing the self protection body diodes of the gate drivers

supposed to switch off. Note that the configuration where the output is pulled down by another HVCMOS stage while both high-voltage devices are turned off (high-impedance state) is also possible. However in both cases, the equivalent small signals circuit of nodes S_H and S_L is shown in Figure 6.23, where R_{on} is the on state resistance of the pMOS transistors pulling up the nodes. Writing the Kirchhoff's law at nodes S_H and S_L yields the same differential equation in both cases. The solution is discussed for node S_H only as it is the most sensitive node. Thus,

$$\frac{dV_{SH}}{dt} + \frac{V_{SH}}{R_{on}C_{mp}} = \frac{C_{mp}}{C_{gp} + C_{mp}} \cdot \frac{dV_{out}}{dt} \tag{6.26}$$

depends on the nature of the output transient, as in the case of the power supply transients. Thus, if the fall time t_f of the output signal is larger than the time constant of the equivalent circuit, it is reasonable to assume that $V_{out}(t)$ is linear during this time, otherwise it should be considered as a step function of amplitude ΔV_{out}. Thus,

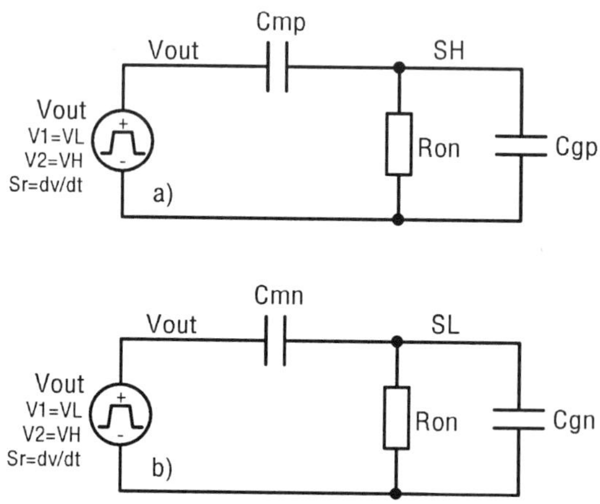

Figure 6.23: Equivalent small signal circuits of nodes a) S_H and b) S_L in the case of the improved static level shifter cell

$$\begin{cases} v_{out}(t) = S_r \cdot t & \left(t_{r,f} > \tau\right) \\ v_{out}(t) = \Delta V_{out} \cdot \varepsilon(t) & \left(t_{r,f} < \tau\right) \end{cases} \tag{6.27}$$

where S_r is the slew-rate of the transient and $\varepsilon(t)$ is the unit step function. As in the case of the power supply transients, the solutions to the differential equations are given by

$$\begin{cases} v_{SH}(t) = S_r \cdot \tau_m\left(1 - e^{-t/\tau}\right) \\ v_{SH}(t) = \Delta V_{out} . e^{-t/\tau} \end{cases} \quad \textbf{(6.28)}$$

where

$$\tau_m = R_{on} C_{mp} \quad and \quad \tau = R_{on}\left(C_{mp} + C_{gp}\right) \quad \textbf{(6.29)}$$

The expression of the voltage spike $v_{sp}(t)$ as a function of time during the output transient is given by Equation (6.28), since the voltage at the gate node S_H is not supposed to change with the output voltage. Therefore, the induced voltage change given by Equation (6.28) yields the amplitude of the spike. Thus,

$$\begin{cases} v_{sp}(t) = S_r \tau_m\left(1 - e^{-t/\tau}\right) & \left(t_{r,f} > \tau\right) \\ v_{sp}(t) = \Delta V_{out} \cdot e^{-t/\tau} & \left(t_{r,f} < \tau\right) \end{cases} \quad \textbf{(6.30)}$$

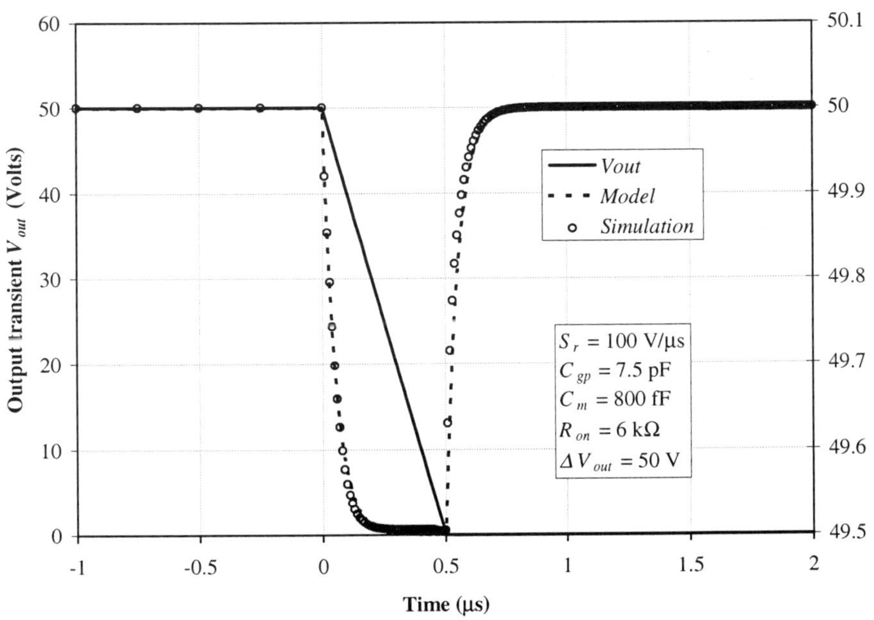

Figure 6.24: Voltage spike resulting from an output stage transient using the improved static level shifter cell

As in the case of the power supply transients, the minimum acceptable rise and fall times should be fixed as a function of the time constant τ of the circuit. If this condition is satisfied, the spike amplitude is fixed by setting the upper limit of the slew-rate S_r as a function of the parasitic time constant τ_m, according to Equation (6.30). Assuming a maximum acceptable spike amplitude value of 0.55V and a parasitic time constant of 5ns, the maximum slew-rate is set to 110V/μs. Figure 6.24 shows a safe output transient where the voltage spike resulting from Equation (6.30) and the one resulting from the simulation data are plotted for comparison.

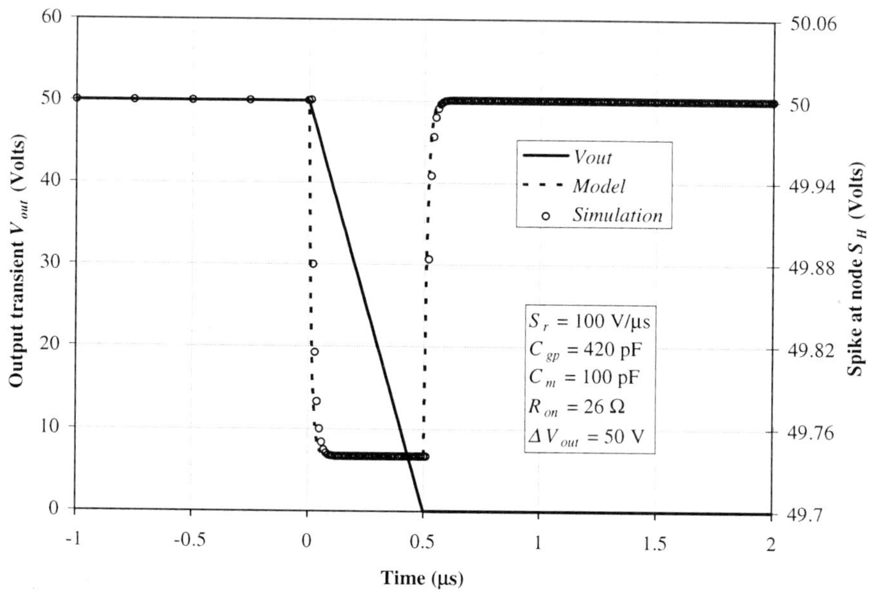

Figure 6.25: Voltage spike resulting from an output stage transient using the low-standby current static level shifter cell

In the previous discussion it was assumed that the HVPMOS transistor is turned off during the falling edge of the output transient. Therefore, the impedance considered in the equivalent small signals circuit R_{on}. The particular case where the HVPMOS transistor is on during the falling edge of the output occurs when two digital interfaces are mounted in a full bridge configuration to drive a floating load. The model shown in Figure 6.23 remains valid except that the circuit time constants become

$$\tau_m = \frac{C_{mp}}{g_m} \qquad and \qquad \tau = \frac{C_{gp} + C_{mp}}{g_m} \tag{6.31}$$

where g_m is the transconductance of the pMOS diode load of the level shifter. As for the case of R_{on}, g_m is scaled to meet the requirements of the gate transients safe operating conditions. Note that the most constraining condition, between the power supply transients and the output stage transients, sets the value of g_m and R_{on}.

Finally, given the topology of the high-voltage buffer stage of the low-standby current level shifter, it can be easily shown that the equivalent small signals circuit during the output transient reduces to that of Figure 6.23. The on state resistance corresponds to the pull-up transistor M_8 of the buffer stage. When the HVPMOS transistor of the output stage is on, during the falling edge of the output, the transconductance g_{m15} of the source follower transistor is considered in place of R_{on8}. Thus, the behaviour of the low-standby current level shifter cell is identical to the basic and improved cells, as shown in Figure 6.25, where the same output transient conditions are used.

6.2 High-voltage analog output interfaces

In high-voltage applications requiring analog signal processing the same approach as for the digital output interfaces should be used. That is, the complex analog functions are implemented with standard low-voltage circuits, and the level shifting of the resulting signal is performed through the high-voltage analog output interfaces discussed in this section. As in the case of the digital interfaces the driving of the gates of the HVPMOS transistors must be performed securely without stressing conditions. This is achieved by the use of analog level shifting techniques based on the same "voltage mirror" circuit block used in the digital level shifters discussed in section 6.1. Given the robustness of this basic block against power supply and output transients, it should be used each time that a HVPMOS transistor is driven. The sensitivity of the block against these transients is a trade-off between the transients constraints and the required circuit performance.

6.2.1 Basic operational transconductance amplifier (OTA)

The simplest high-voltage analog output interface consists of the basic operational transconductance amplifier shown in Figure 6.26, where the analog voltage mirror block is pointed out by the dashed frame. It is used to drive pure capacitive loads. With respect to its original low-voltage topology [14], the amplifier shown in Figure 6.26 was modified and scaled to drive the HVCMOS output stage. Thus, the analog voltage mirroring is performed through the current mirror M_2-M_6 and the pMOS diode load M_9. The equivalent small signals circuit of the voltage mirror is a voltage controlled current source in series with the transconductance of transistor M_9. The voltage variations resulting at node N_2 are mirrored through $1/g_{m9}$ with respect to the high-voltage power supply V_{DDH}, as shown schematically in Figure 6.26. Consequently, the HVPMOS transistor of the output stage is driven as safely as in the case of the digital output interfaces. The behaviour under power supply and output transients is set by the transconductance of transistor M_9, the gate capacitance C_{gp} at node N_3 and the corresponding parasitic capacitance C_p. The transients considerations of this OTA are discussed further in this section. The HVNMOS transistor of the output stage M_8 is driven directly by the complementary output N_1 of the input differential pair. Finally, the current biasing stage uses a high-voltage cascode transistor M_{14} to protect the gate of the pMOS transistors of the high-voltage cascode current mirror M_{15}-M_{16} and M_{17}-M_{18}.

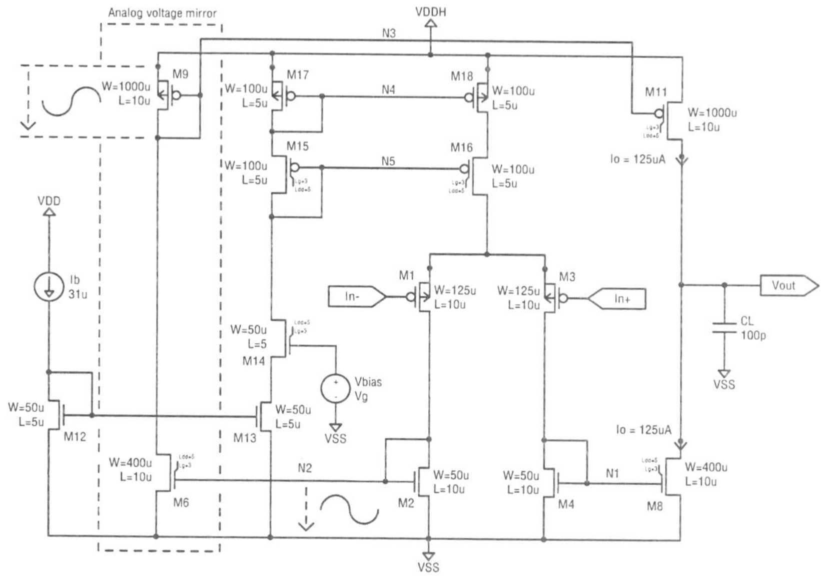

Figure 6.26: Schematic of the basic high-voltage operational transconductance amplifier. The dashed frame points out the analog voltage mirror circuit block

Given the low-to-high voltage output transposition of the input signal, the amplifier is always used in a closed loop configuration, where the gain A is set by the ratio of the DC levels between input and output. Thus, considering a 25V high-voltage power supply and standard 5V analog inputs, A = 5. The measured high-voltage capabilities of the OTA in such a closed loop configuration are shown in Figure 6.27 with and output load C_L of 1nF. The scaling of the amplifier, leading to such a high-voltage performance is similar to that of the low-voltage amplifier except for high-voltage transients. In fact, once the circuit is scaled to meet the required performances, the induced voltage spikes at the "hot nodes" should be checked under maximum transients specifications. The scaling of the OTA shown in Figure 6.26 is given next as an example.

The specifications of the amplifier are, a maximum open loop load of 100pF, a gain bandwidth product (GBW) higher than 1Mhz, and a typical slew-rate higher than 2.5V/μs. Under these conditions [14], the minimum transconductance of the amplifier is given by

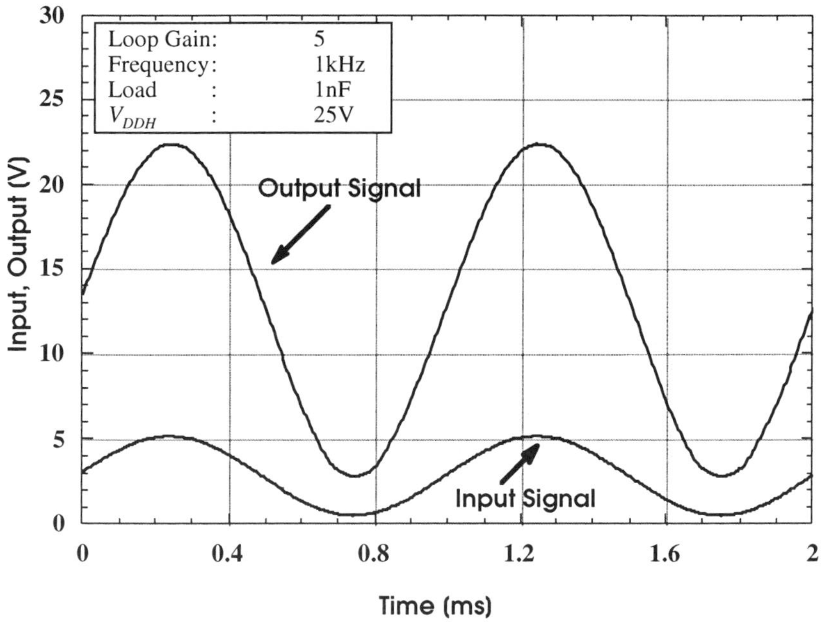

Figure 6.27: low-to-high voltage output transposition of an input signal of 4Vpp. The loop gain is 5 and the high-voltage power supply is 25V

$$g_m = B \cdot g_{m1(3)} = \omega_{GBW} \cdot C_L \approx 630\ \mu A/V \tag{6.32}$$

where B is the current gain between the first and second stage. Considering the minimum slew-rate specification, the maximum output current is

$$I_o = B \cdot I_b = S_{ro} \cdot C_L = 250\,\mu A \tag{6.33}$$

where S_{ro} is the slew rate of the output and I_b is the input biasing current. Thus, the dimensions of the high-voltage output devices is given by

$$S_{11(8)} = \frac{n \cdot g_m^2}{K_{11(8)} \cdot I_o} \tag{6.34}$$

where $S = (W/L)$ and $K = \mu C_{ox}$. For stability considerations the phase margin of the amplifier should be higher than 60°. Therefore, all the parasitic poles must be at least 2.2 times higher than the unity gain bandwidth frequency. Thus, the transconductance of transistors M_4 and M_2 is given by

$$g_{m2(4)} \geq 2.2 \cdot \omega_{GBW} \cdot C_{N1(N2)} = 62\ \mu A/V \tag{6.35}$$

where $C_{N1(N2)}$ are the parasitic capacitances at nodes N_1 and N_2. Considering $S_{2(4)} = S_8/B$ yield

$$B \leq \frac{1}{g_{m2(4)}} \sqrt{\frac{K_8 \cdot I_o \cdot S_8}{n}} = 10 \tag{6.36}$$

Setting $B = 8$, results in

$$\begin{cases} I_b = 31\,\mu A \\ g_{m2(4)} = 78\,\mu A/V \\ g_{m1(3)} = 78\,\mu A/V \end{cases} \tag{6.37}$$

which yields the dimensions of the input devices assuming they are in strong inversion [14]

$$S_{1(3)} = \frac{n \cdot g_{m1(3)}^2}{K_{1(3)} \cdot I_b} \tag{6.38}$$

Finally, assuming a unity current gain between M_9 and M_{11}, the pole associated with node N_3 should be checked to be higher than 2.2 times the unity gain bandwidth frequency. Considering the transconductance given by Equation (6.32), and the parasitic capacitance at node N_3, the third pole is located at about 13Mhz. The Bode diagram resulting from the schematic of Figure 6.26 is shown in Figure 6.28, where the phase margin is higher than 60° at f_{GBW}, and the open loop gain A_0 is about 70dB. The step response to a large signal input is shown in Figure 6.29, where the amplifier is connected in a follower configuration. The slew-rate is about 2.5V/μs and the capacitive load is 100pF. As the input stage is not optimised for a maximum common mode input range, the step starts from the 2.5V, which is the analog ground of CMOS circuits using a 0-5V low-voltage power supply. Finally, the main performances of the OTA are summarised in Table 6.1.

6.2.2 Class-AB output stage operational amplifier

When the output load exhibits a resistive component in addition to the capacitive load, the basic OTA should be replaced by the high-voltage operational amplifier [15] presented in this section. The output stage of this amplifier operates in class AB mode to minimise on chip power consumption. In fact, given the high-voltage power

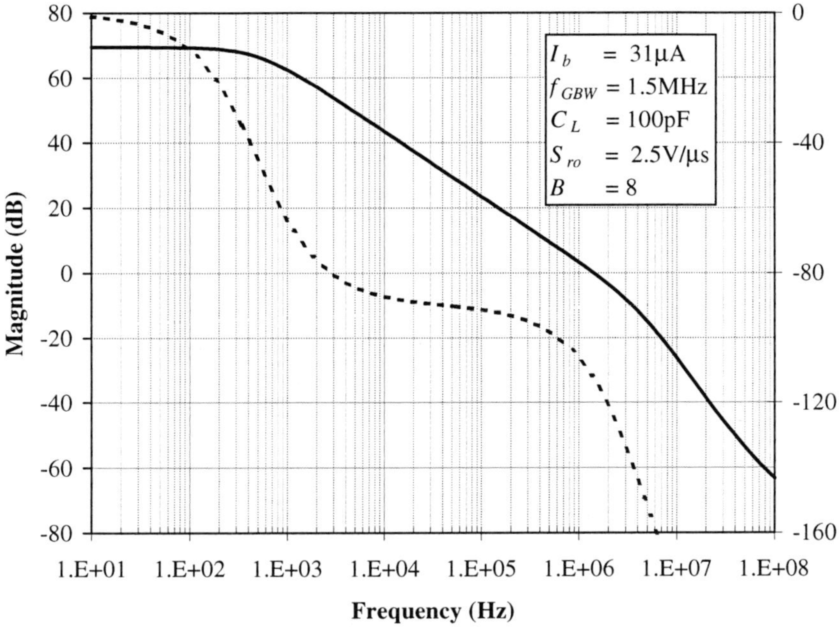

Figure 6.28: Bode diagram of the basic high-voltage OTA shown in Figure 6.26

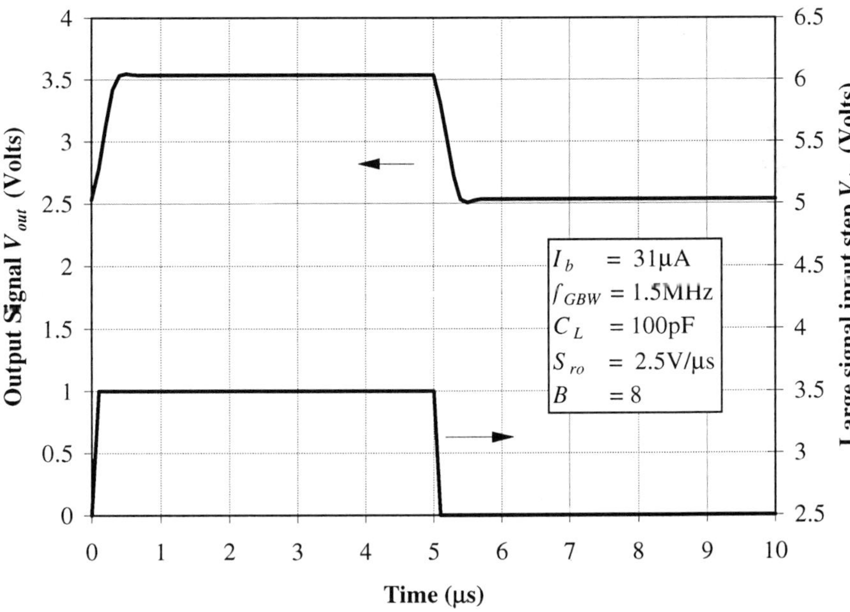

Figure 6.29: Large signal step response of the amplifier showing its slew-rate and stability in a follower configuration

PARAMETER	VALUE
Biasing current I_b :	31 μA
Open loop gain A_0 :	70 dB
Gain bandwidth f_{GBW} :	1.5 MHz
Phase margin (C_L = 100pF) :	60°
Slew-rate S_r (C_L = 100pF) :	2.56 V/μs
HV power supply V_{DDH} :	25V

Table 6.1: Main performances of the high-voltage operational transconductance amplifier

supply range, "normal" low-voltage static currents can lead to excessive power dissipation under V_{DDH}. The main characteristic of the considered class AB amplifier is its ability to source and sink a dynamic current as high as 100mA, starting from a standby current of 100μA. Therefore, it is particularly suited for driving large capacitive loads under high-voltage power supplies. The large output current ratio results from the tight control of the operating point of the HVCMOS stage, under which the transistors are almost in weak inversion in the amplifier's standby mode. In the driving mode, the transistors are forced into strong inversion with their maximum current drive capabilities. This is achieved by the biasing circuit shown in the schematic of Figure 6.30 and controlling the voltage at nodes N_1 and N_2. It is equivalent to a voltage source V_{bias} connected between these nodes. In order to source or sink current at the output, V_{bias} is shifted towards V_{SS} or V_{DDH}. This is performed through the source follower transistors M_2 and M_4. Assuming that M_4 and M_7 have matched biasing conditions (same V_{gs}), the voltage at node N_1 will match that of node N_3, controlling thus the quiescent current of the output transistor M_3. The quiescent current of transistor M_1 is controlled symmetrically. In other words, if I_1 and I_2 are matched to I_b, and M_4 to M_7 and M_3 to M_8, symmetrically M_2 to M_6 and M_1 to M_5, then the quiescent current I_o matches the biasing current I_b in the standby mode. In the dynamic mode, the voltage controlled current source shifts symmetrically the voltage at nodes N_1 and N_2 towards V_{DDH} and V_{SS} through the source follower transistors M_2 and M_4. The transistor diodes M_{13}-M_{14} and M_{11}-M_{12} are used to clamp the full voltage swing of nodes N_1 and N_2 to V_{DD}.

As far as this condition is satisfied they are turned off without altering the circuit performance. Since I_2 is connected to the source of transistor M_4, the cross-coupling of the source followers is necessary to increase the equivalent small signal impedance seen by the input stage of the amplifier. In fact, Figure 6.31 shows the equivalent small signal circuit of the output stage, where the capacitances are neglected. Expressing the admittance y_{N1} of the resulting quadripole, yields

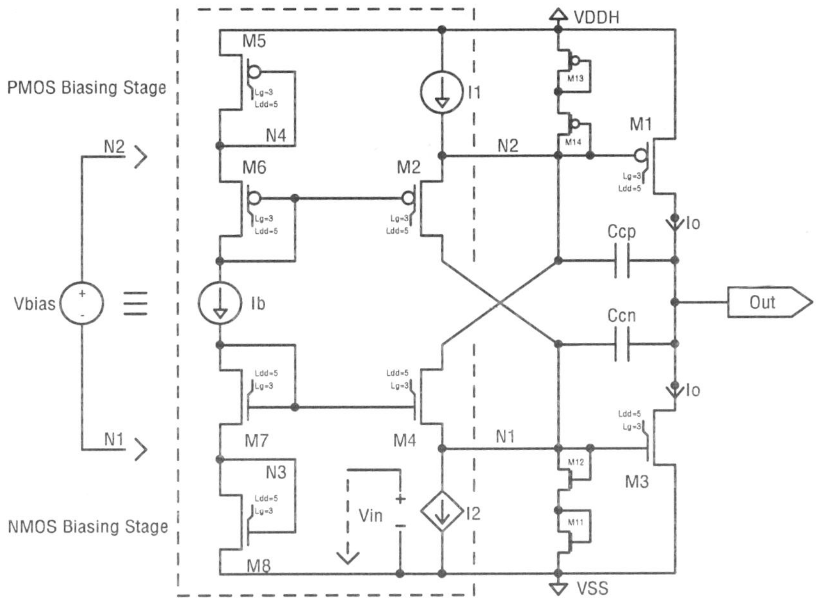

Figure 6.30: High-voltage class AB output stage including the voltage biasing circuit controlling the voltage at nodes N_1 and N_2

$$y_{N1} = \left.\frac{i_{N1}}{v_{N1}}\right|_{v_{out}=0} = g_{02} + g_{01}\frac{g_{ds2} + g_{ds4} - g_{ms}}{g_{02} - g_{ms} + g_{ds2} + g_{ds4}} \qquad (6.39)$$

where the conductances can be reasonably neglected with respect to the source transconductance g_{ms} of the transistors. Thus,

$$y_{N1} = g_{02} + g_{01} \qquad \textbf{(6.40)}$$

where g_{0l} and g_{02} are the output conductances of the current sources I_l and I_2. Consequently, the equivalent impedance seen at node N_l is very high, thus optimising the gain of the input stage. The latter can be a simple low-voltage OTA as shown in the complete schematic of Figure 6.32. Note that the biasing current source I_b is achieved by the same high-voltage cascode technique presented in the basic high-voltage OTA of section 6.2.1. Since the equivalent small signals circuit of the output stage reduces to a simple quadripole of input impedance $z_{in} = 1/(g_{0l}+g_{02})$ and a total output transconductance $g_{mout} = g_{ml}+g_{m2}$, the equivalent small signals circuit

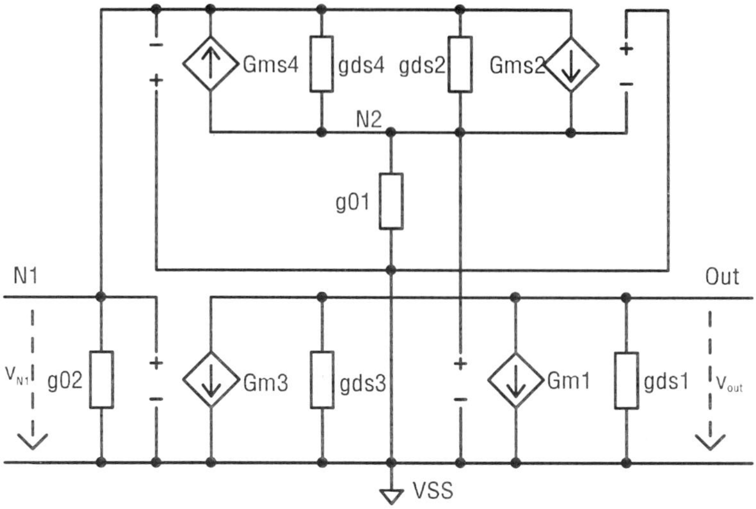

Figure 6.31: Equivalent small signal circuit of the class AB output stage showing the influence of the cross-coupled source followers on the input impedance. The capcitances are neglected.

of the complete amplifier can be considered as a simple two stage operational amplifier (OpAmp), as shown in Figure 6.33. Given the symmetry of the output stage, the compensation capacitor C_c is the sum of C_{cp} and C_{cn}. Note that these capacitors must be made with metal layers over thick oxide since they are connected to the high-voltage part of the amplifier. Finally, the input capacitor of the output stage is $C_{in} = C_{N1}+C_{N2}$ and the output conductance is $g_{dsout} = g_{ds1}+g_{ds3}$. Thus, the scaling of this OpAmp is now a straightforward exercise as considered in the following example.

The maximum open loop load is $C_L = 60$pF and the gain bandwidth product should be higher than 1MHz. The stability is unconditional and the slew-rate of the OpAmp should be about 1V/μs. Such a low value is related to a minimum current consumption requirement, since the high-voltage power supply ($V_{DDH} = 24$V) is generated on chip from a 3V battery with a DC to DC voltage booster. In the final application, the amplifier must drive, in a full bridge closed loop configuration, a floating piezzo-electric transducer with a capacitive component of 10nF. The resistive component is not specified.

Thus, from the unconditional stability specification, the compensation capacitor is given by

$$C_c = 0.22 \cdot C_L \approx 10\,pF \tag{6.41}$$

assuming that the second pole is at least $2.2x f_{GBW}$ and that the right-half-plane zero is

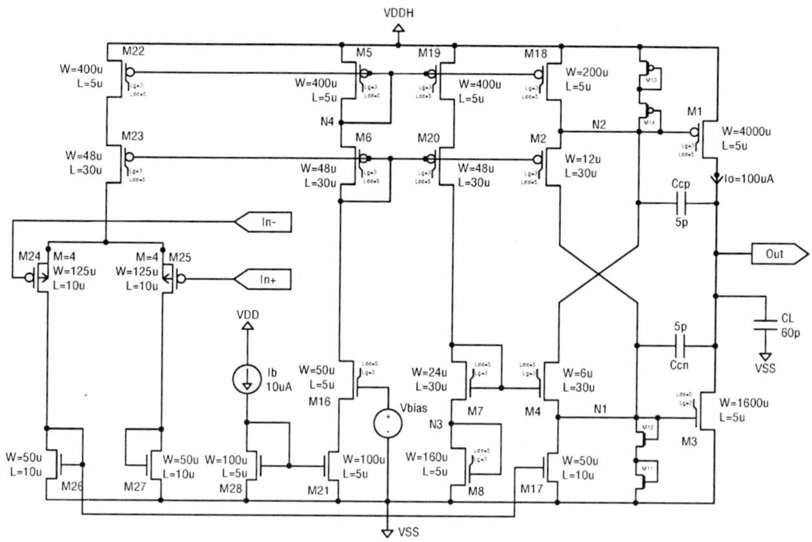

Figure 6.32: Complete schematic of the high-voltage class AB output stage operational amplifier

about $10xf_{GBW}$. From the minimum gain bandwidth product condition, the transconductance of the input stage is given by

$$g_{m,in} \geq C_c \cdot \omega_{GBW} \approx 63\,\mu A/V \tag{6.42}$$

and from the minimum slew-rate condition the biasing current of the input stage is

$$I_b \geq SR \cdot C_c \approx 10\,\mu A \tag{6.43}$$

Assuming the input stage in strong inversion, the dimensions of the input devices are given by

$$S_{24(25)} = \frac{n \cdot g_{m,in}^2}{K_{24(25)} \cdot I_b} \tag{6.44}$$

The actual dimensions of the input devices are twice the theoretical value, since the Miller capacitance of the output devices is about 4pF (40% of C_c). Therefore, to keep the same f_{GBW}, the transconductance should be increased by increasing S. Moreover, to keep the slew-rate constant, the biasing current should be increased as well, which is not performed in this particular case because the total current consumption is more critical. Since no special constraint is specified on the input common mode range, the scaling of M_{26}-M_{27} is not critical.

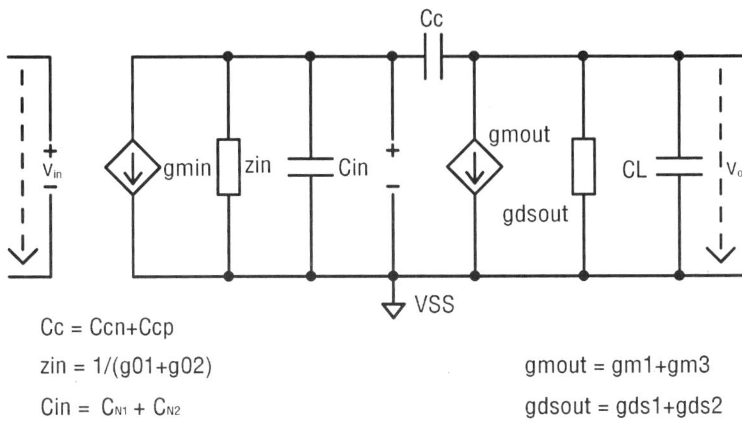

Figure 6.33: Simplified form of the equivalent small signals schematic of the complete amplifier

To source and sink a maximum output current of 100mA, the dimensions of the output transistors are given by

$$S_{1(3)} = \frac{2nI_{o\,\max}}{K_{1(3)}\left(V_{gs} - V_t\right)^2} \tag{6.45}$$

Note that the resulting transconductance of the output stage ($g_{mout} \approx$ 3mA/V) verifies the conditions for which the second pole was assumed to be at least 2.2xf_{GBW} and the right-half-plane zero was assumed to be at10xf_{GBW}. Finally, the scaling of the protection transistors M_{11}-M_{12} and M_{13}-M_{14} is discussed in section 6.2.3 with the transient considerations.

Figure 6.34 shows the linear response of the amplifier when mounted in a full bridge closed loop configuration and loaded with 10nF in parallel with 1kΩ. The high-voltage power supply is 24V leading to a differential full voltage swing at the output of ±24V. The measurement was performed by using a DC input level of 2.5V, leading thus to a closed loop gain A = 4.8. The magnitude Bode diagram, shown in Figure 6.35, exhibits an open loop gain A_0 = 75dB and unity gain bandwidth f_{GBW} = 1MHz, under the maximum open loop load condition (C_L = 60pF). Note that

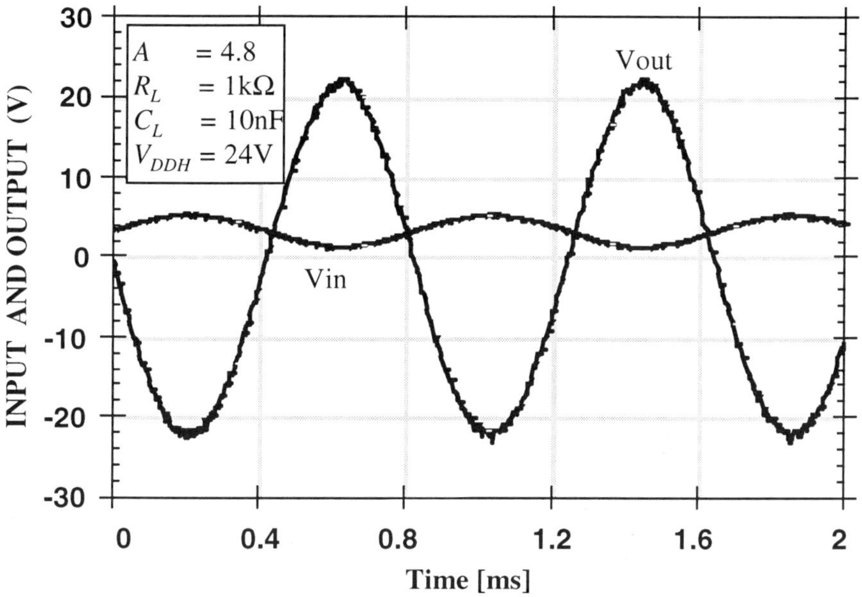

Figure 6.34: Linear high-voltage output capabilities of the class-AB output stage OpAmp when mounted in a full bridge configuration

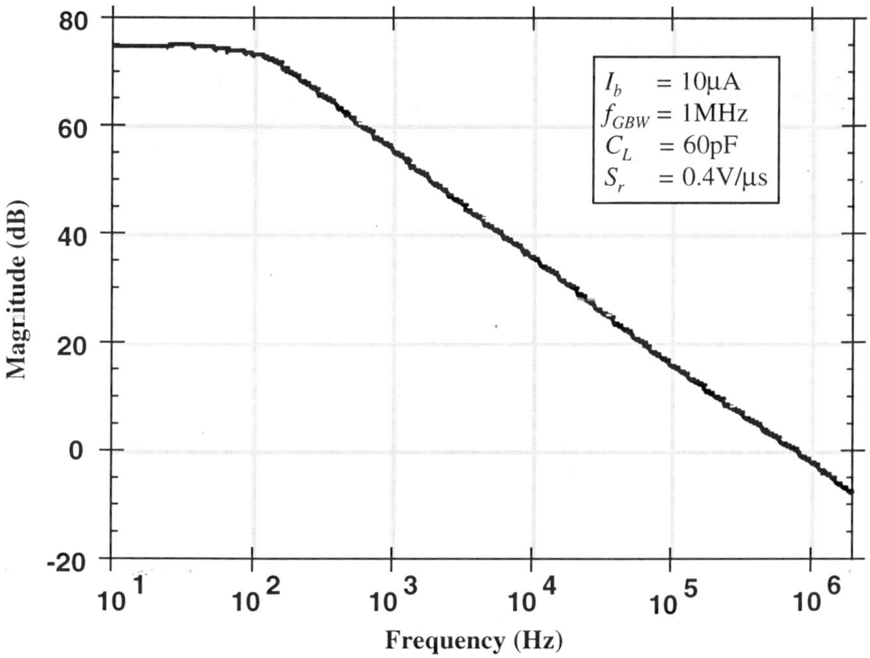

Figure 6.35: Magnitude bode diagram of the class-AB output stage OpAmp

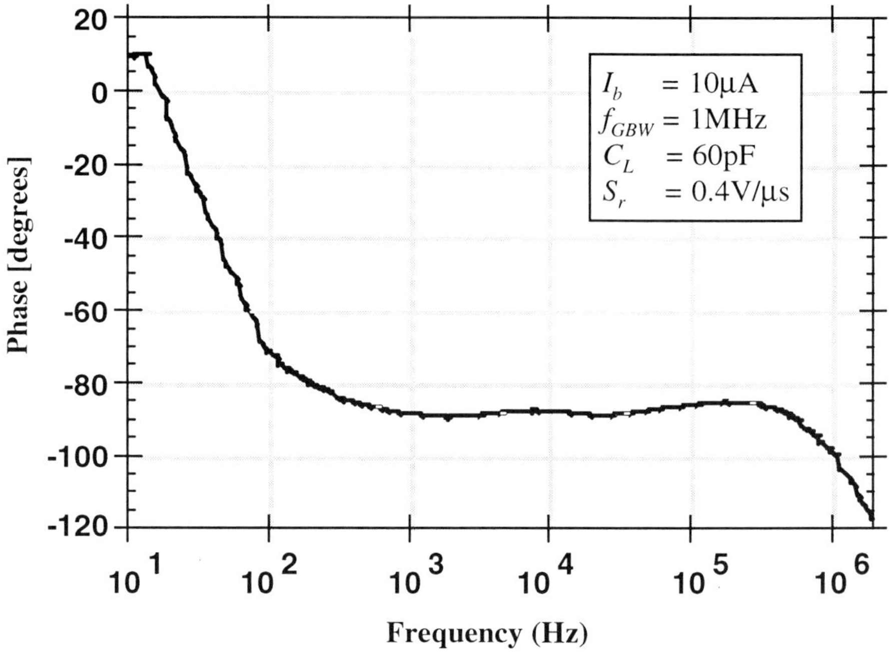

Figure 6.36: Phase bode diagram of the class-AB output stage OpAmp

PARAMETER	VALUE
Biasing current I_b :	10 μA
Standby output current I_o :	100 μA
Peak output current I_p :	100 mA
Open loop gain A_0 :	75 dB
Gain bandwidth f_{GBW} :	1 MHz
Phase margin (C_L = 60pF) :	80°
Slew-rate S_r (C_L = 60pF) :	0.4 V/μs
HV power supply V_{DDH} :	24V
Total standby power P_o	3.6 mW

Table 6.2: Main performances of the high-voltage class-AB output stage OpAmp

the second pole is larger than $2.2xf_{GBW}$. The corresponding phase margin appears in Figure 6.36, where more than 60° are achievable at 1MHz. Finally, the main performances of the amplifier are summarised in Table 6.2.

6.2.3 Transients considerations

1) Basic operational transconductance amplifier: Given the similarities of the level shifting techniques between the basic OTA and the digital output interfaces, the transients considerations developed in section 6.1.4 apply in this case as well. In fact, the *hot nodes* of the amplifier are the gates of the pMOS devices (N_3, N_4 and N_5) against power supply transients, and the gates of the output stage devices (N_1 and N_3) against output transients. Given the slew-rate of the amplifier, the latter case is not critical. In the former, the analog voltage mirror and the biasing current cascode stage have the same topology as the digital voltage mirror. Consequently, the different possibilities considered in the digital interface apply here as well. The first asymptotic case, consists of a large gate capacitance of the considered node C_{gNx} such that the equivalent small signals circuit of the node reduces to a simple capacitive divider. Since C_{gNx} is supposed to be several orders of magnitude higher than the parasitic capacitance of the node C_{pNx}, the induced voltage spike is thus negligible. While this asymptotic case is not applicable in the case of the digital interfaces, it is applicable to the case of the OTA at the static node N_5. The large capacitance value is created artificially by adding an additional filtering capacitor C_{D5} at node N_5, as shown in Figure 6.37. This technique is a very efficient method to protect high-voltage static nodes against power supply transients, at the cost of an increased area. Figure 6.38 shows the simulation data of the induced voltage spike

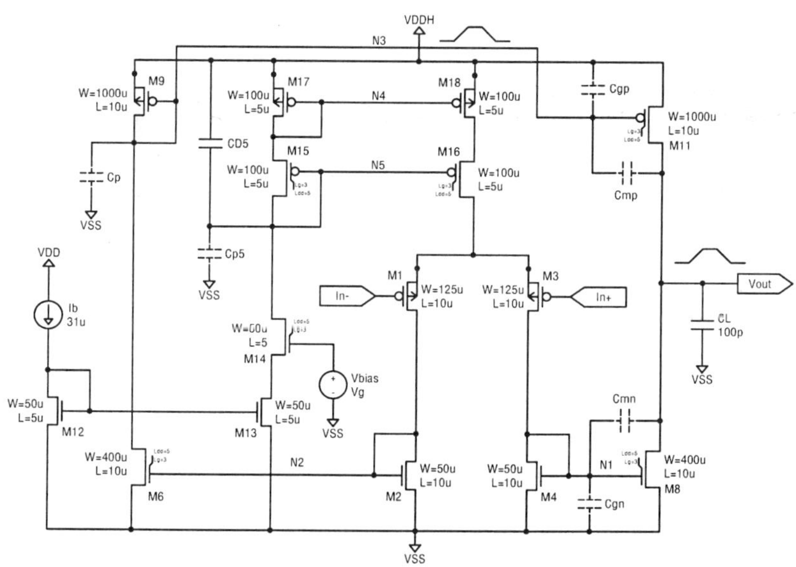

Figure 6.37: Schematic of the OTA including all the parasitic capacitances involved in power supply and output stage transients.

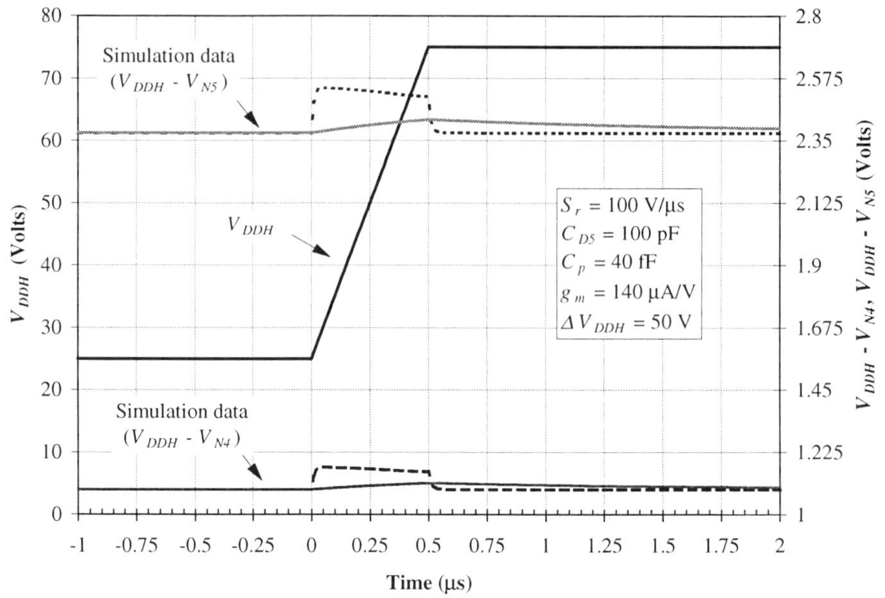

Figure 6.38: Effect of the filtering capacitor on the induced voltage spike at nodes N_4 and N_5.

at nodes N_4 and N_5 before and after adding a filtering capacitor of 100pF. The transient condition is identical to that of the digital output interface, and the corresponding equivalent impedance of the capacitor is about 800Ω.

While the filtering of the power supply transient is possible at the static nodes, it is unfortunately not the case at the driving node N_3. At this node, the situation is identical to the output S_H of the improved static level shifter driving the gate of the HVPMOS transistor. The equivalent small signals circuit of node N_3 during a power supply transient is shown in Figure 6.39. When the capacitive load is large with respect to the Miller capacitance C_{mp}, the total parasitic capacitance to ground is given by

$$C_{pt} = C_p + C_{mp} = C_p + C_{gdo} \qquad for \qquad C_L >> C_{mp} \tag{6.46}$$

and when the output load is negligible with respect to C_{mp}, the total parasitic capacitance to ground is given by

$$C_{pt} = C_p \qquad for \qquad C_L << C_{mp} \tag{6.47}$$

In both cases, the expression of the voltage spike at node N_3 is given by Equation (6.21). Note that C_{mp} is given by the gate overlap capacitance C_{gdo} of transistor M_{11}, since the latter is operating in the saturation mode. The simulation data of the voltage

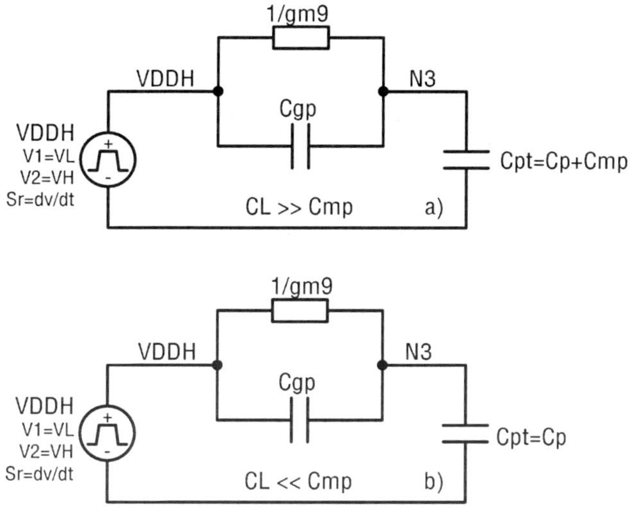

Figure 6.39: Equivalent small signals circuit of node N_3 when the OTA is loaded a) with a large capacitive load b) negligible load

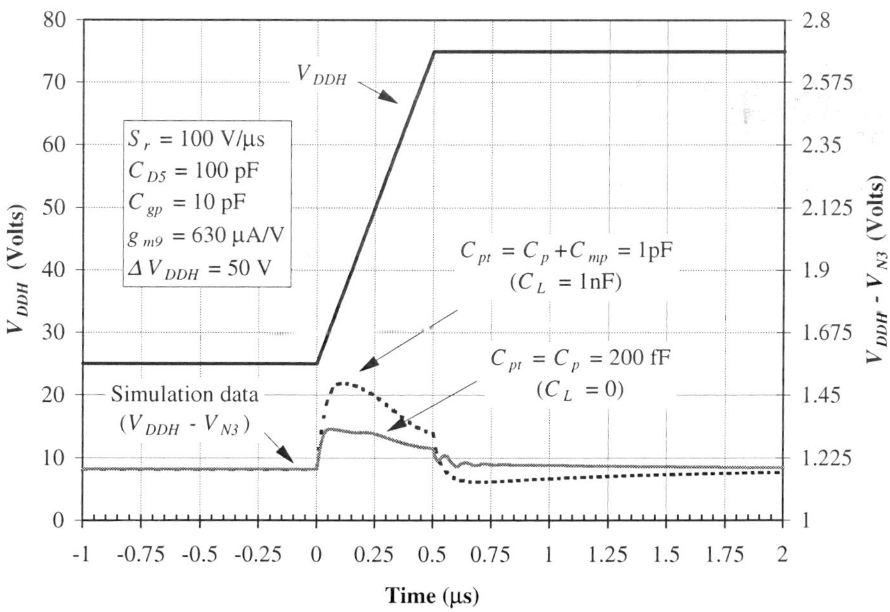

Figure 6.40: Effect of the load capacitor on the induced voltage spike at nodes N_3. The OTA is mounted in a follower configuration.

spike at node N_3 is shown in Figure 6.40, where the same power supply transient condition is used. Note that the gate of the HVPMOS transistor is safely driven under the worst case condition of large capacitive loads. Since the transconductance of transistor M_9 is higher than that of the level shifter, the time constant of the spike is significantly smaller.

Finally, the time constant of the amplifier influences the induced voltage spike as well. The faster the time constant, with respect to the rise and fall times of the transients, the higher the spike amplitude. In fact, when the time constant of the feedback loop of the amplifier is fast, the output is less sensitive to the power supply transient. Consequently, it acts as in the case where the capacitive load is larger than the Miller capacitance C_{mp}. Actually, the time constant of the amplifier is directly related to the output load. Fast amplifiers assume small loads, while slow ones usually drive large capacitive loads. Therefore, on the basis of the asymptotic small signals circuits shown in Figure 6.39 and the performance of the amplifier, the sensitivity of node N_3 to the power supply transients should be performed by simulation, as shown in Figure 6.40.

2) Class-AB output stage operational amplifier: As was discussed in section 6.2.2, the main objective of the cross-coupled source followers is to increase the impedance of nodes N_1 and N_2. This is contradictory with the transients considerations, where the analysis had pointed out that the lower the impedance of the considered node the less critical are the induced voltage spikes. In fact, Figure 6.41 shows an induced voltage spike at node N_2 when no protection is used and under the same power supply transient condition as in the previous cases. The spike reaches an amplitude of 12V assuming an output load of 10nF and a compensation capacitor of 5pF. The amplifier is mounted in a follower configuration. This is obviously not acceptable and leads to the use of protection devices. The ideal protection case would have been the use of a zener diode clamping the maximum voltage swing to the low-voltage safe operating conditions. Since the SVX approach is essentially based on the standard CMOS technologies, such devices are not available. Therefore, to cope with this constraint, their electrical characteristic should be reproduced. This is achieved by the protection diode transistors, M_{11}-M_{12} and M_{13}-M_{14}, as shown in Figure 6.42. Actually the number of transistors in series is a trade off between surface and circuit performance. In fact, the higher the number of protection transistors, the less the amplifier performance will be affected. The only drawback is the additional silicon surface occupied by these transistors. This is discussed in this section for the class-AB output stage OpAmp. Thus, as far as the induced voltage spike amplitude is lower than the turn on voltage V_{on}, the impedance of nodes N_1 and N_2 is high. When the protection transistors are turned on, the equivalent small signals circuit reduces to the one shown in Figure 6.43, where the transconductance g_m of the protection transistors is assumed higher than the source transconductance g_{ms} of the source follower transistors. Note that α/g_m assumes that all the protection diodes have the same dimensions, where α is the total number of transistors in series.

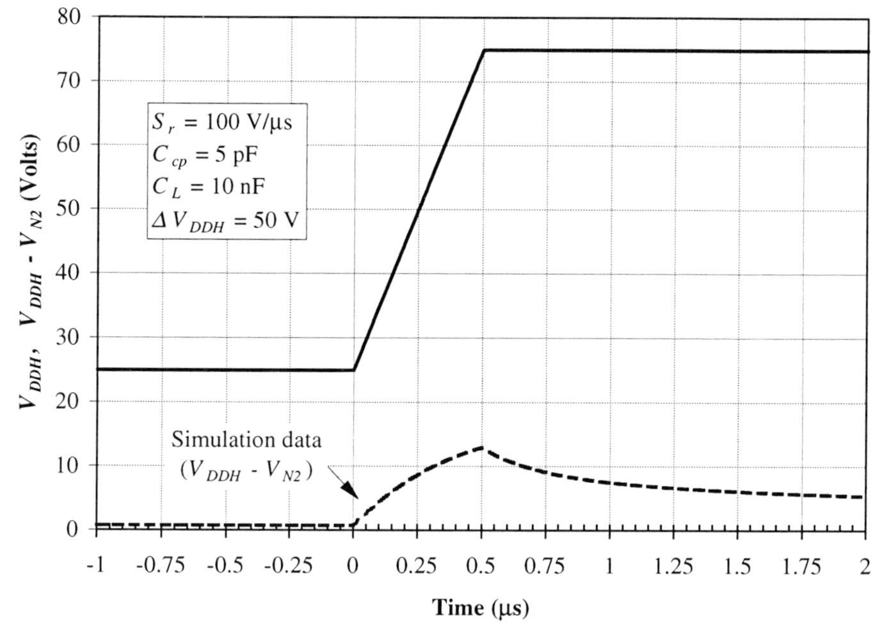

Figure 6.41: Induced voltage spike at node N_2 of the amplifier when no protection is used to clamp the voltage

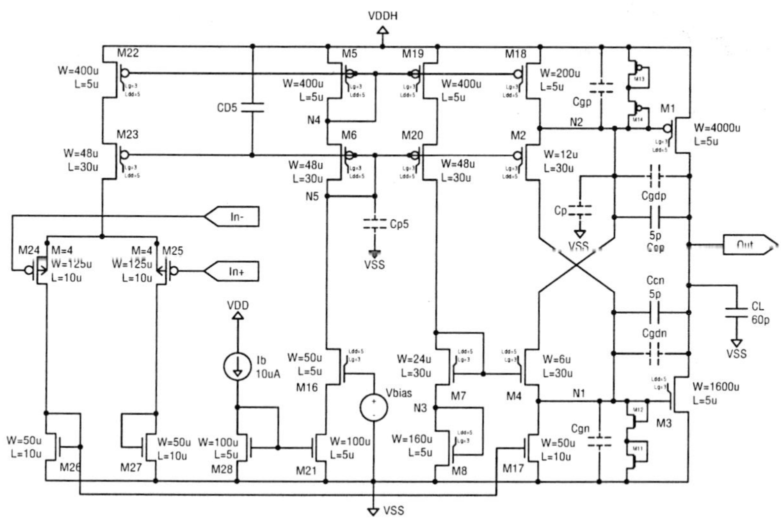

Figure 6.42: Schematic of the class-AB output stage OpAmp including all the parasitic capacitances involved in power supply and output stage transients

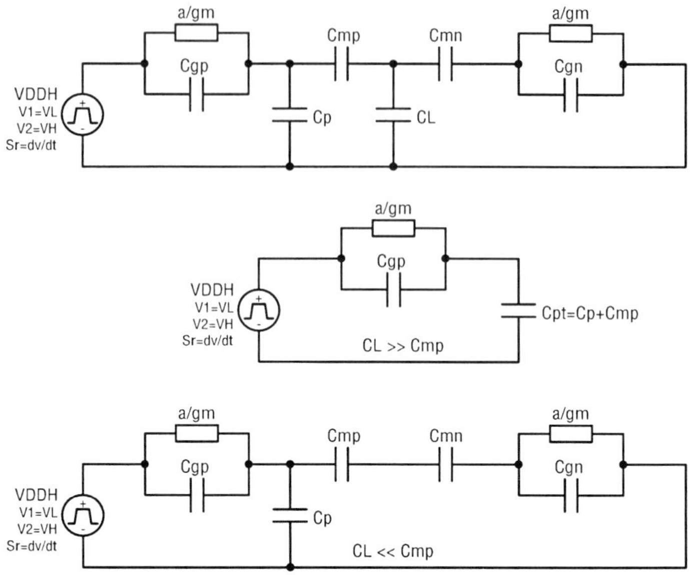

Figure 6.43: Equivalent small signals circuit of nodes N_1 and N_2 during the power supply transient and when the protection diodes are turned on

Thus according to Figure 6.43, the induced voltage spike at node N_1 in this case, also depends on the output load condition. When the capacitive load is large with respect to the Miller capacitance C_{mp}, the total parasitic capacitance to ground is given by

$$C_{pt} = C_p + C_{mp} = C_p + C_{cp} + C_{gdo} \qquad for \qquad C_L >> C_{mp} \tag{6.48}$$

In this case, the expression of the voltage spike at node N_1 is given by Equation (6.21). Consequently, the transconductance g_m of each transistor is given by

$$g_m = \alpha \cdot S_r \cdot \frac{C_{pt}}{v_{sp(t)}}\left(1-e^{-t/\tau}\right) \approx \alpha \cdot S_r \cdot \frac{C_{pt}}{V_{sp\max}} \qquad (t >> \tau) \tag{6.49}$$

assuming a maximum acceptable voltage spike amplitude V_{spmax}. Thus, on the one hand, the higher the number of protection transistors, the higher will be the transconductance value and the larger are the devices. On the other hand, the increase of the number α results in a higher turn-on voltage. Consequently, the circuit performance will be less affected. Considering $\alpha = 4$, a maximum voltage spike $V_{spmax} = V_{DD} + V_{bi} - V_{t0}$ (V_{t0} assumes the output transistor almost in weak inversion), $C_{pt} \approx 8$pF and $S_r = 100$V/µs, the resulting transconductance is about 700µA/V. Assuming that all the transistor diodes are in the same well and have the

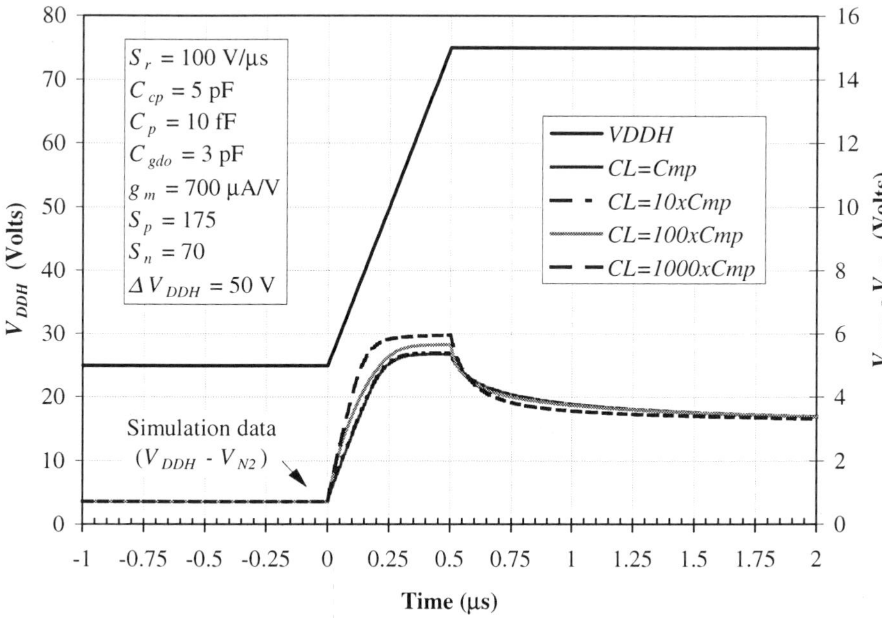

Figure 6.44: Induced voltage spike at node N_2, where the effect of the output load is shown

same dimensions, it can be shown that the total voltage drop across the transistors is given by

$$V_{DD} + V_{bi} = V_{gs1} \cdot \sum_{i=1}^{\alpha} n_i^{i-1} \tag{6.50}$$

where n is the slope factor, which is assumed constant as a function of the bulk-to-source voltage, and V_{gs1} the gate-to-source voltage of the first transistor. Thus, Equations (6.49) and (6.50) yield the dimensions of the transistors

$$S = \frac{n_1 \cdot g_m}{K\left(V_{gs1} - V_{t0}\right)} \tag{6.51}$$

Figure 6.44 shows the simulation data of different voltage spikes amplitudes as a function of the load value. The asymptotic worst case occurs when the Miller capacitance becomes negligible with respect to the load. In this case, the total parasitic capacitance is given by Equation (6.48) and the protection transistors clamp the full voltage swing of node N_2 to $V_{DD} + V_{bi}$. Note that the other asymptotic case is reached when the load capacitor is almost equal to the Miller capacitance. In this case, the total parasitic capacitance is given by the series connection of C_{mp}, C_{mn} and

C_{gn}, which are set by the circuit performance. Thus, the gates of the pMOS devices are driven safely even if the impedance of node N_2 was initially very high. Finally, as in the case of the operational transconductance amplifier, the output transients are not critical in terms of induced voltage spikes due to the low slew-rate value.

6.3 References

[12] M. J. Declercq, M. Shubert, F. Clement, "5V-to-75V CMOS Output Interface Circuits" *IEEE Int. Solid-State Circuits Conference*, 1993, pp. 162-163

[13] Patent

[14] M. G. Degrauwe, J. Rijmenants, E. A. Vittoz, H. J. De Man, "Adaptive Biasing CMOS Amplifiers" *IEEE Journal of Solid-State Circuits*, Vol. SC-17, No. 3, June 1982, pp. 522-528

[15] H. Ballan, M. J. Declercq, F. Krummenacher, "Design and Optimization of High-Voltage analog and digital circuits built in a standard 5V CMOS technology" *IEEE CICC proceedings,* paper 25.6, May 1994

7 12V DELTA-SIGMA CLASS-D AUDIO AMPLIFIER

Abstract:
Among the various existing medium- and low-power system applications, where a full high-voltage integrated circuit solution can be used, there is a limit beyond which the SVX technique loses its advantages with respect to dedicated technologies. The class-D audio amplifier presented in this chapter is a typical medium power application suited for this purpose. In addition to the power limit considerations, this chapter points out the flexibility of this approach in combining low-voltage signal processing and high-voltage features. In fact, the low-voltage part of the amplifier consists of a delta-sigma (Δ-Σ) modulator and some protection features, such as over-temperature and over-current detection cells.

Since a fully integrated solution is chosen for the amplifier [27], the most critical part in the design of the high-voltage integrated circuit is the output power stage. In fact, the power levels imposed by the low impedance value of the speaker load are the most constraining requirements. But, before considering these requirements, section 7.1 first presents the general system aspects such as the existing modulation techniques, and the maximum achievable output power under the typical supply voltage value of a car battery. The motivations for choosing a Δ-Σ modulator are also discussed. The switched capacitor implementation of the modulation stage is presented in section 7.2. Then, section 7.3 moves to the actual power constraints of a fully integrated class-D audio amplifier solution. The surface density of the high-voltage devices is first calculated, leading to the minimum on state resistance achievable within the available silicon surface. Then, the driving circuits required for the power devices are discussed. Given the efficiency of the amplifier, an over-temperature detection cell is presented. An over-current protection cell is presented as well, since the maximum current handling capabilities of the metal wires is much lower than that of the power devices. Finally, the measured performances and the possible system improvements are presented in sections 7.4 and 7.5.

7.1 General system considerations

The most commonly used configuration in class-D audio amplifiers implements a Pulse-Width-Modulator (PWM) and a power output stage, as shown in Figure 7.1. The PWM stage produces a high-frequency rectangular wave whose average value tracks that of the input signal. A minimum distortion rule of thumb, consists of choosing a switching frequency at least five times that of the highest frequency component in the audio signal. The rectangular wave signal results from the comparison of a high-frequency triangular waveform to the input audio signal. The amplitude of both signals must be matched in order to avoid output distortion. The level of output distortion will also depend on the linearity of the triangular waveform. The latter is generated either by the use of digital techniques such as counters [28], [29] for an all-digital audio power amplifier solution, or by the use of analog integrators [30] for classic analog solutions. The power output stage employs complementary MOSFET devices mounted in a full bridge configuration, which is used because of the asymmetrical nature of the power supply (0-V_{DDH}). Moreover, it facilitates the output filtering since the odd harmonic components of the modulated signal are cancelled, and it quadruples the output power in comparison to a half-bridge solution. The output *LC* filter is required to remove the switching frequency components from the output signal.

The pulse-width-modulation technique used in class-D amplifiers, keeps the frequency of the modulated signal constant, and adapts its duty cycle to the input amplitude. Consequently, if the frequency of the modulated signal is high-enough, the total harmonic distortion (*THD*) is minimised. Actual implementations were reported [31],[32] with switching frequencies ranging between 150kHz and 500kHz.

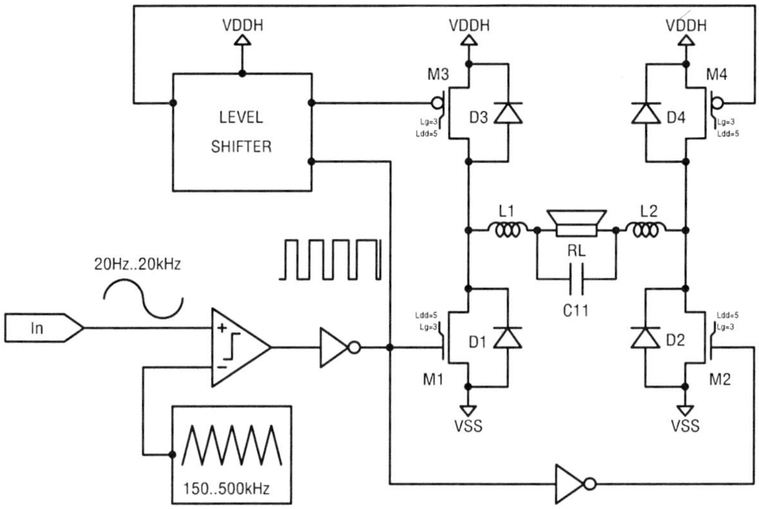

Figure 7.1: Schematic of a classic PWM implementation of a class-D audio amplifier

A higher switching frequency would, in theory, lead to a better quality signal, since the higher the ratio of switching frequency to modulating frequency, the smaller are the amplitudes of the unwanted intermodulation products in the output signal. However, a higher switching frequency results in an increased distortion due to the finite rise and fall times of the rectangular signal with respect to its pulse period. Therefore, the choice of the switching frequency depends on the power devices and their characteristics.

The typical spectrum of a PWM signal, corresponding to a sinewave input and resulting from a full bridge output stage, contains harmonic components at the even multiples of the switching frequency, as shown schematically in Figure 7.2a. The noise floor is assumed constant at about -60 dB. Another interesting modulation approach consists of shaping the noise floor of the modulated signal. This can be achieved by the use of a Δ-Σ modulator, where the noise at the output of the modulator is shaped by a high-pass transfer function, as shown schematically in Figure 7.2b. The order of the transfer function is given by the order of the Δ-Σ modulator. Note that the noise shaping characteristic is valid under the assumption of a uniform power spectral density of the noise at the input. The main advantage of this modulation technique is the higher base-band signal-to-noise ratio (*SNR*) that can be achieved in comparison to the previous modulation case. The choice of the order of the modulator and its actual implementation is discussed in more detail in section 7.2. Finally, although the modulation type is different from the case of a PWM modulator, the average value of a Δ-Σ modulated signal tracks the amplitude of the input as well. In fact, while the PWM technique keeps a constant switching

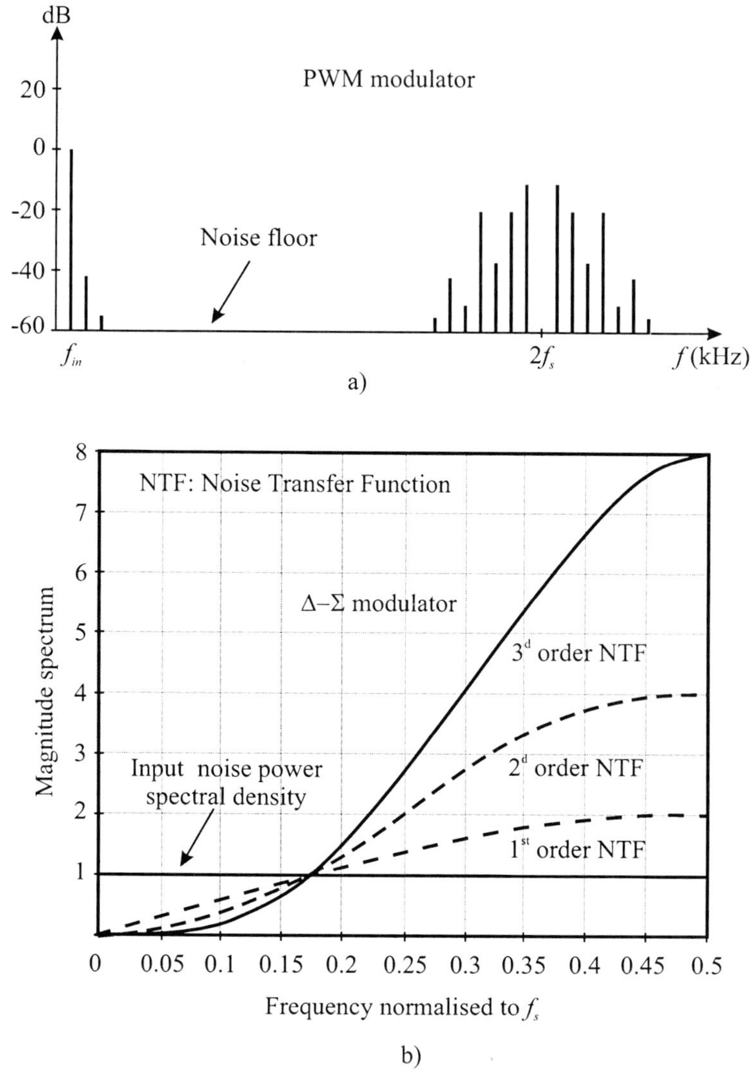

Figure 7.2: a) Typical spectrum of a PWM modulated signal. b) noise shaping characteristic of Δ-Σ modulators

frequency by adapting the duty cycle of the signal to the input amplitude, the signal resulting from a Δ-Σ modulator is frequency modulated. However, assuming a high switching frequency with respect to the modulating input, the average of both rectangular waveforms is identical over the complete cycle of the input signal.

The last point of the system considerations consists of the estimation of the maximum output power ratings. Thus, considering a 12V high-voltage power supply value, resulting typically from a car battery, the maximum output power P_{omax} is a function of the speaker load. Assuming an 8Ω load, P_{omax} is computed on the basis of a sinewave input signal. Thus, considering ideal switches ($R_{dson} = 0\ \Omega$)

$$P_{o\max} = \frac{U_{rms}^2}{R_L} = \frac{\left(12/\sqrt{2}\right)^2}{8} = 9\ W \tag{7.1}$$

where U_{rms} is the root-mean-square value of the output sinewave. As will be pointed out in the following sections, such a theoretical value is reduced by the finite on state resistance of the power devices.

7.2 Modulation stage

As was mentioned in the previous section, the choice of a Δ-Σ modulator is motivated by its noise shaping characteristic, leading to a higher base-band signal-to-noise ratio. Moreover, in comparison to the PWM technique, it exhibits an increased linearity given the closed loop nature of the modulator. In fact, the spectrum of a Δ-Σ modulated signal is free of harmonic contents up to the overload limit. An overview of these considerations in addition to the actual implementation of the modulator are the aim of this section.

7.2.1 Theoretical considerations

Among the variety of existing Δ-Σ modulators, the first order architecture shown in Figure 7.3a is the simplest. It consists of an integrator, a one bit analog-to-digital (A/D) converter, which is usually a simple comparator, and a digital-to-analog (D/A) converter in the feedback path, which is usually a simple switch comutating between V_{DD} and ground. The difference between the input and the analog representation of the quantised output are integrated. It is basically a non-linear system formed by a high base-band gain block in the feedforward path, which is linearised by the feedback loop. Thus, assuming that the loop noise is mainly due to the quantisation operation, it is removed from the base-band by the integrator gain. As the gain of the integrator decreases as a function of the frequency, the noise is amplified outside the signal band. It results in the high-pass noise shaping transfer function shown in Figure 7.2. In the feedforward path the signal is kept unchanged. The linear Z domain model of such a modulator is shown in Figure 7.3b, where the quantiser is replaced by a noise source $E(z)$, the integrator by the transfer function $H(z)\cdot z^{-1}$ and the D/A converter by the unity transfer function as it is assumed ideal. It can be shown [16],[17] that the quantised output V_{out} is given by

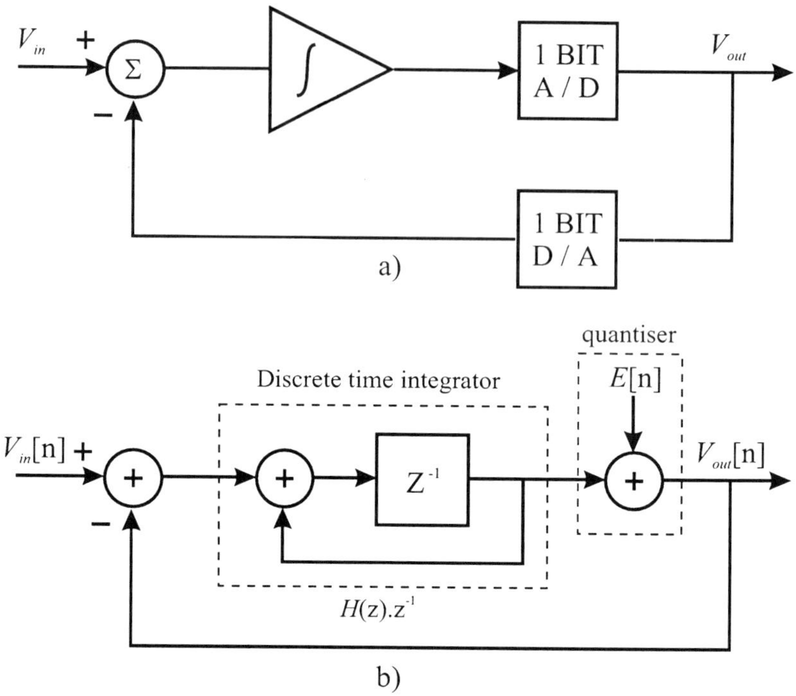

Figure 7.3: Architecture of a first order Δ-Σ modulator a) Continuous time representation b) Linear Z domain model

$$V_{out}(z) = V_{in}(z) \cdot z^{-1} + \frac{E(z)}{1 + H(z) \cdot z^{-1}} \tag{7.2}$$

where V_{in} is the sampled input and $H(z)$ is the gain of the integrator given by

$$H(z) = \frac{1}{1 - z^{-1}} \tag{7.3}$$

For base-band frequencies ($z^{-1} \approx 1$ and $H(z) >> 1$)

$$V_{out}(z) \approx V_{in}(z) + \frac{E(z)}{1 + H(z)} \tag{7.4}$$

Consequently, noise in the modulated signal is suppressed by the integrator gain $H(z)$. The main disadvantage of the first order Δ-Σ modulator is its limit cycle oscillations resulting in the presence of periodic components in the output. They are more commonly known as tones since they manifest themselves in the output spectrum as tones present at frequency *f/T* and its harmonics, where *T* is the period of the limit cycle. Detailed analysis on the presence of these tones for DC and sinusoidal inputs are provided in [18],[19],[20],[21]. Thus, the use of a first order Δ-Σ modulators in audio applications is not recommended because of the presence of such tones in the output signal spectrum. This limitation is overcome by the use of higher order Δ-Σ modulators, where more noise power is pushed outside the base-band, as shown in Figure 7.2. However, it is difficult to maintain feedback loop stability when more than second order integration is used. Moreover, it is useless to significantly increase the *SNR* of the modulation stage by increasing its complexity, since the distortions introduced by the power stage of the amplifier become dominant. Therefore, the use of a second order Δ-Σ modulator is the best trade-off between design complexity and performance in comparison to the classic PWM technique. The equivalent time domain representation as well as the corresponding linear *Z* domain model are shown in Figure 7.4. It can be shown [16],[17] that the quantised output V_{out} is given by

$$V_{out}(z) = V_{in}(z)\cdot z^{-1} + \frac{E(z)}{1 + H_1(z)\cdot H_2(z)\cdot z^{-1} + H_2(z)\cdot z^{-1}} \tag{7.5}$$

For base-band frequencies ($z^{-1} \approx 1$ and $H_1(z)$ and $H_2(z) >> 1$)

$$V_{out}(z) \approx V_{in}(z) + \frac{E(z)}{1 + H_1(z)\cdot H_2(z)} \tag{7.6}$$

Consequently, noise in the modulated signal is suppressed by the product of the integrator gain of each stage. Note that for higher orders the noise is attenuated by the number of integrators. But, as was mentioned previously the stability of the feedback loop becomes critical.

Assuming that the modulator output is filtered by an ideal low-pass filter, the base-band *SNR* is given by

$$SNR = 10\log\frac{P_x}{P_n} \tag{7.7}$$

where P_x and P_n are the powers of the signal and noise at the output of the modulator. The base-band power noise P_n of a second order Δ-Σ modulator is given by

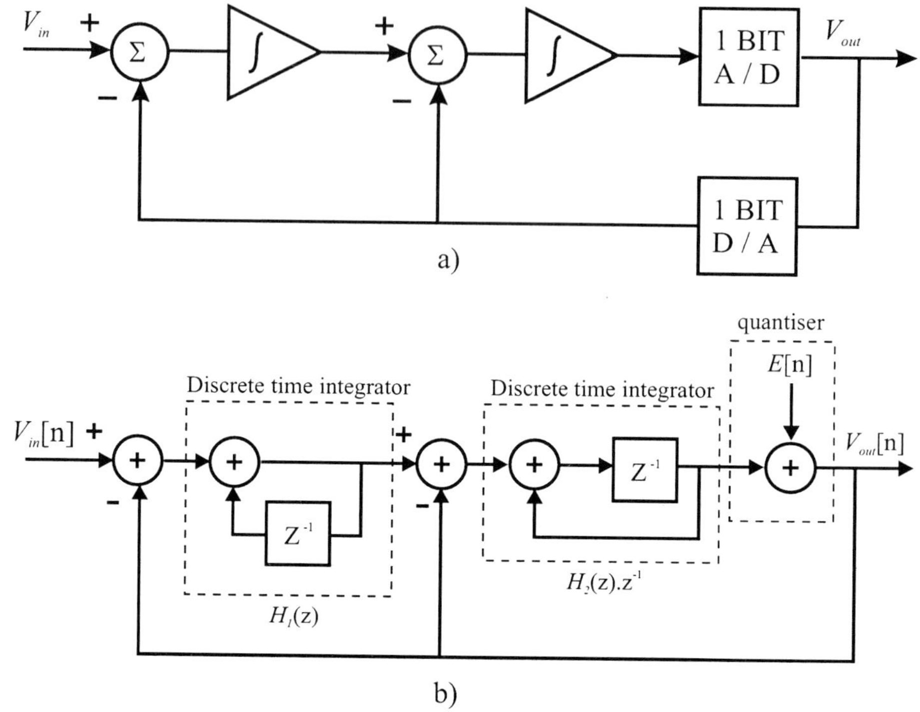

Figure 7.4: Architecture of a second order Δ-Σ modulator a) Continuous time representation b) Linear Z domain model

$$P_n = P_e \cdot \frac{\pi^4}{5} \cdot \left(2 \cdot \frac{f_b}{f_s}\right)^5 \tag{7.8}$$

where P_e is the quantisation power of the noise $E(z)$ injected by the comparator, f_b is the base-band width and f_s is the sampling frequency. Consequently, the signal-to-noise ratio of a second order Δ-Σ is given by

$$SNR = 10\log\left(P_x\right) - 10\log\left(P_e\right) - 10\log\left(\frac{\pi^4}{5}\right) - 15.05 + 10\log\left(\frac{f_s}{f_b}\right)^5 \tag{7.9}$$

Note that since $10\log(2)^5 = 15.05$, the SNR improves by 15dB for every doubling of the oversampling ratio (f_s/f_b). Therefore, it must be maximised in order to achieve the best audio performance. However, in order to minimise distortion and dynamic switching losses, the sampling frequency for this design is set to 500kHz. Thus, considering a typical class-D audio base-band of 15kHz, the maximum theoretical signal-to-noise ratio is given by

$$SNR_{\max} = 56\,dB \qquad \begin{bmatrix} f_b = 15\,kHz \\ f_s = 500\,kHz \\ P_e = 1/12\,(1\,bit\,A/D) \\ P_x = 0.5\,(V_{in} = V_{DD}) \end{bmatrix} \tag{7.10}$$

The *SNR* value computed in Equation (7.10) is an absolute maximum theoretical value. In fact, it results from a full swing sinewave input. Actually, full swing signals at the input of second order Δ-Σ modulators result in a significant decrease of the *SNR* value due to the overload limit. The overload of the modulator is caused by the saturation of the output of the second integrator, when the input signal approaches the full swing range. It manifests itself by a significant increase of harmonic distortion tones, assuming a sinusoidal input, leading to the decrease of the *SNR* value. This is represented schematically in Figure 7.5, where the *SNR* decreases for input amplitudes greater than about -5dB. For lower amplitude values the *SNR* decreases linearly with signal power P_x. The overload limit is generally characterised by simulation, since no theoretical considerations are yet reported on the characterisation of its spectral noise properties. However, for well designed modulators this limit usually occurs for input amplitudes in the range of -5dB. Thus,

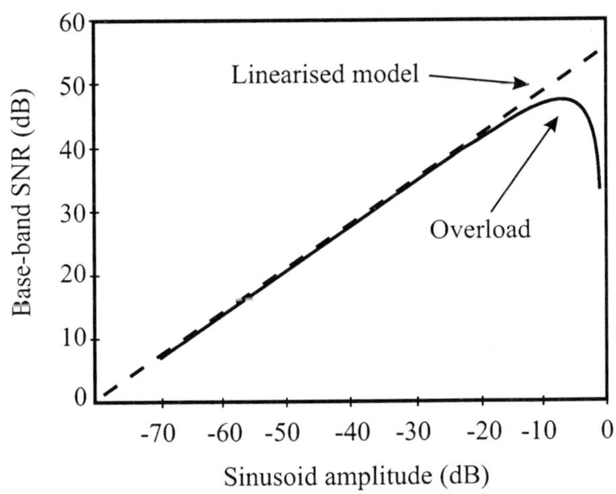

Figure 7.5: Base-band *SNR* of a second order Δ-Σ modulator as a function of a sinusoidal input amplitude

the actual *SNR* value is expected to be lower than that given by the linearised model of Equation (7.9). In the considered audio application, the maximum input signal amplitude is about -5dB, assuming a 5V low-voltage power supply and a typical compact disc player output (1 V_{rms}). For higher input amplitudes, the *SNR* starts decreasing significantly. The use of dithering techniques [22] can improve the linear dependence of the *SNR* on the signal power P_x by breaking up the tones structures resulting from overload. However, it does not eliminate the overload behaviour of the modulator. Moreover, in most of the power audio systems, the total harmonic distortion of the amplifiers at full output power is always lower than the typical sinewave input one. Therefore, given the additional complexity introduced by these techniques, in comparison to the expected slight improvements, their use is not justified in the considered class-D audio amplifier.

7.2.2 Switched capacitor implementation

The second order Δ-Σ modulator uses a classic switched capacitor implementation, as shown in Figure 7.6. It is formed by two non-inverting integrator blocks, a comparator with a D flip-flop as an A/D converter, and a synchronising logic in the feedback loop controlling the switching sequence of the reference source V_{ref} in order to form the D/A converter. In order to minimise the charge injection, the integrators are switched by two non-overlapping clock phases, Φ_1 and Φ_2, and two

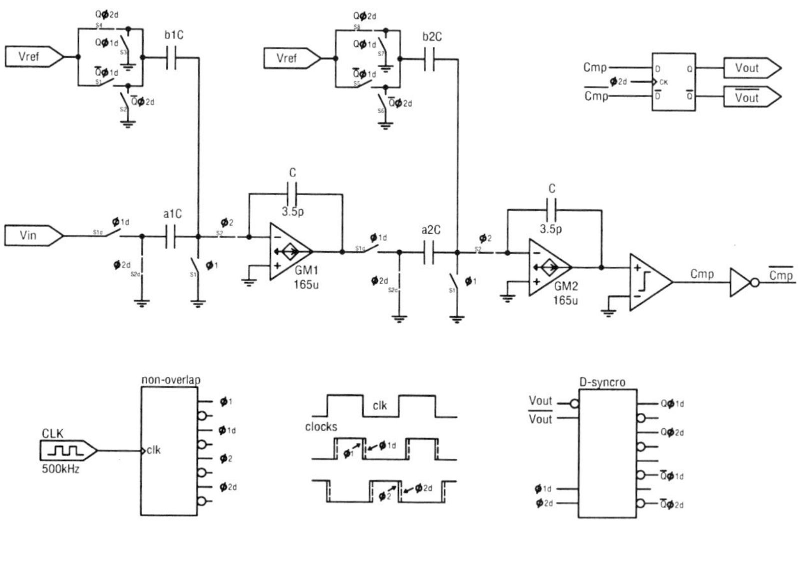

Figure 7.6: Switched capacitor implementation of the second order Δ-Σ modulator

additional delayed clock phases, Φ_{1d} and Φ_{2d}. In fact, it is shown in [23], [24] that channel charge injection and clock feedthrough caused by S_{1d}, S_{2d}, S_1 and S_2 can be reduced if S_{1d}, S_{2d} are turned-off after the turn-on of S_1 and S_2. The use of more elaborate techniques reducing the charge injection, such as two dummy switches for each active switch or a fully differential integrator structure, are not suited for our application as they are surface consuming. Moreover, the charge injection is not very critical in this design, since the resolution resulting from the low oversampling ratio is about 8 bits only.

The switching sequence of the reference source V_{ref} is such that positive charges are transferred to the output of the integrators when the modulated signal is running with a "0" logic level, and negative charges are transferred in the opposite case. In other words, while the sum between V_{in} and V_{ref} is integrated in the former case, the difference is integrated in the latter. Note that for the stability of the loop, only a fraction of these voltages is sampled. The fraction is given by the ratio between the integration capacitance C and the summing capacitances at the inputs of the integrators ($a \cdot C$ and $b \cdot C$). The optimisation of the coefficients of each integrator

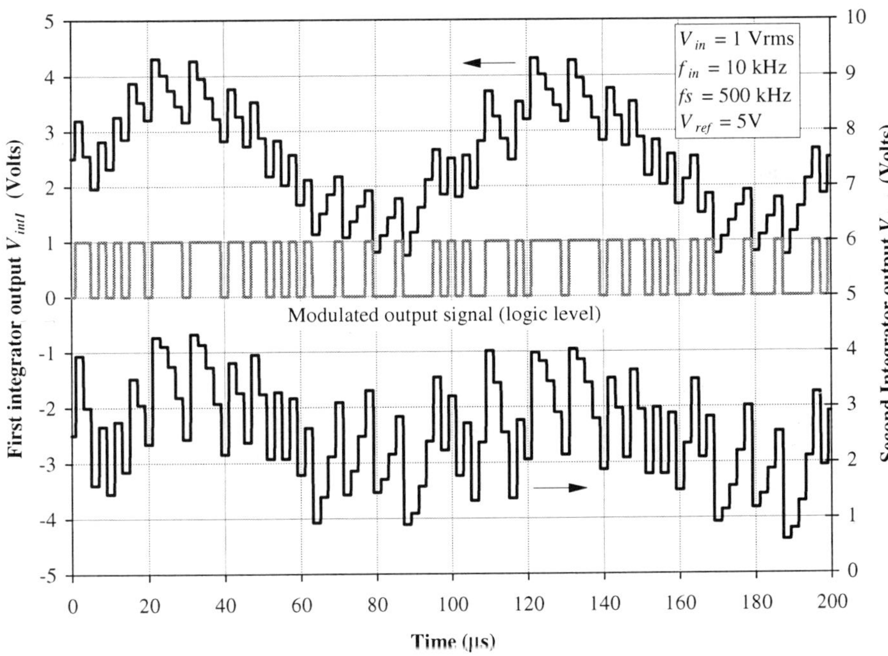

Figure 7.7: Waveforms at the outputs of the integrators when a 1V_{rms} input sinewave is applied

block, i.e. a_1,b_1 and a_2,b_2, was performed by simulation. The corresponding waveforms are shown in Figure 7.7, where the following optimum values are used

$$\begin{bmatrix} a_1 = 0.28 & b_1 = 0.28 \\ a_2 = 0.71 & b_2 = 0.57 \end{bmatrix} \quad (7.11)$$

Note that the use of an analog behavioral modelling approach is required to perform such a simulation. In fact, a full transistor level simulation requires about 48 hours to carry out the results of one cycle of a 10kHz input sinewave. Therefore, a C based analog behavioral model of the Δ-Σ modulator shown in Figure 7.6 was implemented in a SPICE type simulator [25]. This approach is particularly suited to perform fast-fourrier-transforms (*FFT*), where a large number of periods is required. Moreover, it helps to simulate the complete class-D amplifier in a reasonable amount of time. The reader interested in the code of the modulator model can refer to appendix ***B***.

Except for the *OTA*s of the integrators, the remaining blocks of the modulator are not critical. Therefore, given the simple CMOS circuit techniques used for these blocks, their schematics are attached in appendix ***B*** without being

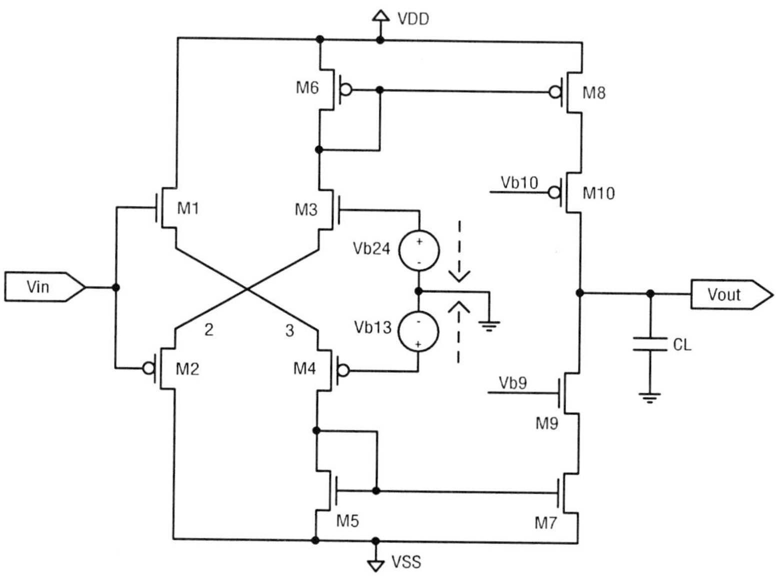

Figure 7.8: Simple implementation of a class-AB operational-transconductance amplifier

discussed here. Thus, this section continuous with the design of the *OTA*s only. As will be discussed in section 7.3, the power consumption and efficiency of the complete audio amplifier are the most critical constraints. Therefore, the architecture of the *OTA* uses the class-AB principle shown in Figure 7.8, where the current flowing through the cross coupled branches is identical in the quiescent mode. As V_{in} is shifted towards V_{DD} or V_{SS}, the current increases through M_1 or M_2. This current is supplied to the output load through the current-mirrors M_5-M_7 and M_6-M_8. The maximum current drive capability is reached when transistors M_3 or M_4 leave the saturation mode, due to the increasing gate-to-source voltage. Note that the output current can be several orders of magnitudes higher than the bias current.

The *I-V* transfer characteristics of a class-A and class-AB *OTA*s are plotted in Figure 7.9, assuming the same transconductance in both cases. While the current supplied by the class-A circuit saturates at I_0, the one supplied by the class-AB circuit is several orders of magnitudes higher. Note that in the class-AB case, a slight assymmetry appears in the positive and negative output saturation currents due to the lack of a perfect symmetry in the circuit. However, such a dissymmetry is negligible as long as the current supplied by the amplifier is larger than that of a linear source. In fact, the settling time of the output voltage will always be smaller than the linear case, where $\tau = g_m/C_L$, until the current supplied by the linear source becomes higher. Consequently, the settling time condition is valid for a limited input voltage range $[V_{in}^- .. V_{in}^+]$, which must be as large as possible to avoid distortion.

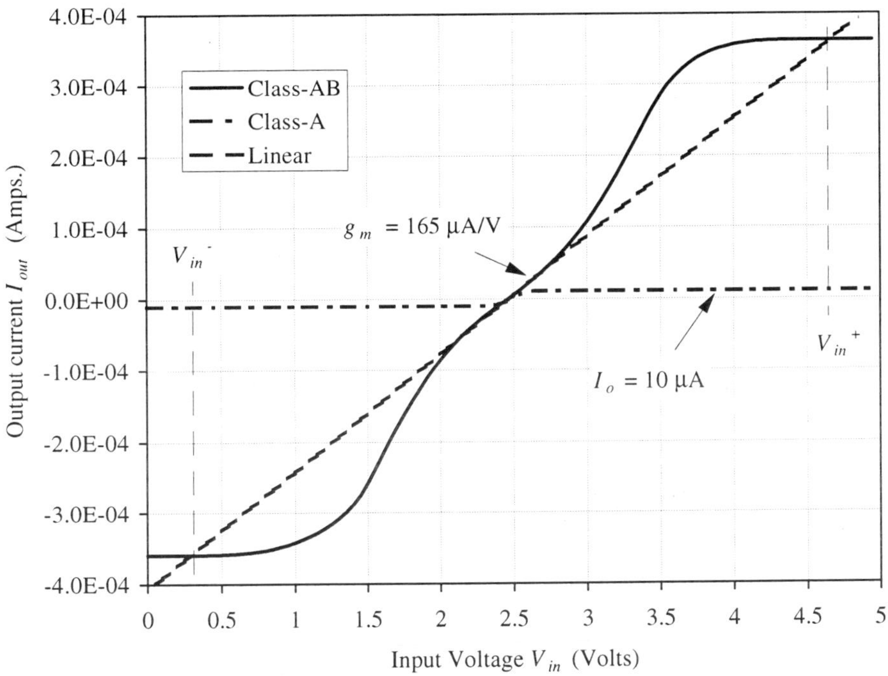

Figure 7.9: *I-V* characteristics of a class-A and class-AB OTAs

Two main drawbacks arise from the use of the structure shown in Figure 7.8. On the one hand, the number of stacked threshold voltages in each branch is about $2V_t$, limiting thus the current drive capabilities of the amplifier. On the other hand, the output cascode stage M_9-M_{10} limits the full voltage swing at the output. These limitations are overcome by the use of the improved implementation shown in Figure 7.10, where the number of stacked threshold voltages is reduced to $1V_t$. Moreover, the cascode stage M_9-M_{10} uses a dynamic biasing scheme [26] in which the bias voltages V_{b9} and V_{b10} vary with the current flowing through the transistors. It maintains the output transistors M_7-M_8 at the limit of the saturation mode for all values of the output current. In order to maximise the internal voltage swings, the quiescent voltages at nodes N_2 and N_3 are set to ground by the use of the input voltage followers M_{1p}-M_{1n} to drive the gates of the transistors of the input stage M_{2n}-

M_{2p}. The gates of the opposite cross-coupled transistors of the input stage, i.e. M_{3n}-M_{3p}, are biased symmetrically by the circuit shown in Figure 7.10b, such that the same current flows in both branches. Any mismatch between these currents increases the input offset.

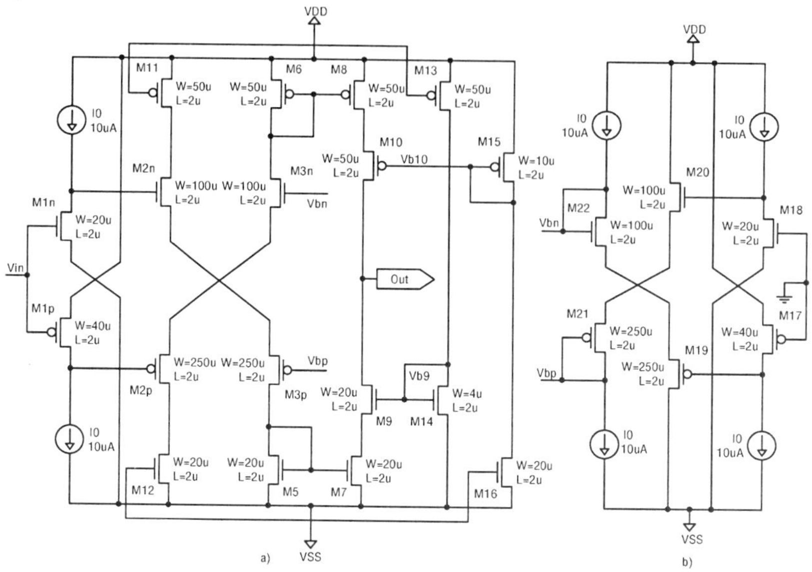

Figure 7.10: Improved implementation of the class-AB *OTA*

The total transconductance of the amplifier is given by the sum of the transconductance of each branch of the input stage. Thus

$$g_{mt} = g_{m23} + g_{m32} \tag{7.12}$$

where

$$g_{m23} = g_{m32} = \frac{g_{mn} \cdot g_{mp}}{n_n \cdot g_{mn} + n_p \cdot g_{mp}} \tag{7.13}$$

where g_{mn} and g_{mp} are the tranconductances of the nMOS and pMOS transistors. In order to maximise the transconductance of the amplifier, the transistor must operate in the weak inversion mode. Thus, assuming $n_n = n_p = n$ and $g_{mn} = g_{mp} = I_0/(nU_t)$, the resulting total transconductance value is given by

$$g_{mt} = 2 \cdot \frac{g_m}{2n} = \frac{I_0}{n^2 \cdot U_t} \tag{7.14}$$

This allows the scaling of the amplifier on the basis of the required performances. Considering the sampling frequency set previously f_s = 500kHz, the time constant of

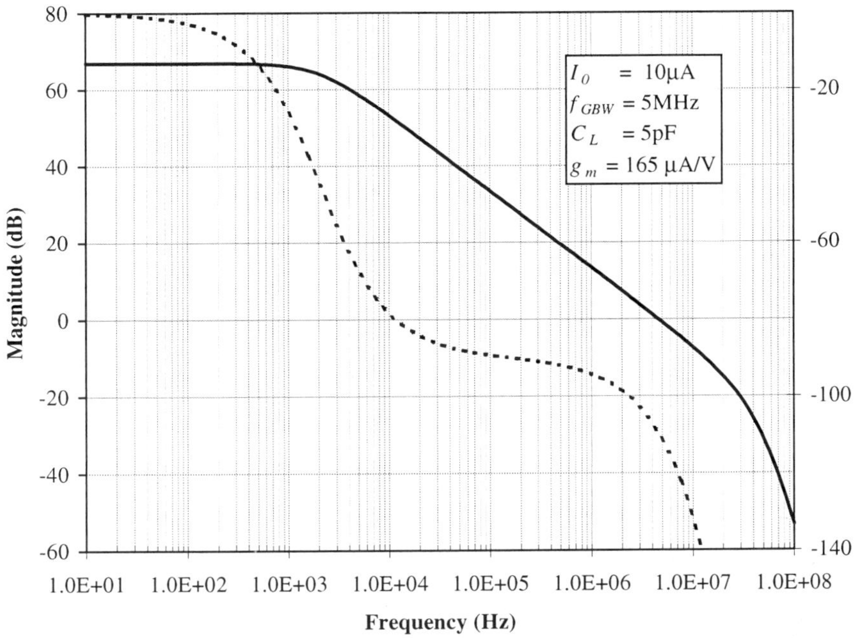

Figure 7.11: Bode diagram of the class-AB OTA under the maximum load condition

the amplifier should be at least 10 times smaller than the sampling period in order to avoid distortion. Consequently, the gain bandwidth product should be $GBW \geq$ 5MHz. Assuming $C_L \leq$ 5pF, the resulting total transconductance must be $g_m \geq$ 160µA/V. The *I-V* transfer characteristic shown in Figure 7.9 results from the simulation of the implemented *OTA*. As it appears, the total transconductance achieved by this amplifier is about 165µA/V. The corresponding Bode diagram under the maximum load condition is shown in Figure 7.11, where GBW = 5MHz and A_0 = 70dB. Note that the phase margin is higher than 60°.

The measured performance of the complete Δ-Σ modulator, using such a class-AB *OTA* structure and the optimised stability coefficients given by Equation (7.11), is presented in section 7.4.

7.3 Output stage

It is obvious that the power efficiency η of class-D audio amplifiers, using discrete power devices to drive the output load, approaches the theoretical value of 100%. In fact, typical power efficiencies achievable with such a solution [31], [32] are in the range of 90%. This is mainly due to the very low on state resistance available with discrete power devices. It ranges between few tenths and few hundreds of milliohms depending on the power requirements. In the case of a fully integrated solution, a trade-off must be considered between the chip surface and the system performances.

7.3.1 High-voltage power devices

To achieve a maximum power efficiency the total on state resistance of the full-bridge switches must be minimised within the available chip surface. The latter is limited by the production yield, which is difficult to estimate a priori for such a high-voltage integrated circuit. However, considering a full digital integrated circuit of $30mm^2$ fabricated in the considered technology, the production yield is typically of 80%. Therefore, the upper limit of the die will be set by this value. In order to estimate the active silicon area, the pad ring should be subtracted from the total chip surface. At this level several iterative floor-planing considerations have led to a total number of 72 pads of which 48 are required for the multiple parallel drain/source outputs forming the power devices, 12 are required for the inputs and power supplies and 12 are used for the testability of the chip. Consequently, assuming an elementary pad surface of $0.125mm^2$, 30% of the total silicon surface is occupied by the pads. From the remaining $20mm^2$, $3mm^2$ are used for the connectivity and the guard rings. Thus, only $17mm^2$ are available to implement the complete class-D audio amplifier. Considering that about $3mm^2$ are necessary for the low-voltage part of the circuit, which consists of the Δ-Σ modulator, the protection features and the driving circuits, the power devices must fit within $14mm^2$. Consequently, the minimum total on state resistance $R_{dson,min}$ can now be calculated for each device, assuming that their surface density is known. This information is computed in the following paragraphs on the basis of the measurement data and layout rules presented in chapter 5.

In order to optimise the surface density of the power devices their layout topology is slightly modified with respect to that presented in Figures 5.50 and 5.51. In fact, the surface occupied by these devices is minimised when the transistor is designed with a common lightly doped sharing two channel regions, as shown in Figure 7.12 for both devices. Thus, the surface density s_n of the HVNMOS transistor is given by

$$s_n = \frac{W}{S_n} = \frac{50 \cdot 10^{-3}}{2.52 \cdot 10^{-3}} = 19.85 \left(mm / mm^2 \right) \qquad \textbf{(7.15)}$$

where W is the width of the transistor in mm and S_n is the surface in mm^2 including the majority carrier guard. The surface density s_p of the HVPMOS transistor is given by

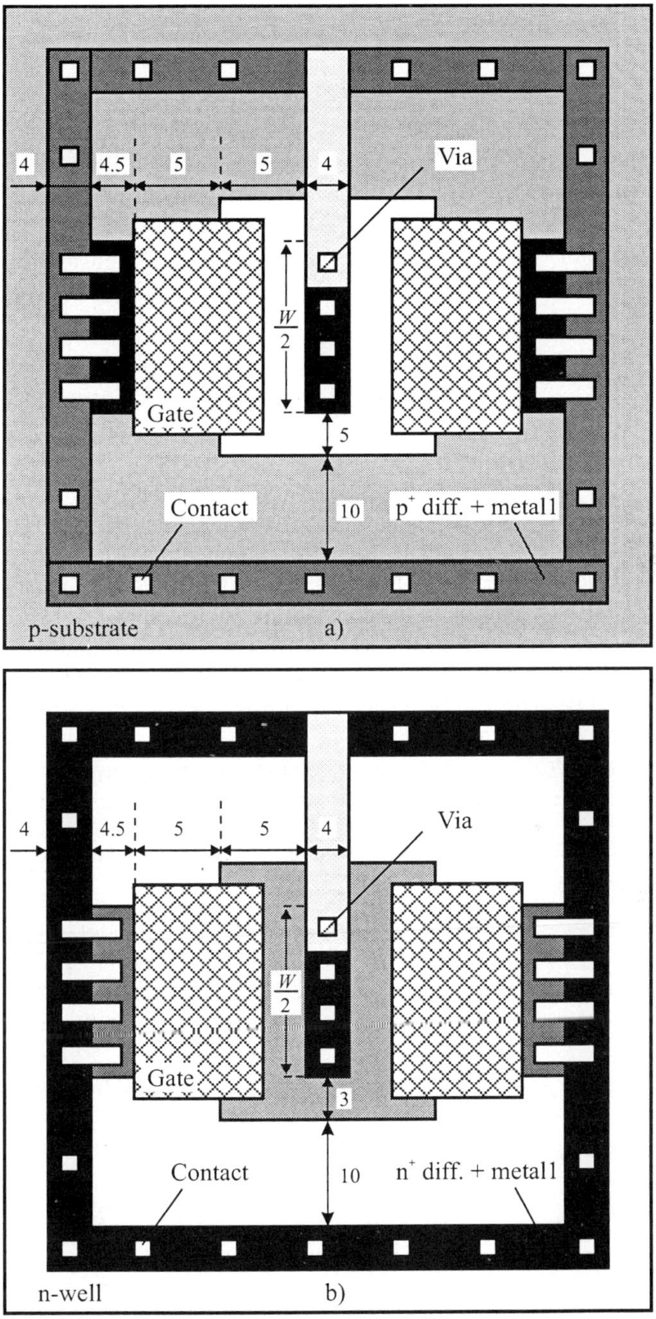

Figure 7.12: Condensed layout version of the high-voltage devices in order to optimise their surface density a) HVNMOS b) HVPMOS

$$s_p = \frac{W}{S_p} = \frac{50 \cdot 10^{-3}}{2.36 \cdot 10^{-3}} = 21.2\left(mm / mm^2\right) \tag{7.16}$$

According to the measurement data presented in chapter 5, the total on state resistance of the HVNMOS transistor is $R_{dson,n} = 60\ \Omega$ per 1mm width and that of the HVPMOS transistor is $R_{dson,p} = 110\ \Omega$ per 1mm width. Considering the surface density achieved by each transistor, the resulting on state resistance per 1mm^2 surface is

$$\begin{cases} R_{dson,n} = 3\Omega.mm^2 \\ R_{dson,p} = 5.2\,\Omega.mm^2 \end{cases} \tag{7.17}$$

In order to achieve equal resistance values for minimum distortion, the ratio between the surface occupied by the HVNMOS and HVPMOS transistors must equal that of the resistance values. Consequently, considering two identical half-bridge outputs, the surface occupied by one HVNMOS transistor is 2.56 mm^2 and the one required by the HVPMOS transistor is 4.44mm^2. The resulting minimum total on state resistance that can be achieved by each transistor is

$$R_{dson,\min} = 1.17\Omega \qquad \text{(per transistor and for } S_{tot} = 14\text{mm}^2\text{)} \tag{7.18}$$

where S_{tot} is the total available surface value. The resulting characteristics of the power devices are summarised in Table 7.3.

	HVNMOS	HVPMOS	Unit
Total Width (W_{tot})	51000	94000	µm
Gate capacitance (C_{gb})	250	430	pF
Saturation current (I_{max})	4	3	A
$R_{dson,\min}$	1.17	1.17	Ω

Table 7.3: Main electrical characteristics of the full-bridge power devices

Given the non-negligible on state resistance value of the power devices, the actual maximum output power achievable with the full-bridge is set by the voltage resulting

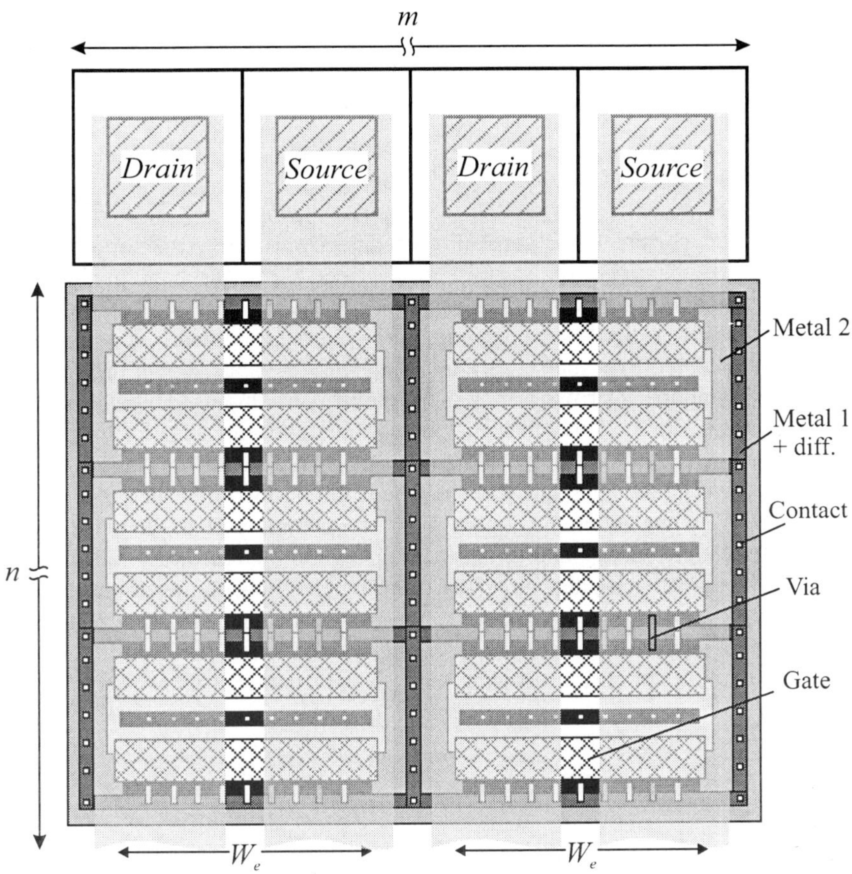

Figure 7.13: Layout topology of the power devices maximising the surface of the metal 2 wires

from the resistive divider formed by $R_{dson,\min}$ and the load. Thus, P_{omax} can be expressed as

$$P_{o\max} = \frac{\left(0.77 \cdot U_{rms}\right)^2}{R_L} = \frac{\left(0.77 \cdot 12/\sqrt{2}\right)^2}{R_L} = 5.4\,W \tag{7.19}$$

and the corresponding power P_{diss} dissipated in the chip is given by

$$P_{diss} = \frac{\left(0.23 \cdot U_{rms}\right)^2}{2 \cdot R_{dson,\min}} = \frac{\left(0.23 \cdot 12/\sqrt{2}\right)^2}{2.34} = 1.6\,W \tag{7.20}$$

Finally, neglecting the switching losses, the maximum power efficiency achievable with such a fully integrated amplifier is given by

$$\eta = \frac{P_{o\max}}{P_{tot}} = \frac{P_{o\max}}{P_{o\max} + P_{diss}} = 77\% \tag{7.21}$$

Considering the current drive capability of the power stage, the topology of the transistors must maximise the surface of the metal 2 wires handling the output current in order to avoid electromigration problems and metal evaporation. Therefore, the power devices are designed such that the total width of a high-voltage transistor is achieved by a matrix of *nxm* elementary cells, as shown in Figure 7.13, where n is the number of rows and m is thenumber of columns. The width W_e of an elementary transistor cell matches that of two juxtaposed pads, forming thus the drain and source access of column m_i. Consequently the surface of the metal 2 wires is maximised within the available surface. The resulting layout of the complete class-D amplifier is shown in Figure 7.14, where the central part implements the low-voltage features, the drivers and the guard rings, while the half-bridges of the output stage are placed at each side. Note that the juxtaposition of the drain and source allows a better connectivity in comparison to the case where the drain pad is facing the source one. In fact, in the former case, only metal 2 wires are used in the vertical direction to handle the current flowing through the drain an source terminals, while in the latter case, metal 2 wires are required for one terminal and metal 1 wires are required for the other terminal.

7.3.2 Output drivers

In order to drive the huge gate capacitances of the power devices, dynamic drivers are required to save static power consumption. In the case of the HVPMOS transistors, the low-standby current static level shifter cell discussed in chapter 6 symmetrically drives the devices of each half-bridge. The use of high-voltage buffer stages, between the classic static level shifter cell and the gate of the output pMOSFETs, reduces the static consumption by a factor of 40 leading to a total static power dissipation of 20mW instead of the 550mW required by the classic static cell. In the case of the HVNMOS transistors, standard CMOS buffers are used between the modulation stage and the output stage. They control the signal driving the HVNMOS transistors and the signal at the input of the level-shifter. Given the current drive capability of the power devices, an overlap cancellation logic is used in addition to the drivers in order to avoid transient short-circuits between the high-voltage power supply and ground. The complete schematic of the drivers and the overlap cancellation logic is shown in Figure 7.15. Since no overlap cancellation is achievable at the complementary outputs of the level shifter cell, the digital function implemented by the low-voltage logic is given by

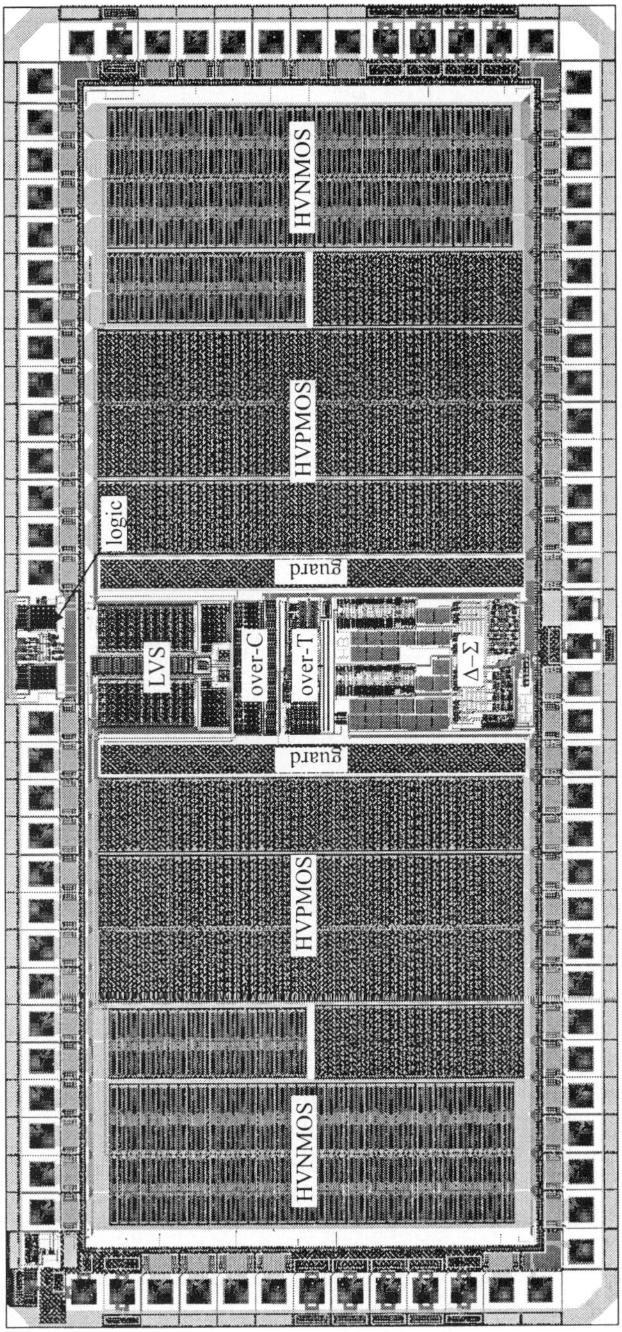

Figure 7.14: Complete layout of the Δ-Σ class-D audio amplifier

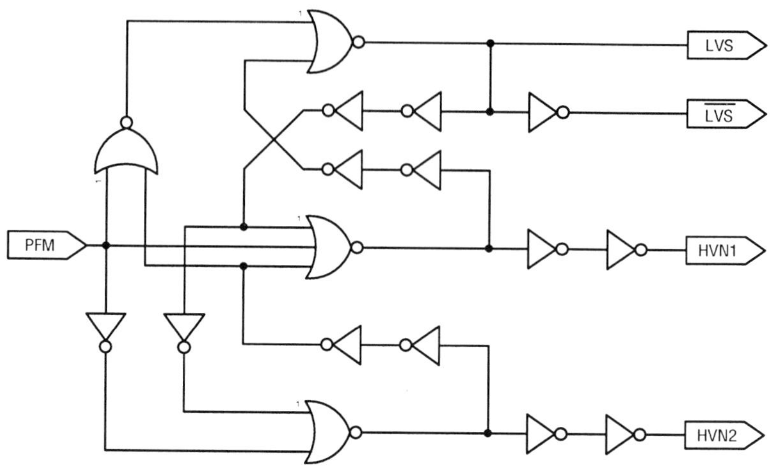

Figure 7.15: Overlap cancellation logic avoiding transient power supplies short-circuit

$$\begin{cases} HVN2 = \overline{PFM + LVS + HVN1} \\ HVN1 = \overline{\overline{PFM} + \overline{LVS}} \\ LVS = \overline{\overline{PFM + HVN1} + HVN2} \end{cases} \tag{7.22}$$

which consists of switching off the HVNMOS devices while the HVPMOS devices are turned on and off complementarily. The measured driving signals at the gates of the power devices are shown in Figure 7.16. Note that the rise and fall times of these signals are about 100ns. Consequently, the total delay time during which no current is supplied to the load is about 300ns. It represents less than 10% of the minimum period of the modulated signal, assuming a modulator clock frequency of 500kHz. It is a good trade-off achieved for this design between power consumption and distortion, since the dynamic losses are minimised without altering noticeably the total harmonic distortion and the power efficiency.

7.3.3 Protection features

Given the current drive capabilities of the power devices with respect to that of the metal 2 wires, special precautions must be taken to avoid electromigration problems or metal evaporation under short circuit conditions. Moreover, since the upper limit of the power efficiency is about 77% and considering the power dissipated in the

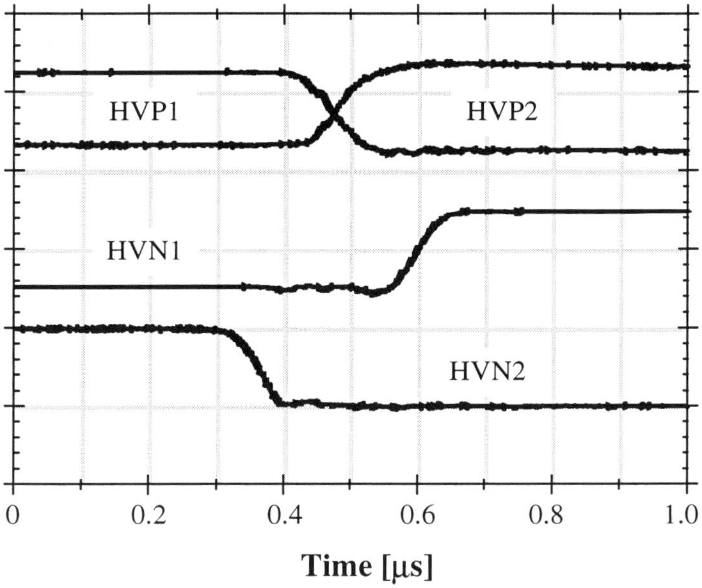

Figure 7.16: Switching sequence of the power devices

chip, additional precautions must be taken to avoid an excessive heating of the die. These features are implemented by the use of an over-current detection cell, sensing the instantaneous output current value, and by the use of an over-temperature detection cell sensing the temperature of the die.

1) Over-temperature detection cell: The over-temperature condition is detected by the comparison of the base-emitter junction voltage of a bipolar transistor V_{BE}, which has a negative thermal coefficient, to a voltage proportional to the absolute temperature (*PTAT*). This is implemented by the circuit shown in Figure 7.17, using a compatible lateral bipolar transistor (*CLBT*), available in any standard CMOS process, and a *PTAT* current source. This current source is formed by transistors M_1-M_2, M_3-M_4, M_7-M_8 and resistor R_1. Assuming that M_1-M_2 are operating in weak inversion, the voltage drop across R_1 is proportional to the absolute temperature according to

$$V_{R1} = U_T \ln\left(\frac{S_4}{S_3} \cdot \frac{S_1}{S_2}\right) \tag{7.23}$$

where $S = (W/L)$. Assuming that $S_4 = S_3$ and that the hysteresis switch M_9 is in the on state, the voltage drop across R_2 is *PTAT* as well according to

$$V_{PTAT} = K \cdot U_T \ln\left(\frac{S_1}{S_2}\right) \tag{7.24}$$

where

$$K = \frac{S_6}{S_3} \cdot \frac{R_2}{R_1} \tag{7.25}$$

is the slope factor of the voltage V_{PTAT}. It is controlled by the ratio of resistors R_2 and R_1 and the gain of the current mirror formed by M_6 and M_1. Consequently, since the voltage V_{BE} has a negative thermal coefficient, it is possible to detect a given temperature by choosing an appropriate slope factor K. The maximum operating

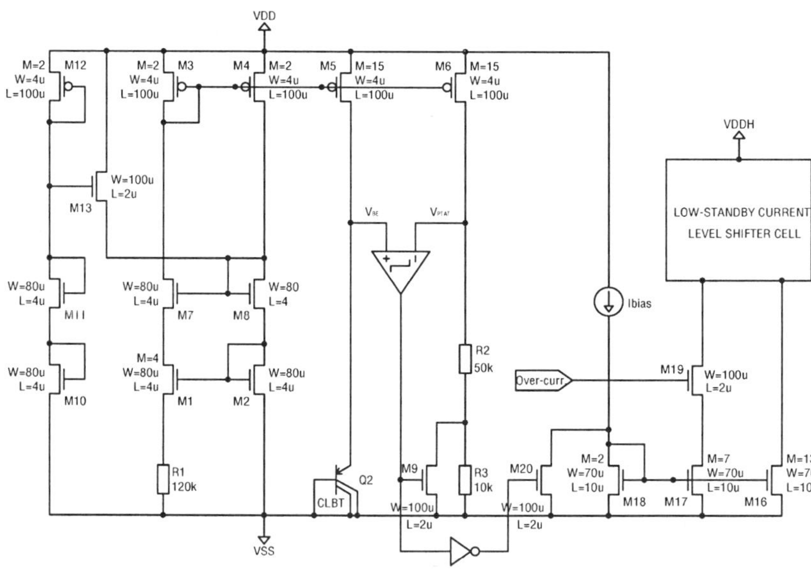

Figure 7.17: Complete schematic of the over-temperature protection cell

temperature of the chip is fixed at 125 °C. Resistor R_3, in parallel with transistor M_9, introduces a hysteresis between the turn-off and turn-on temperature. The turn-on temperature is set to 100 °C, as shown in Figure 7.18, where the voltage of the base-emitter junction of Q_1 and the voltage across R_2 are compared. When an over-temperature is detected, the output of the comparator inhibits the biasing current source of the level shifter in order to switch-off the HVPMOS transistors of the full-bridge. Thus, the resulting waveforms at the output of the level shifter after an over-temperature on and off states are compared in Figure 7.19. In the former case, the transistor of the conducting branch of the full-bridge switches-off within 150μs. This time delay is due to the time constant required to discharge the huge gate-capacitance, when the biasing current of the level shifter is switched-off. In contrast to this case, when an over-temperature off condition is detected, the HVPMOS transistors of the full bridge are turned on without any delay since the current source of the level shifter is switched-on again. The precision of the temperature sensing depends on the offset of the comparator, the precision of the

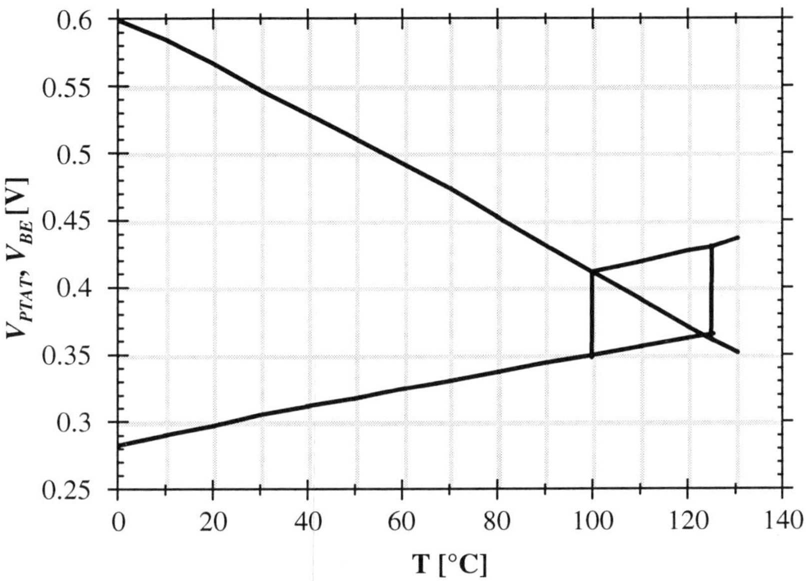

Figure 7.18: Measured behaviour of the voltages V_{BE} and V_{PTAT} as a function of the temperature

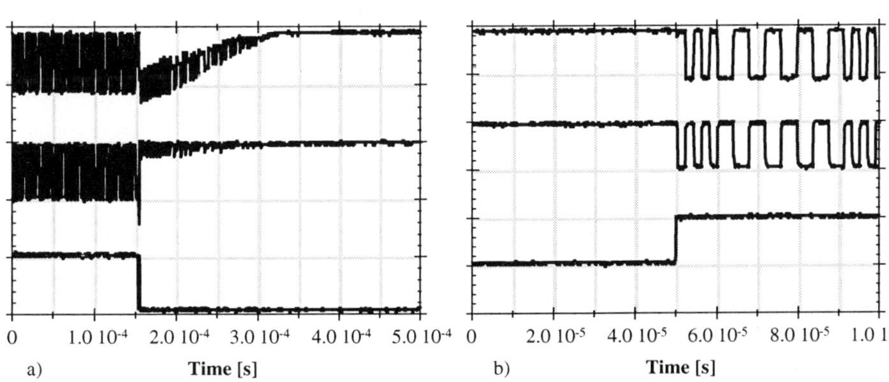

Figure 7.19: Measured waveforms at the output of the level shifter after an a) over-temperature on condition b) over-temperature off condition

PTAT source and particularly on the circuit location on the chip. When placed in the middle of the die between the half-bridges, the turn on and off temperatures are 125°C and 100°C. Note that resistor R_2 can be trimmed externally in order to precisely tune these conditions.

2) Over-current detection cell: Consider the worst case condition where one of the outputs of the full bridge is short circuited to ground. The continuous current that can flow through the corresponding HVPMOS transistor is given by the ratio of the power supply value (12V) to the on state resistance of the transistor (~1.17Ω). Consequently, a peak current of about 10 amperes can suddenly flow under such an extreme condition. The same reasoning applies for short-circuit of the HVNMOS transistor, since R_{dson} is identical for all the output devices. It is obvious that the metal wires cannot handle such current conditions. Therefore, the output current must be limited to the maximum value allowed by the smallest metal surface, without altering the circuit performance. This is the case of the metal wires of the HVNMOS transistor. According to the topology shown in Figure 7.13, 2.5 columns of elementary cells are required to achieve the on state resistance calculated in section 7.3.1, as it appears in Figure 7.14. The corresponding maximum continuous current that can be handled by the metal wires is 1.8A, assuming a maximum current density of 1.5 mA/μm. In fact, since a pad width is about 250μm, the maximum continuous current of column m_i is I_{cmax} = 750 mA (there are two drain terminals per pad width). The worst case short circuit condition occurs when a maximum DC input signal is applied. In fact, in this case, the current flowing through the output transistor is almost continuous, whether a PWM or a Δ-Σ modulator is used. Therefore, the maximum acceptable short circuit current value must be limited to

$$I_{short} = 1.8\ A \tag{7.26}$$

Note that the circuit performance is not altered, since the peak current value under normal operating conditions is given by

$$I_{peak}(R_L) = \frac{\sqrt{2} \cdot U_{rms}}{R_L} = \frac{0.77 \cdot 12}{8} \approx 1.2\ A \tag{7.27}$$

The filtered output signal and current flowing into the speaker load is shown in Figure 7.20, where a full swing input sinewave is used as a modulating signal.

The detection of an over-current condition is achieved by sensing the voltage drop created by the output current across the metal wires. The resulting voltage is compared to a reference value setting the short-circuit condition.

As soon as this condition is detected, the output current is clamped by reducing the biasing current of the level shifter in order to reduce the gate-to-source voltage of the HVPMOS transistors. The schematic of the circuit implementing this protection feature is shown in Figure 7.21. It consists of a shifted offset operational amplifier,

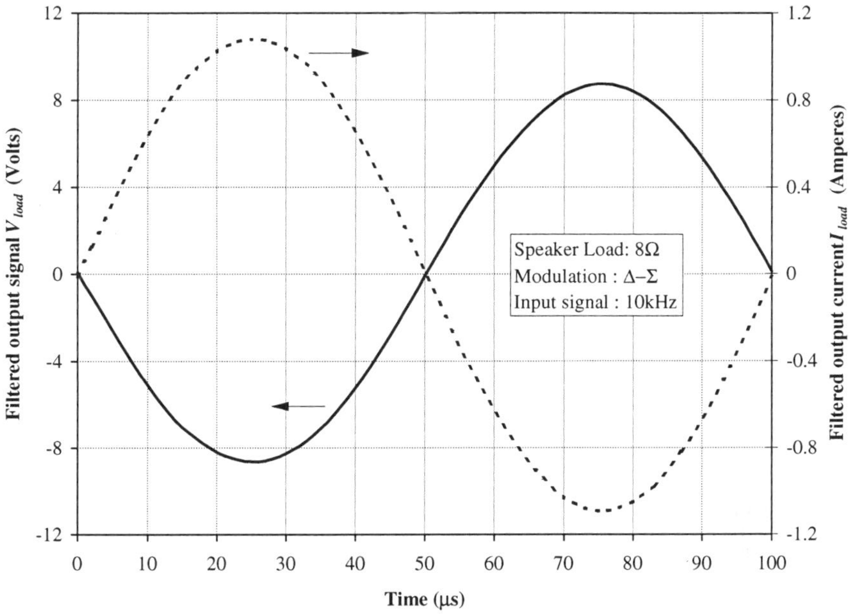

Figure 7.20: Filtered output signal and current flowing into the speaker load under maximum modulating input amplitude

where the offset shift V_{shift} is created by the reference current source I_{ref}. The value of V_{shift} is set by the value of the short circuit current I_{short} flowing into the metal wire resistance R_{M2} of the corresponding HVNMOS transistor. Consequently one protection cell is required for each half-bridge. Considering the current value given by Equation (7.26) and a total resistance R_{M2} ≈27.5 mΩ, the resulting offset shift is given by

$$V_{shift} = R_{M2} \cdot I_{short} = 27.5 \cdot 10^{-3} \cdot 1.8 \approx 50\,mV \tag{7.28}$$

The simulation data of the short circuit detection is shown in Figure 7.22 for all possible combinations of the electrical parameters deviations. The maximum deviation around the typical value of 50mV is ±5% in the worst case conditions. The typical deviation of the offset of the amplifier and the spreading of the sheet resistance of the metal 2 layer must be added as well, leading thus to a total deviation of ±20%. The corresponding deviation of the short circuit condition is about ±350mA around the typical value of 1.8A, which is obviously not acceptable. Therefore, the possibility of tuning the short circuit condition externally should not be excluded. This is achieved either by trimming the reference current value I_{ref} or by the use of an external precision shunt (typical precision of ±0.5%..2%) to perform the sensing operation. The former solution was chosen in the considered

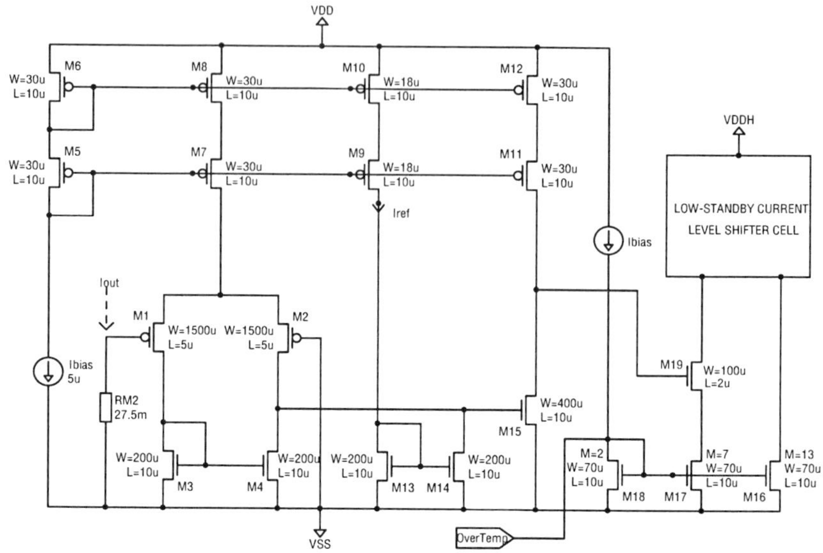

Figure 7.21: Complete schematic of the over-current protection cell

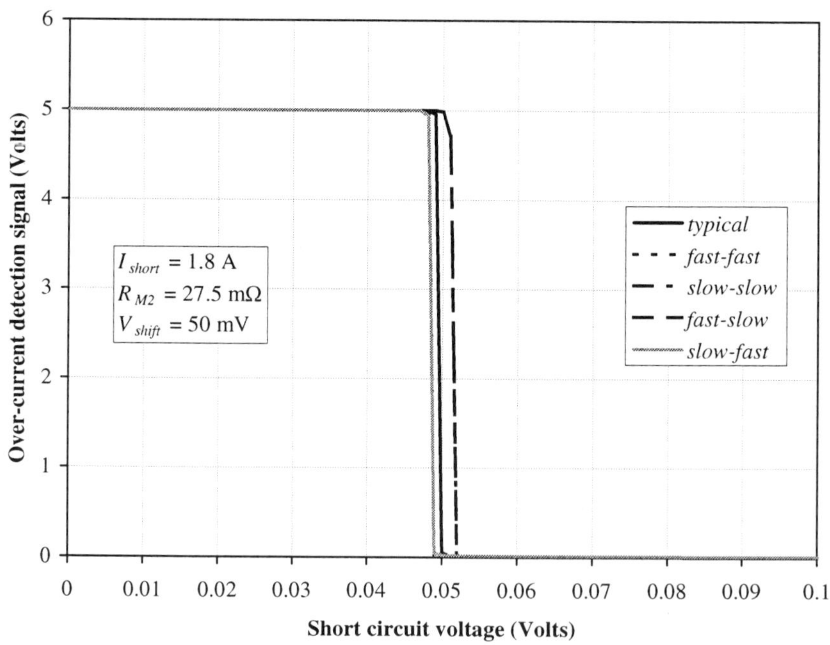

Figure 7.22: Simulation data of the short circuit detection condition

implementation. Unfortunately, the trimming range of the reference source is not wide enough to compensate for the different deviations discussed above. Moreover, the choice of limiting the maximum current supplied by the HVPMOS transistors is an efficient solution against the particular case where the load is in a short circuit condition. The case where one of the outputs of the full-bridge is short circuited to ground or to V_{DDH} , is not covered by the use of two over-current protection cells only, at the level of the HVNMOS transistors. Therefore, the detection and the limiting action must be performed on each power device of the full bridge. Such optimisation considerations are currently under-taken for a second implementation of the class-D amplifier.

7.4 Measured performances

The system performances are considered at the level of the Δ-Σ modulator and the level of the complete class-D audio amplifier. Such a distinction is made because of the non negligible additional distortions introduced by the full-bridge power stage. In fact, during the switching sequence of the power devices the amplitude of the modulated signal at the output of the amplifier is not constant. This is due to the finite on state resistance of the devices when they are turned on, and to the conduction of their body diodes when they are turned off. The comparison between the signals at the output of the Δ-Σ modulator and the output of the amplifier is shown in Figure 7.23. The additional amplitude "modulation" of the resulting signal decreases the *SNR* of the Δ-Σ modulator and increases the distortion of the filtered signal. These different aspects in addition to the power efficiency of the amplifier are the aim of this section.

7.4.1 Signal-to-noise ratio

The measured *FFT* spectrum of the filtered signals at the outputs of the Δ-Σ modulator and the class-D amplifier are shown in Figure 7.24. A fourth order Butterworth function is used in both cases with a cut-off frequency of about 15kHz. When the *FFT* measurement is performed on the output of the modulator the power stage is completely turned-off such that the characteristics of the modulator are not altered. The noise floor of the resulting spectrum is at about -70dB, and the peak of the signal is at 0dB since the input reference is set to $1V_{rms}$. When the power stage is turned-on and the *FFT* is performed on the output signal, the noise floor is increased by an amount of 20dB. Moreover, the third harmonic of the input signal is strongly present in the spectrum as it is located at about -30dB. This is due to the effect of the finite R_{dson} of the power transistors and to the conduction of their "free wheel" diodes. Note that the third harmonic is much higher than the second one, since the distortion resulting from the diodes and the on state resistance is symmetrical in the positive and negative alternation of the output signal, as it appears in Figure 7.23. The increase of the noise floor is first related to the amplification of the input signal by the gain of the amplifier which is given by

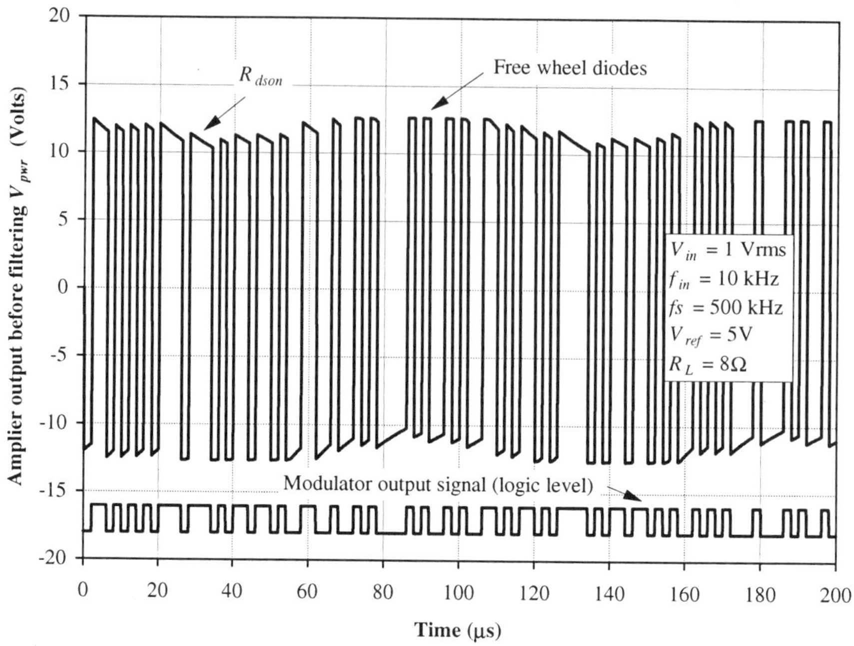

Figure 7.23: Difference between the modulated signals at the output of the Δ-Σ modulator and at the output of the power stage

$$A = 20\log\left(\frac{R_L}{R_L + 2R_{dson}} \cdot \frac{2V_{DDH}}{V_{DD}}\right) \approx 10\,dB \tag{7.29}$$

The remaining 10dB are mainly due to the thermal noise induced by the power devices, given the efficiency of the amplifier. In fact, when the *FFT* is performed on the output of the modulator when the power devices are turned-on without being loaded, the resulting noise floor is located between the two boundaries shown in Figure 7.24.

A first order computing of the *SNR* of the signals shown in Figure 7.24 can be performed on the basis of the total number of samples available for the entire spectrum. In fact, since 1598 samples are used for a total spectrum width of 40kHz, the spacing between two successive spectrum-lines is 25Hz. Thus, the noise level in the considered base-band is given by

$$L_e = L_{nf} + 20\log\left(\frac{\sqrt{f_b}}{\sqrt{25}}\right) \tag{7.30}$$

where L_{nf} is the noise floor level and f_b is the base-band width. Similarly, since the signal power is not concentrated in a unique spectrum-line, its level is given by

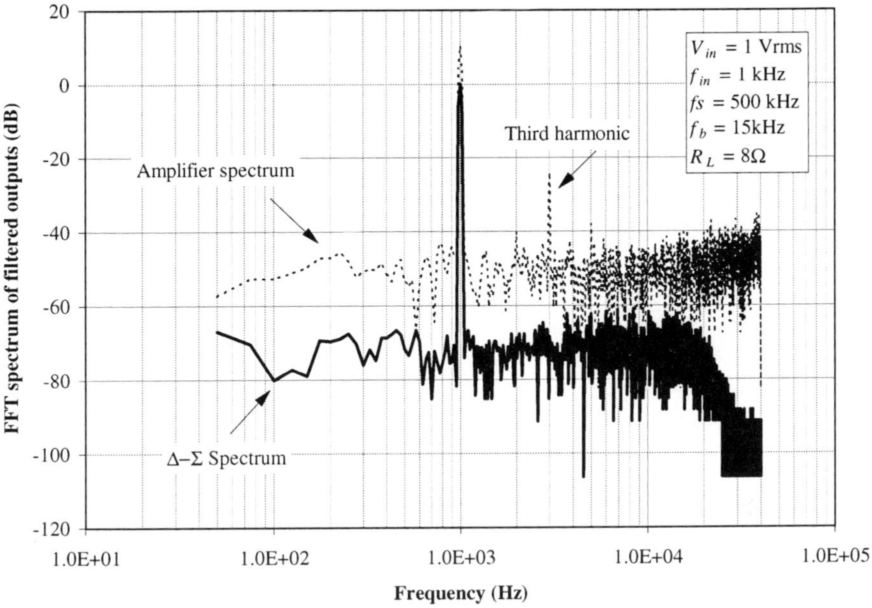

Figure 7.24: FFT spectrum of the signals at the output of the Δ-Σ (filtered) and the output of the amplifier

$$L_x = L_{peak} + 20\log\left(\sqrt{n_x}\right) \tag{7.31}$$

where L_{peak} is the peak level and n_x is the number of signal spectrum-lines. Consequently, the *SNR* can be expressed as

$$SNR = L_x - L_e = L_{peak} + 20\log\left(\sqrt{n_x}\right) - L_{nf} - 20\log\left(\frac{\sqrt{f_b}}{\sqrt{25}}\right) \tag{7.32}$$

Considering a base-bandwidth of 15kHz and 4 signal spectrum-lines, the *SNR* value, estimated on the basis of the measurement data shown in Figure 7.24, is given by

$$\begin{bmatrix} SNR_{(\Delta-\Sigma)} \approx 48\,dB \\ SNR_{(Amp)} \approx 38\,dB \end{bmatrix} \tag{7.33}$$

Thus, according to the maximum theoretical value given by Equation (7.10), the measured signal to noise ratio is in general agreement with the theoretical considerations. Moreover, the *SNR* value given by Equation (7.33) is actually the maximum achievable value since the applied input signal amplitude is in the overload limit range. However, since the distortions introduced by the power stage are more dominant than the quantisation noise, it is more convenient at this level to express the performances of the amplifier as a function of the signal distortion.

7.4.2 Total harmonic distortion

The most commonly used measurement data characterising the non linearity of audio amplifiers, is the total harmonic distortion (*THD*). It consists of applying a sinewave

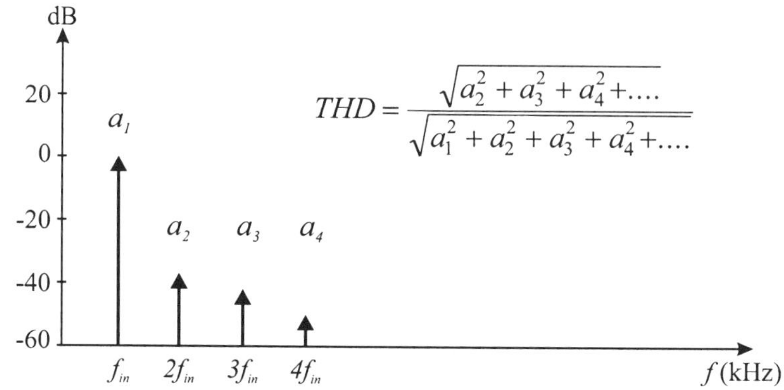

Figure 7.25: Total harmonic distortion definition

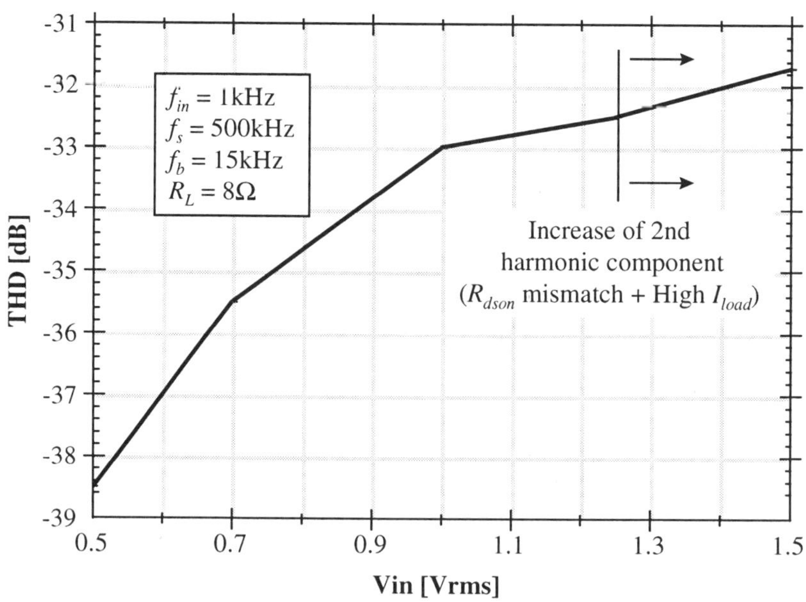

Figure 7.26: Measured total harmonic distortion of the amplifier as a function of the input amplitude

input and measuring the resulting harmonic components at the output of the amplifier. The *THD* is defined as being the ratio of the root-mean-square value of these harmonic components to the total components, as shown schematically in Figure 7.25. This ratio is measured in Figure 7.26 as a function of the input sinewave amplitude where the frequency is set to 1kHz. Note that the ratio is limited to the third harmonic since the level of higher order harmonics is lower than the noise floor. The chosen input signal range covers all standard audio players specifications. The corresponding *THD* ranges between 1% at 0.5Vrms and 2.5% at 1.5Vrms, which is in the typical range of medium quality class-D audio amplifiers. As will be shown in section 7.5, this range can be significantly improved by the use of a fully closed-loop configuration. Note that the sudden increase of the *THD* value for input amplitudes higher than $1.25V_{rms}$, is due to the increasing second harmonic component. This increase is due to the assymmetry between the on state resistance of the p-channel and n-channel devices. It becomes important with the input signal amplitude because the output current supplied to the load is significant. Thus the dissymmetry between the resistance values has an increasing effect on the second harmonic component.

7.4.3 Intermodulation distortion

When the input frequency of the sinewave increases, the amplitude of the harmonic components is attenuated by the output filter. Thus, the *THD* is limited by the cut-off frequency of the filter. Another and more convenient method used generally to cope with this limitation is the intermodulation distortion. It consists of measuring the non-linearity of the amplifier to an input signal resulting from the modulation of a high-frequency input sinewave f_2 by a low-frequency one f_1 of amplitude quadruple (12dB), as shown schematically in Figure 7.27. The typical spectrum of such a modulation consists of lateral components around f_2 at frequencies, $f_2 \pm f_1$, $f_2 \pm 2f_1$, $f_2 \pm 3f_1$, etc. The total intermodulation distortion is defined as the ratio of the root-mean-square value of these lateral components to the fundamental one at f_2.

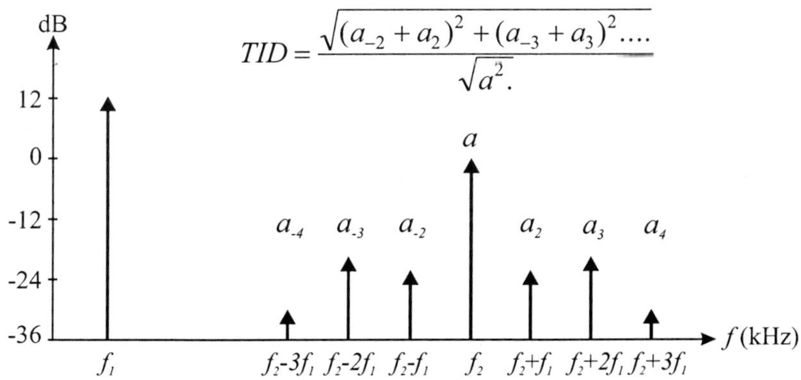

Figure 7.27: Definition of the intermodulation distortion

The measured total intermodulation distortion, limited to the third order lateral components, is shown in Figure 7.28, where the high-frequency signal is at 10kHz and the low-frequency one is at 100Hz. While keeping the 12dB of level difference between the input sinewave signals, the highest amplitude signal is swept from -9dB to 0dB. The 0dB reference level is set to $1V_{rms}$. The resulting total intermodulation distortion ranges between -20dB and -10dB. Note that such a measurement result is not always specified in the data sheets of audio amplifiers, as it is a more constraining measurement condition.

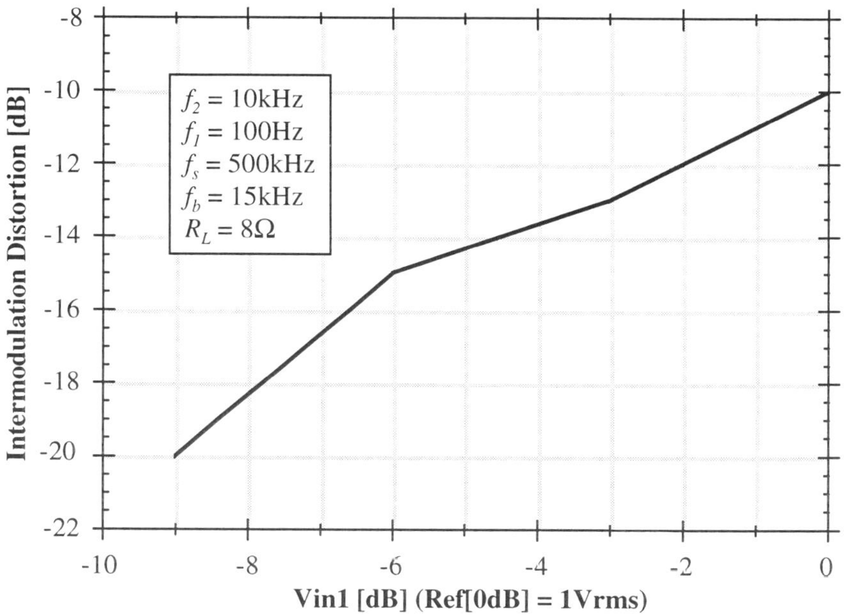

Figure 7.28: Measured total intermodulation distortion as a function of the input amplitude

7.4.4 Amplifier efficiency

The last interesting measurement data characterising audio amplifiers is the power efficiency of these circuits. The resulting curve is shown in Figure 7.29 as a function of the output power P_{out}. The efficiency is given by the ratio of P_{out} to the total power supplied to the amplifier P_{tot}. The latter is determined from the measurement of the power supply current, which increases as a function of the input amplitude. The output power is obtained by measuring the amplitude of the output voltage and the value of the current flowing through an 8Ω resistor, at 1kHz input frequency. The resistance of the load is kept as constant as possible, by cooling the load, in order to avoid inaccuracy due to temperature drift. The discrepancies with the maximum theoretical value (~77%), given by Equation (7.21), are mainly due to the switching losses.

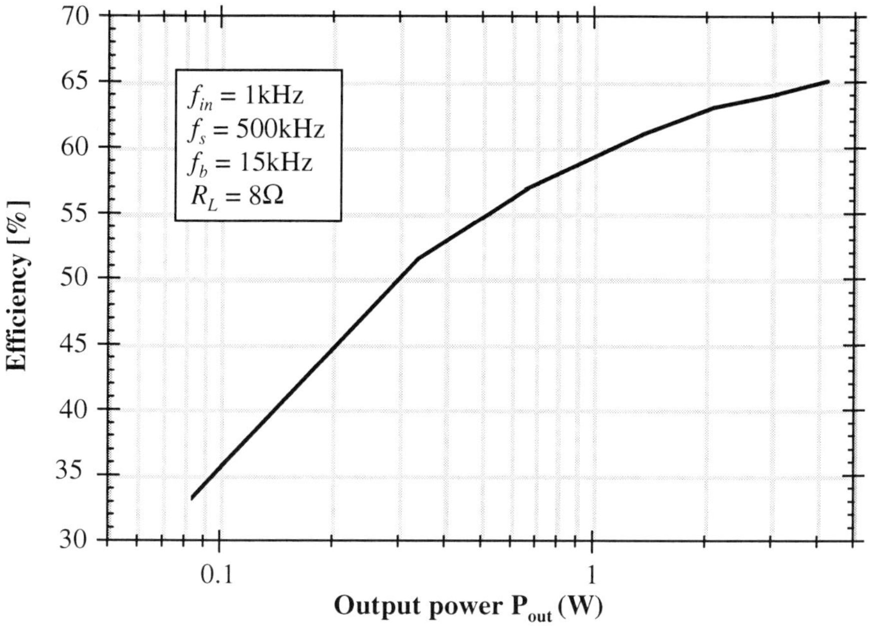

Figure 7.29: Efficiency as a function of the output power

7.5 System improvements

The performances achieved by the fully integrated class-D amplifier presented in the previous sections can be improved significantly by optimising at least one of the following parameters:

1. Reducing the total on state resistance value. This can be achieved only by the use of an advanced high-voltage technology such as the I^2T presented in chapter 3, where the total on state resistance per mm^2 is optimised. Note that the increase of the surface of the power devices using the SVX technique is not possible since the production yield of the die can be seriously affected

2. Increasing the sampling frequency of the Δ-Σ modulator in order to achieve a better signal-to-noise ratio. Care must be taken when choosing such an option. In fact, increasing the sampling frequency implies that the switching rise and fall times of the power devices should be decreased to avoid additional distortions and conduction losses. However, the reduction of the rise and fall times is limited by the increase of the losses due to the recovery time of the "free wheel" diodes as well as the switching losses.

3. Choosing a closed loop configuration of the amplifier, where the feedback loop of the Δ-Σ modulator is performed at the level of the output stage and not at the level of the modulation stage. In fact, in order to compensate the additional

distortions shown in Figure 7.23, they must be taken into account by the modulator.

The last two optimisation options are currently undertaken for a second implementation of the amplifier. The main improvement results and possible degradation of the power efficiency are discussed here. It is obvious that the use of these options with an advanced technology would result in additional improvements.

As was mentioned in the previous sections, the upper limit of the sampling frequency of the Δ-Σ modulator is limited by the power stage. Therefore, f_s was set to 500kHz in the implemented version of the class-D amplifier. Actually, in comparison to a classic PWM solution operating at 500kHz switching frequency, the sampling

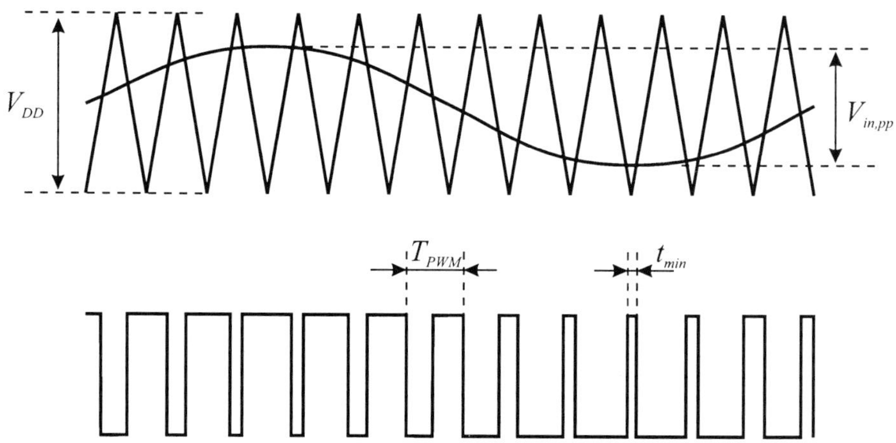

Figure 7.30: Minimum pulse width of a PWM signal modulated with a sinewave input

frequency of the Δ-Σ modulator can be increased up to 2.5MHz, assuming the same input modulating condition. In fact, since the modulation technique in both cases is basically different, the power stage is under the same switching constraints by using a Δ-Σ modulator running at 2.5MHz rather than a PWM modulator at 500kHz with $V_{in} = 1V_{rms}$ and $V_{DD} = 5V$. This is shown schematically in Figure 7.30 and can be explained as follows. Since the PWM technique keeps a constant switching frequency by adapting the duty cycle of the modulated signal to the input amplitude, the minimum pulse width duration t_{min} is given by the modulation ratio. Thus, t_{min} is given by

$$t_{\min} = \frac{T_{PWM}}{2} \cdot \left(1 - \frac{V_{in,pp}}{V_{DD}}\right) \approx 400\,ns \qquad (7.34)$$

where T_{PWM} is the period of the unmodulated *PWM* signal, and where the high-frequency triangular waveform is assumed rail-to-rail. Since the signal

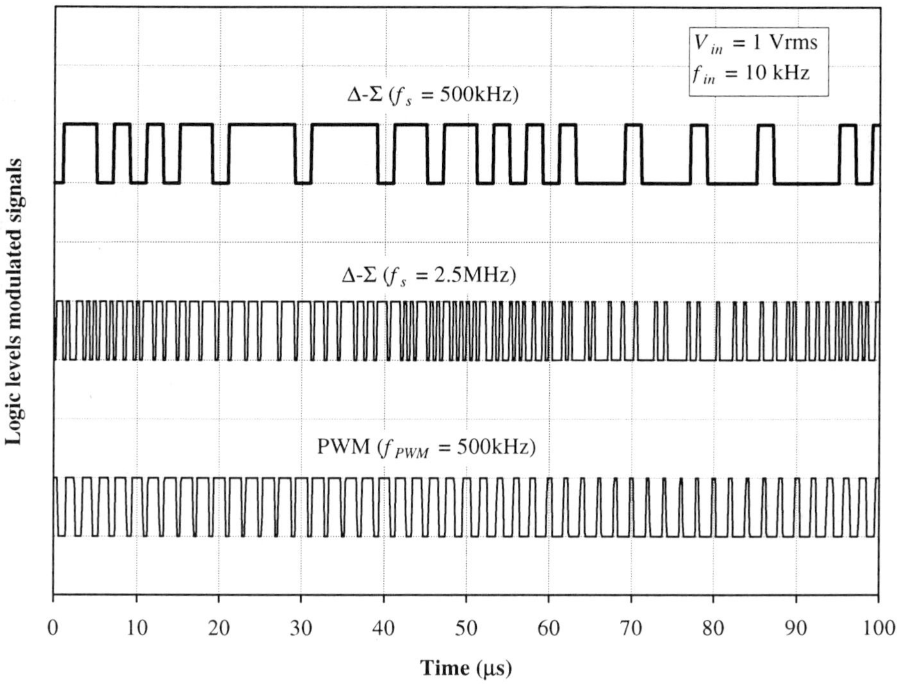

Figure 7.31: Comparison between a PWM and Δ-Σ modulated signals using the same input modulating sinewave

resulting from a Δ-Σ modulator is frequency modulated, the minimum pulse width is given by one clock period. Consequently, the equivalent sampling frequency of the Δ-Σ modulator is given by

$$f_s = \frac{1}{t_{\min}} = \frac{1}{400 \cdot 10^{-9}} = 2.5\ MHz \tag{7.35}$$

The comparison between a 500kHz PWM signal and a Δ-Σ modulated signal using sampling frequencies of 500kHz and 2.5MHz is shown in Figure 7.31 where the same input modulating sinewave is used. It appears clearly that while the modulated signal resulting from a sampling frequency of 500kHz is rich in low-frequency harmonic tones for high-input sinewave amplitudes, the harmonic contents of the modulated signal at 2.5MHz are equivalent to that of the PWM technique. Moreover, the minimum pulse width duration of the Δ-Σ modulator is identical to the PWM case. Consequently, under the same transient constraints applied to the power stage, the maximum theoretical *SNR* value achievable by the new sampling frequency is given by Equation (7.9)

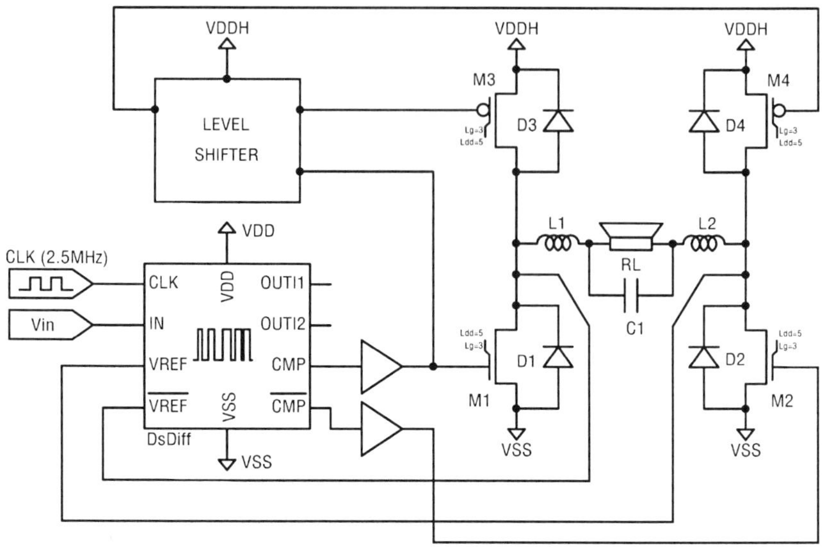

Figure 7.32: Closed loop configuration of the amplifier

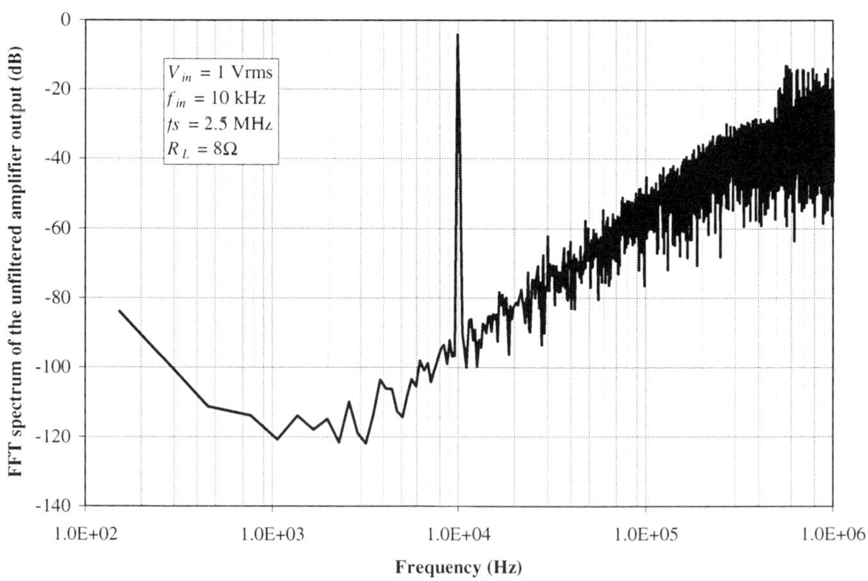

Figure 7.33: Simulation data of the typical noise shaping performance expected for the improved amplifier

$$SNR_{max} = 90dB \quad \begin{bmatrix} f_b = 15\,kHz \\ f_s = 2.5\,MHz \\ P_e = 1/12\,(1bit\ A/D) \\ P_x = 0.5\,(V_{in} = V_{DD}) \end{bmatrix} \tag{7.36}$$

In addition to an increase of the sampling frequency, a closed loop configuration is chosen in order to compensate the distortions of the output stage, as shown in Figure 7.32. The main difficulty with such a solution is the processing of the differential high voltage signal resulting from the output stage. The preliminary simulation data of the performances achieved by the improved option are shown in Figure 7.33. Note that the amplitude of the input signal is identical to that of the open loop solution, and the R_{dson} values are unchanged. The improvement of the performances is very interesting since the linearity of the amplifier is increased for high input amplitudes. Moreover, the increase of the signal-to-noise ratio is significant.

Finally, the analysis is focused now on the effect of a higher sampling frequency on the power efficiency. The use of external "free-wheel" diodes is probably obligatory given the high slew-rate values involved during the switching sequence of the power stage. In fact, simulations were performed using external "free-wheel" diodes dedicated for switched power applications. They have shown that the use of a fast recovery diode (MR811) specified for recovery times ranging between 350ns and 750ns, induces current peaks of 8A at the switching instants. These peaks are eliminated by the use of an ultra-fast recovery diode (MUR110) specified for recovery times ranging between 75ns and 100ns.

7.6 References

[16] J. C. Candy, G. C. Themes "Oversampling Delta-Digma Data Converters" *IEEE Press*, 1992

[17] P. M. Aziz, K. V. Sorensen, J. V. D. Spiegel "An overview of Sigma-Delta Converters" *IEEE Sig. Process. Magazine.*, pp. 61-84, January 1996

[18] V. Friedman, "The Structure of Limit Cycles in Sigma Delta modulation", *IEEE Trans. On Communications*, pp. 972-979, August,1988

[19] J. Candy, O. Benjamin, "The Structure of Quantization noise from Sigma-Delta Modulation", *IEEE Trans. On Communications*, pp. 1316-1323, September,1981

[20] R. Gray, "Spectral analysis of quantization noise in a single-loop sigma-delta modulator with DC input", *IEEE Trans. On Communications*, pp. 588-599, June,1989

[21] R. Gray, "Quantization noise spectra", *IEEE Trans. On Information Theory*, pp. 1220-1244, November,1990

[22] S. Norsworthy, D. Rich "Idle channel tones and dithering in delta sigma modulators" *95th Convention of the Audio Engineering Society*, pre-print 3711, October 1993

[23] D. G. Haigh, B. Singh "Switching scheme for switched capacitor filters which reduces the effect of parasitic capacitances associated with switch control terminals", *IEEE ISCAS proceedings*, pp. 586-589, April 1983

[24] K. Lee, R. G. Mayer "Low-distortion switched capacitor filter design techniques" *IEEE Journal of Solid State Circuits*, Vol. SC-20, No. 6, pp. 1103-1113, December 1985
[25] SMASH simulator, Dolphin Integration, Meylan Cedex, France
[26] R. Castello, P. R. Gray "A high performance micropower switched capacitor filter" *IEEE Journal of Solid State Circuits*, Vol. SC-20, No. 6, pp. 1122-1132, December 1985
[27] H. Ballan, M. Declercq, "12V Δ-Σ Class-D Amplifier in 5V CMOS Technology", *CICC* proceedings, paper 26.5, May 1995
[28] J. M. Goldberg, M. B. Sandler " Noise Shaping and Pulse Width Modulation for an All-Digital Audio Power Amplifier", *Journal of Audio Eng. Soc.,* Vol. 39, No. 6, June 1991
[29] M. O. J. Hawksford " Dynamic Model-Based Linearization of Quantized Pulse-Width Modulation for Applications in Digital-to-Analog Conversion and Digital Power Amplifier System", *Journal of Audio Eng. Soc.,* Vol. 40, No. 4, April 1992
[30] J. Hancock " A Class D Amplifier Using MOSFETs with Reduced Minority Carrier lifetime", *Journal of Audio Eng. Soc.,* Vol. 39, No. 9, September 1991
[31] D. A. Grant, J. Gowar, "Power MOSFETs Theory and Applications", John Wiley & Sons, pp. 313-319, 1989
[32] B. E. Attwood, "Design Parameters Important For Optimization of very High Fidelity (Class D) Audio Amplifiers", *Proceedings Conv. Audio Eng. Soc.,71st* (1982)

8 CONCLUSIONS

1) High-voltage technologies and devices

The first part of this book was dedicated to high-voltage CMOS devices and related technologies. If the most common and the most straightforward approach consists in selecting a CMOS technology with dedicated high-voltage options, it has been shown that very interesting results can be obtained as well on the basis of a standard low-voltage CMOS technology, without any modification of the process steps. Owing to the economical and technical advantages offered by this approach, a special chapter has been dedicated to the design and optimization of high-voltage devices based on this SVX approach. The technology selected to illustrate this methodology is a standard 2μm n-well CMOS process, because an in-depth reliability analysis exists for the high-voltage extension of this process, and because many high-voltage circuits have been designed and mass-produced on its basis.

After the examination of the potential problems arising from excessive drain voltage conditions, detailed analysis of the process steps have lead to the most reliable solution for implementing complementary high-voltage devices. For the HVNMOS transistor, the n-well layer was used as the lightly doped drain (LDD), while the "p-TUB" and "channel stop" implants were used for the LDD of the HVPMOS. In the latter case, only the logic equations generating the implant masks was modified. These actions do not change the process flow under any circumstances. For both devices, the field oxide layer was used in order to implement the necessary field-plates.

Measurement results have shown that reliable and reproducible high-voltage capabilities as high as 120V and 75V can be achieved by the HVNMOS and HVPMOS transistors. While the use of floating field rings and additional field-plates can improve the high-voltage capabilities of the HVNMOS up to 170V, those of the HVPMOS remain unchanged. Two dimensional simulations have shown that this limit is related to the depth of the n-well containing the device. In fact, beyond 75V, punch-through breakdown occurs between the lightly doped drain junction and the n-well junction. However, since most of the HVICs developed with the SVX technique operate under a maximum power supply voltage of 30V, the scaling of the devices was optimized to meet this specification with the best surface efficiency. Reliability considerations have shown that the optimized version can operate safely up to this power supply limit.

On the basis of exhaustive measurements of the characteristics of the devices, two macro-models were developed to allow the simulation of the HVICs. They both use a low-voltage MOS transistor and other discrete components in order to model the low-voltage and high-voltage sections of the devices. In the case of the HVPMOS transistor, it was shown that a simple resistor and diode correctly model the electrical characteristics of the lightly doped drain. In the case of the HVNMOS transistor, a JFET in series with a variable resistance were necessary to carry out a correct model. The critical dynamic behaviors of both devices are roughly modeled by the extrinsic capacitances of the low-voltage MOS transistors.

The final step, which is as important as the previous ones, was the definition of the necessary layout rules and precautions that avoid parasitic effects. It was performed at a device and circuit levels, where the importance of using majority and minority carrier guards was pointed out. Although they are surface consuming, it was shown that minority carrier guards using the n-well layer are the most efficient layout precautions against latchup triggering.

In order to appreciate the performances achieved by the SVX devices, their main high-voltage features have been compared to several dedicated technologies. Results have shown that the initial motivation for extending a low-voltage CMOS process with high-voltage features was justified. On the one hand, no additional masks or processing steps are required in this approach, which optimises the total number of masks in comparison to dedicated technologies. As a matter of fact, usually 2 to 5 masks are necessary to add specific high-voltage layers to a standard low-voltage CMOS process, while in the SVX approach the total number of masks remains unchanged, that is 12 to 13 masks. On the other hand, the high-voltage capabilities of the SVX devices are comparable or even better than those of dedicated technologies. Particularly, given the thinner gate oxide, the surface density is better, for a given transconductance and total ON state resistance R_{dson} of the devices. This appears clearly in the power application developed in chapter 7, where the R_{dson} achieved by the power devices are $3\Omega.mm^2$ in the case of the HVNMOS and $5.2\Omega.mm^2$ in the case of the HVPMOS. Typical values in available equivalent 2μm dedicated high-voltage technologies range between 5 and 8Ω.mm^2 in the case of the n-channel devices, and 9 and 20Ω.mm^2 in the case of the p-channel devices. Lower values can be found in advanced high-voltage submicron CMOS technologies, where 2 and 4Ω.mm^2 are typical.

It is obvious that the SVX approach is not the unique solution in high-voltage technologies. Actually, the selection of a high-voltage process depends on the application type. Given the wide field of high-voltage applications, different technological solutions are available, with or without modification of the process steps. These solutions have been presented and investigated in the different chapters of this book, with an emphasis on the SVX technique, such that the main particularities of existing high-voltage technologies are covered. This should help in the selection of the appropriate solution for a given application. Typically, for high-voltage power applications involving high-current levels, power device technologies using vertical DMOSFETs are more convenient than SVX or any dedicated high-voltage technology implementing lateral devices. On the other hand, when the application requires a complex signal processing with high-voltage output capabilities, technological solutions implementing lateral DMOSFETs should be considered. The choice between the SVX technique or a modified CMOS process is application specific.

When using a submicron technology, it becomes more difficult to reach high-voltages with an unmodified process, and an extension of the technology based on the addition of masking steps and implantation steps, may be the best approach for reaching voltages in the range of 50V or more. On the other hand, the SVX approach remains the most straightforward for creating 5V devices in a low-voltage technology, making it very useful for creating standard interfaces.

2) About the design of HVICs

The second part of this book adresses the design of high-voltage integrated circuits. As explained in the text, most of the existing HV devices obtained either by a dedicated process or by the SVX technique can only sustain the nominal voltage between the source and drain terminals, while the gate-to-source voltage is limited to much lower values. To cope with this problem, reliable level shifting techniques have therefore been discussed for both digital and analog cells.

When a chip with high-voltage output drivers involves an intensive signal processing part, experience has shown that the most favorable approach usually consists of performing the signal processing at low voltage, followed by a level-shift for driving the high-voltage output interfaces. Multiple supply voltage can be solved by using on chip step-up or step-down converters.

Two digital output interfaces have been detailed for low- and medium-power applications. The sensitivity of these interfaces to power supply and output transients was analyzed. The critical parameters of the transients, such as the slew-rate and the rise and fall times, were characterized as a function of the load condition and circuits performance. On the basis of concrete circuit examples, typical safe values were up to 100V/μs for the slew-rate condition and 500ns for the rise and fall times. Two analog output interfaces were developed as well: an operational transconductance amplifier introducing the concept of analog level shifting and a class-AB output stage operational amplifier. The latter is currently used to drive a piezo-electric transducer in an industrial high-voltage application. Detailed transient analysis has shown that the amplifiers behave safely under the same transient conditions as those used in the digital interfaces.

Among the various existing medium-power applications, for which a full integrated circuit solution can be used, the limits of the SVX technique were set by a typical medium-power application example, the class-D audio amplifier. A Δ-Σ modulator was chosen as a typical low-voltage signal processing block suited for CMOS implementation, in addition to some protection features such as over-temperature and over-current detection cells. Given the power efficiency achieved to drive an 8Ω speaker load (about 70% @ P_{out} = 5.6 Watts) with respect to the surface occupied by the power devices, the use of an advanced technology such as the I2T is probably more suitable, since the on state resistance of the power devices is almost 3 times lower. However, it should be recalled that the processing cost of the I2T is higher than the SVX. Therefore, from an economical point of view, the power limit beyond which a dedicated technology is more appropriate, depends on the surface of the chip, its processing cost and its production yield. From the technical point of view, the power limit of SVX can be set by the maximum current handling capabilities of the metal wires, as discussed in chapter 7.

APPENDIX A

This appendix presents the macro-models of the high-voltage devices using the SVX technique. The simulator used to implement these models is SMASH [1]. The SPICE parameters of the discrete components modelling the transistors are attached as well. Finally, the main high-voltage layout rules are presented for a maximum V_{DDH} value of 30V.

1) Macro-model of the HVNMOS transistor

```
.SUBCKT HVNMOS D D2 D1 G S B PARAMS: W=50E-6 L=5E-6 LG=4.25E-6
   + LDD=4.5E-6 M=1 AD=840E-12 PD=148E-6 AS=200E-12 PS=108E-6

GLDD D D2 <behavioral>
RLDD D D2 10E9
GFP D2 D1 <behavioral>
RFP D2 D1 10E9
MLV D1 G S B NHV W='W' L='L-1.0E-6' M='M'
+ AD=0 PD=0 AS='AS' PS='PS'
RLV D1 S 10E9
DHV B D DNHV 'AD'

[static]
double        imax, rldd, rfp, V0;
double        ids1, ids2, did2, i2g, i2b;
double        vd, vd2x, vd1, vgds;
double        vdsat, v2sat, v2gsat;
double        v1, v2, v2g, v1g;
```

```
[state]
double          VD2;

[header]
#include        <math.h>
#define q       1.6E-19
#define e0      8.854E-12
#define esi     (11.7*e0)
#define eox     (3.9*e0)
#define vsat    1.0E5
#define nd      (2.5E15*1.0E6)
#define na      (7.0E14*1.0E6)
#define LD      (LDD-LG)
#define λHV     0.01
#define Esat    3.0E6
#define tox     1E-6
#define xj      5.5E-6
#define K       10
#define K1      2

#define ln(A) ( (A)<=0 ? -1E-21 : 2.3*log(A) )
#define myexp(A) ( exp(A)<=1E-21 ? 1E-21 : exp(A) )
#define myfabs2(A,B) ( (A)-(B)<=0 ? 1E-21 : (A)-(B) )

[initial]
imax    =       nd*q*vsat*W*M;
rldd    =       sqrt(2*esi*na/(q*nd*(nd+na)));
rfp     =       sqrt(2*esi/(q*nd));
V0      =       (q*esi*nd/2)*pow(tox/eox,2);

[behavior]
vdsat   =       myfabs2(S(VD2),V(B));
v2sat   =       myfabs2(S(VD2),V(D1));
v2gsat  =       V0+0.5*((S(VD2)-V(G))+sqrt(pow((S(VD2)
                -V(G)),2)+pow(K1,2)));
vd2x    =       vdsat*(1-(1/ln(1+myexp(K)))*ln(1+myexp(K*(1-(V(D2)
                -V(B))/vdsat))));
vd1     =       myfabs2(V(D1),V(B));
v2      =       myfabs2(vd2x,V(D1));
v2g     =       V0+0.5*((vd2x-V(G))+sqrt(pow((vd2x-V(G)),2)+pow(K1,2)));
v1g     =       V0+0.5*((V(D1)-V(G))+sqrt(pow((V(D1)-V(G)),2)+pow(K1,2)));
vd      =       myfabs2(V(D),V(B));
v1      =       myfabs2(V(D),V(D2));
vgds    =       0.5*((V(D)-vdsat)+sqrt(pow((V(D)-vdsat),2)+pow(K1,2)));
```

```
did2    =       -(imax/pow(Esat*LG+v2sat,2))*((xj + esi*tox/eox)*v2sat
                -(2*rldd/3)*(pow(vdsat,1.5) - pow(vd1,1.5))
                -(2*rfp/3)*(pow(v2gsat,1.5) - pow(v1g,1.5)))
                +(imax/(v2sat+Esat*LG))
                *(xj + esi*tox/eox - rldd*(sqrt(vdsat)) - rfp*(sqrt(v2gsat)));

make_zero(did2);

if ( (vd>=vd2x) && (vd2x>=vd1) )
        {
        ids1    =       (imax/(Esat*LD+v1)) * (xj*v1-(2*rldd/3)*(pow(vd,1.5)
                        -pow(vd2x,1.5)));

        ids2    =       (1+λHV*vgds)*(imax/(Esat*LG+2))
                        *((xj+esi*tox/eox)*v2
                        - (2*rldd/3)*(pow(vd2x,1.5) - pow(vd1,1.5))
                        - (2*rfp/3)*(pow(v2g,1.5) - pow(v1g,1.5)));
        }
else
        {
        ids1    =       0.0;
        ids2    =       0.0;
        }

GLDD~val = ids1;
GFP~val = ids2;

.ENDS HVNMOS
```

2) SPICE parameters for the HVNMOS transistor

```
.MODEL DNHV D IS=190E-6 RS=3.85E-6 N=1.047 M=0.4
+ FC=0.8 CJ0=200E-6 VJ=0.35

.MODEL          NHV     NMOS
+       LEVEL           =       3
+       LD              =       2.5E-7
+       TOX             =       4E-8
+       NSUB            =       1.16E16
+       VTO             =       0.7
+       GAMMA           =       0.7
+       UO              =       621
+       THETA           =       0.06
+       DELTA           =       1.7
```

```
+        KAPPA     =    1
+        RSH       =    30
+        XJ        =    3.6E-7
+        VMAX      =    1.75E5
+        NFS       =    1.12E11
+        JS        =    0.0
+        CJ        =    2.63E-4
+        MJ        =    0.56
+        CJSW      =    4.82E-10
+        MJSW      =    0.36
+        PB        =    0.95
+        CGSO      =    2.95E-10
+        CGDO      =    4.00E-10
+        PHI       =    0.65
```

3) Macro-model of the HVPMOS transistor

```
.SUBCKT HVPMOS D G S B PARAMS: W='50u' L='5u' M='1'
+ AD='840E-12' PD='148E-6' AS='200E-12' PS='108E-6'

MLV D2 G S B PHV W='W' L='L-1E-6' M='M'
+ AD=0 PD=0 AS='AS' PS='PS'
Rldd D D2 '50E-3/(W*M)'
DHV D B DPHV 'AD'

.ENDS HVPMOS
```

4) SPICE parameters for the HVPMOS transistor

```
.MODEL DPHV D IS=22E-6 RS=50E-9 N=1.06 M=0.65
+ FC=0.8 CJ0=200E-6 VJ=0.85

.MODEL          PHV     PMOS
+        LEVEL          =    3
+        LD             =    2E-7
+        TOX            =    4E-8
+        NSUB           =    1.2E16
+        VTO            =    -0.7
+        GAMMA          =    0.53
+        UO             =    245.5
+        THETA          =    0.15
+        DELTA          =    0.93
+        ETA            =    0.11
```

+	KAPPA	=	10
+	XJ	=	3E-7
+	VMAX	=	5.76E5
+	NFS	=	3.55E10
+	RSH	=	95
+	JS	=	0
+	CJ	=	2.7E-4
+	MJ	=	0.44
+	CJSW	=	4.41E-10
+	MJSW	=	0.31
+	PB	=	0.81
+	CGSO	=	2.65E-10
+	CGDO	=	8.00E-10
+	PHI	=	0.6

5) high-voltage layout rules ($V_{DDH} \leq 30V$)

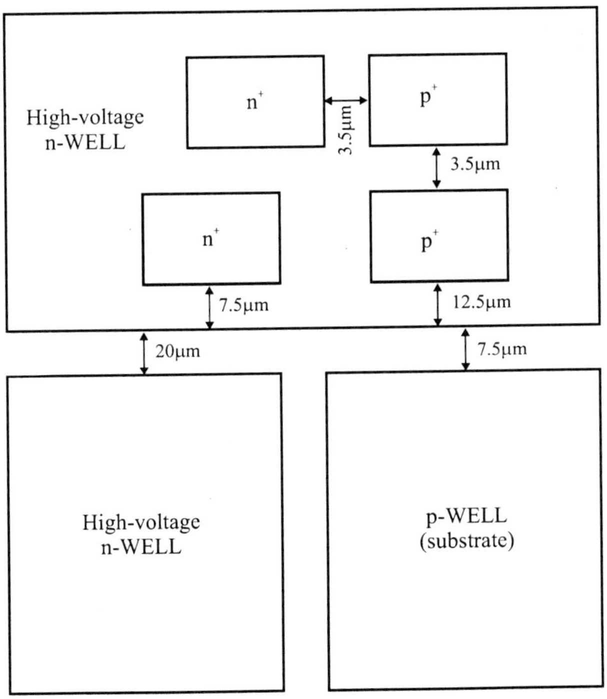

6) References

[1] Dolphin integration, Meylan Cedex, France

APPENDIX B

This appendix presents the code of the analog behavioral model of the second order Δ-Σ modulator implemented in SMASH [1]. The schematics of the general purpose blocks of the class-D audio amplifier are attached as well.

1) Analog behavioral model of the second order Δ-Σ modulator

```
DECLARATIONS:

INPUTS: IN VREF CLK VDD VSS
OUTPUTS: OUTI1 OUTI2 OUTCMP
PARAMS: A1 A2 B1 B2
GLOBAL_DOUBLE: XN1 XN2 YN1 YN2 OUT1 OUT2 CMP Q NQ

BEHAVIOR:
{
 double IN1,VREF1,VDD1,VSS1;

 IN1        =   IN - (VDD+VSS)/2;
 VREF1      =   VREF - (VDD+VSS)/2;
 VDD1       =   VDD - (VDD+VSS)/2;
 VSS1       =   VSS - (VDD+VSS)/2;

   if (simtime==0)
{
        OUTCMP = 0.0;
```

```
        OUTI1  = (VDD+VSS)/2;
        OUTI2  = (VDD+VSS)/2;
        OUT1   = 0.0;
        OUT2   = 0.0;
        CMP    = VSS1;
        Q      = 0.0;
        NQ     = 1.0;
        XN1    = 0.0;
        XN2    = 0.0;
        YN1    = 0.0;
        YN2    = 0.0;
        return;
  }
  if (POSEDGE(CLK_Node,2.5,0.0))
{
        OUT1   = A1*XN1 + XN2 + NQ*B1*VREF1 - Q*B1*VREF1;
        OUT2   = A2*YN1 + YN2 + NQ*B2*VREF1 - Q*B2*VREF1;
        if (OUT2 > 0.0)
                CMP = VDD1;
        else
                CMP = -VDD1;
         }
  if (NEGEDGE(CLK_Node,2.5,0.0))
{
        XN1    = IN1;
        XN2    = OUT1;
        YN1    = OUT1;
        YN2    = OUT2;
        Q      = (CMP+VDD1)/(2*VDD1);
        NQ     = (VDD1-CMP)/(2*VDD1);
        }
    if (OUT1 > VDD1)
        OUT1   = VDD1;
    if (OUT2 > VDD1)
        OUT2   = VDD1;
    if (OUT1 < VSS1)
        OUT1   = VSS1;
    if (OUT2 < VSS1)
        OUT2   = VSS1;

        OUTI1  = ((VDD+VSS)/2) + OUT1;
        OUTI2  = ((VDD+VSS)/2) + OUT2;
        OUTCMP = ((VDD+VSS)/2) + CMP;
}
```

2) Schematics of the general purpose blocks of the class-D amplifier

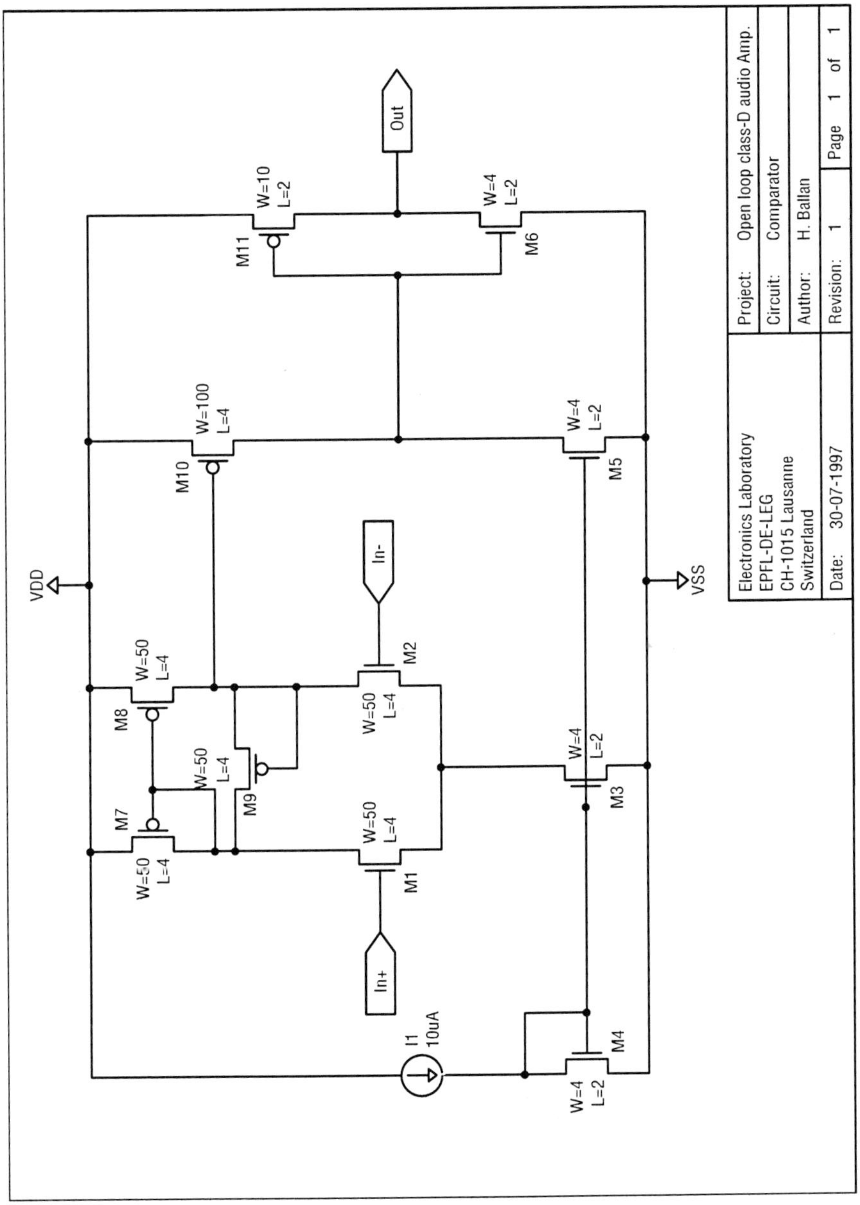

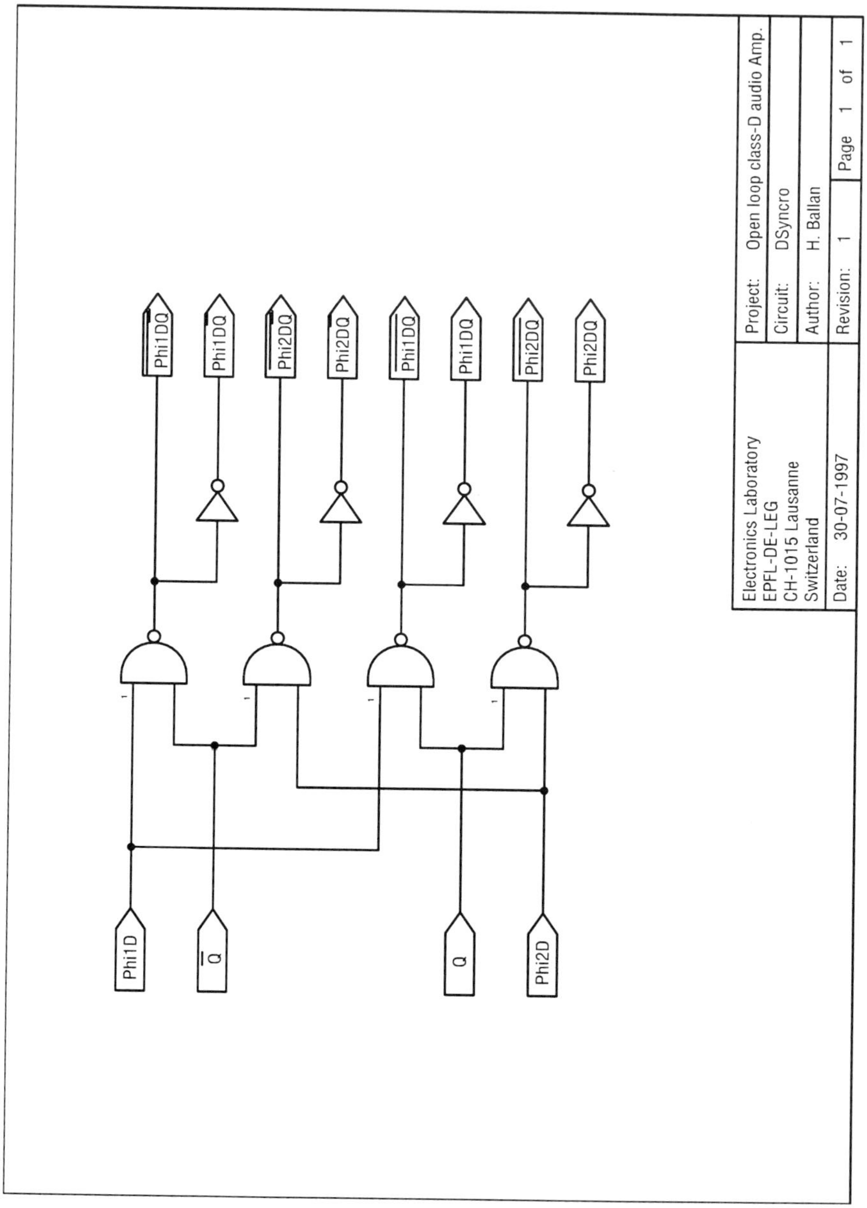
Phi1D
Q
Q
Phi2D
Phi1DQ
Phi1DQ
Phi2DQ
Phi2DQ
Phi1DQ
Phi1DQ
Phi2DQ
Phi2DQ
Electronics Laboratory
EPFL-DE-LEG
CH-1015 Lausanne
Switzerland
Date: 30-07-1997
Project: Open loop class-D audio Amp.
Circuit: DSyncro
Author: H. Ballan
Revision: 1
Page 1 of 1

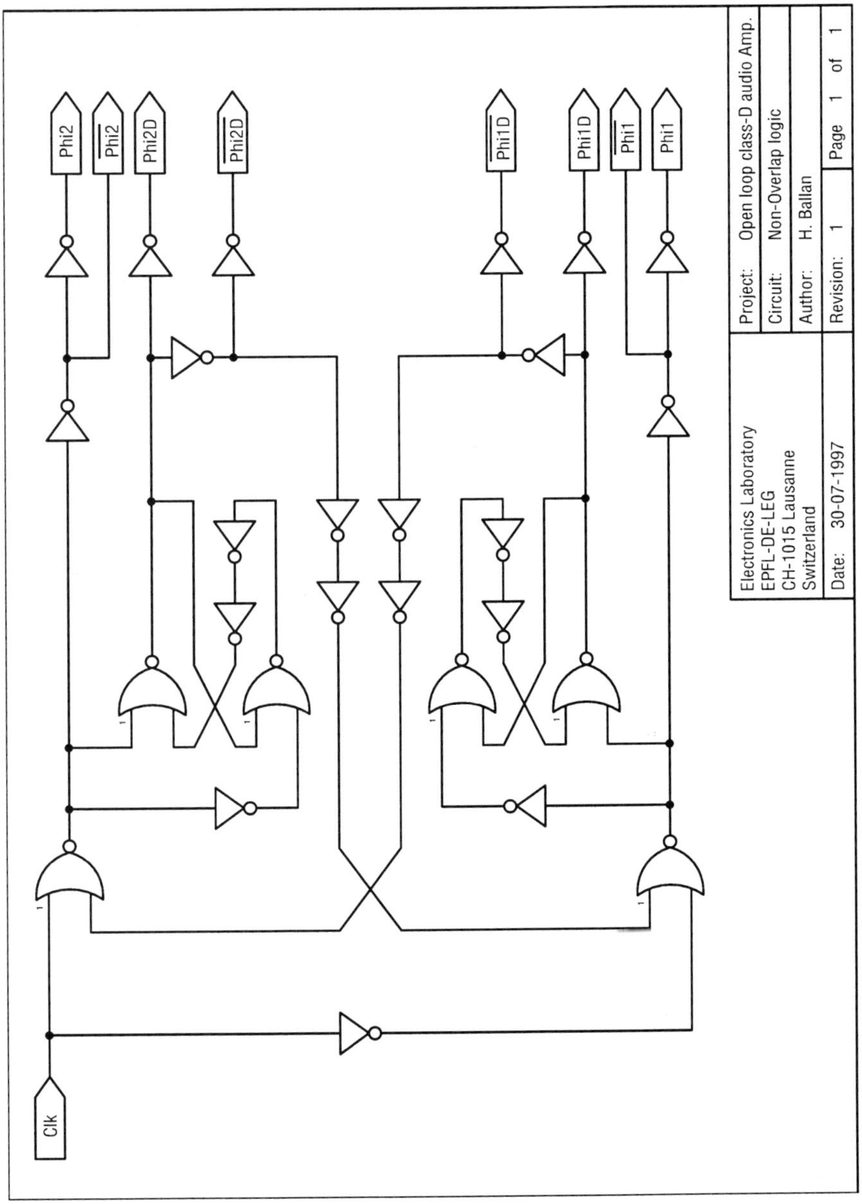
Phi2
Phi2
Phi2D
Phi2D
Phi1D
Phi1D
Phi1
Phi1
Clk
Electronics Laboratory
EPFL-DE-LEG
CH-1015 Lausanne
Switzerland
Project: Open loop class-D audio Amp.
Circuit: Non-Overlap logic
Author: H. Ballan
Date: 30-07-1997
Revision: 1
Page 1 of 1

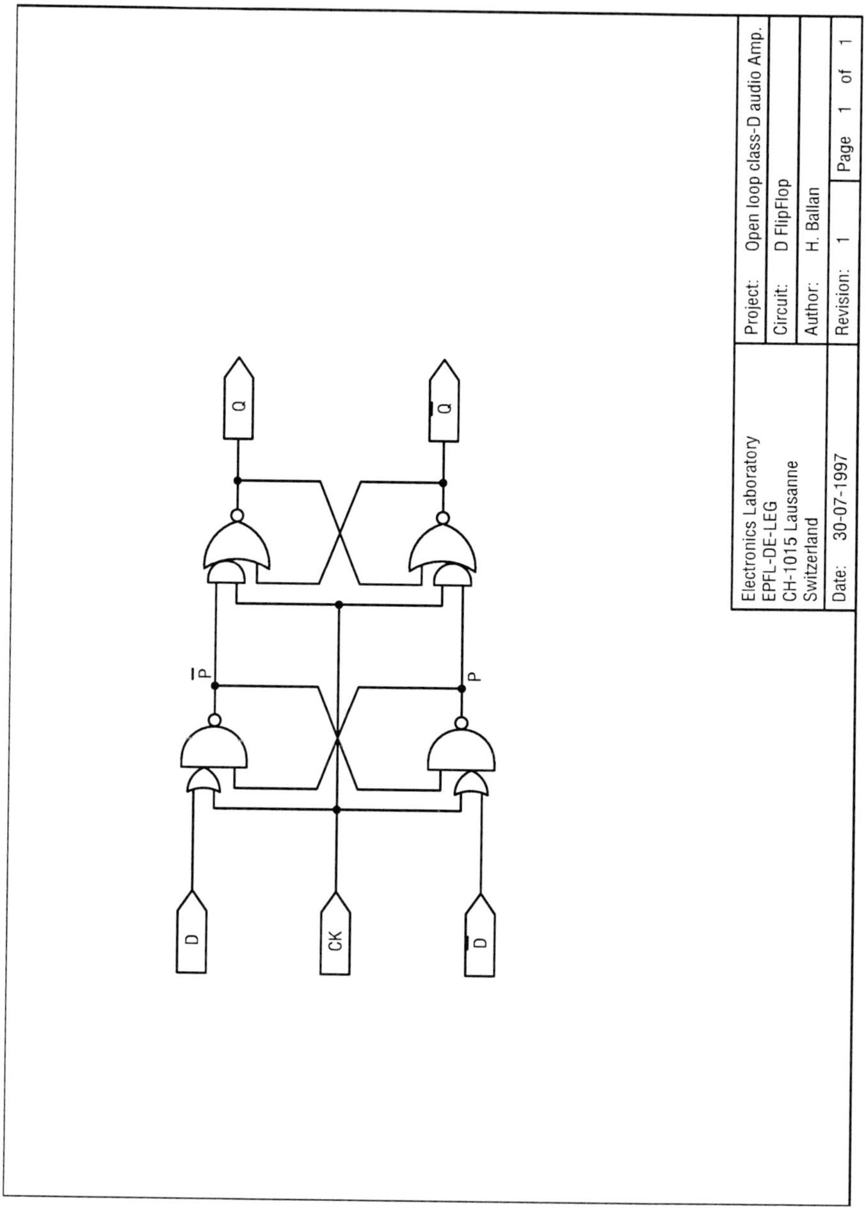

3) References

[1] Dolphin integration, Meylan Cedex, France

INDEX

Abrupt-junction 25; 26; 28; 29; 30
Accumulation... 37; 128-131;140; 164
Aluminum 80
Amplifier ..4; 125; 178; 188; 207-265
Automotive 1; 52; 53
Avalanche
  breakdown 5; 14; 23-43; 54-67
  ....; 91-118
  hot-electrons 15; 19
  multiplication....................... 48; 67

Ballasts 52
Barrier
  energy 7; 8; 12; 15; 16; 43-46
  lowering......................... 6; 44; 110
  potential..................... 44; 104; 110
Base current 40; 41
Base resistance...................... 171; 172
Base-emitter junction..... 42; 249; 250
Bipolar transistor
  open-base 40
  parasitic 40-42; 175
Breakdown
  avalanche 5; 14; 23-43; 54-67
  ....; 91-118
  cylindrical 104
  gate-oxide 42
  snapback 5; 7; 23; 40; 43
  spherical.....................................34
  surface.................................... 5; 23

Buffer region 53
Built-in potential 46; 131; 194
Bulk current 133; 134; 172
Buried layer............52-56; 68; 69; 72
Burn-in 43

Capacitance
  drain-to-bulk 142; 145; 164; 165
  field-plate ..128-130; 140; 142; 164
  gate-oxide................................ 140
  gate-to-bulk 128; 140
  gate-to-drain 130; 142; 145; 163
  ....; 164; 197; 198; 200
  gate-to-source 129; 140
Carrier guard
  majority 171; 173; 177; 178
  minority 174; 178
Cascode 208; 213; 219; 239
Charge injection 236
Current flow lines......... 108; 110; 114

Delta-sigma 227
Depletion layer............ 23-67; 97-110
  ....; 120-148; 156; 167
Device
  degradation.........................6; 7; 16
  lifetime 8; 20; 21; 22
DIBL 43-47; 104; 122; 134
Diffusion
  lateral 58; 86; 91-107; 122; 132-157

window 30; 33; 35; 97
Diode
abrupt-junction 25; 26
body 125; 131; 203; 255
diffused-junction 35
fast-recovery 265
graded-junction 27
DMOSFET
lateral 51-71; 85; 101-103; 146; 167
vertical 53-56; 69
Dopant 30
Drain
drift region 54
lightly doped . 20; 38; 53-58; 63; 67
.... ; 85-167

Efficiency
field-plate 66; 68; 98
power 242-248; 255; 260-265
Electric field
concentration 37; 38; 67; 107
lines 30-36; 63; 66; 86; 98
.... ; 107-113; 133; 156
strength 42; 43
Electromigration 246; 248
Electron injection....... 15; 19; 40; 172
Emitter . 40-42; 73; 171-172; 249-250
Energy barrier........7; 8; 12-16; 43-46
Epitaxial layer. 38; 54-71; 85; 97; 103
Equipotential lines 107; 112; 113
Etching 78-80; 86-87; 97-98; 132
.... ; 139; 142
Extrinsic capacitance 127; 164

Factor
ideality 125; 136
mobility degradation 157
multiplication 10; 41; 42; 134
Failure
B-mode 43
Field-oxide 79; 86; 87; 93
Field-plate .. 36-39; 49; 63-68; 85-164
Filp-Flop
high-voltage 182; 236
Flat band 65
Follower . 57; 188; 190; 207; 210-212
.... ; 221; 222
Gate
current 7; 12-21; 112; 121; 134
field-plate 63; 87
overlap 63; 164; 220
Gauss' law 64
Gaussian profile 98; 104
Graded-junction 24; 27-30; 62; 170
Grooves 69

Half-bridge ...228; 244; 246; 252; 253
Harmonic
components 228; 229; 259
distortion 235; 236; 248; 258
tones 263
High-pass 229; 231
High-temperature 29
High-voltage
analog output interface 207
buffer188; 190; 198; 200; 207; 246
cascode 208; 213
class-AB output 215
digital output interface 181; 192
DMOS 73
flip-flop 182
integrated circuits (HVICs) .. 6; 117
.... ; 143; 179; 188; 192
nMOS .. 38; 85-90; 97-99; 106; 107
.... ; 118; 121; 127; 138; 174
OPAMP 210
OTA 211; 213
output stage 192
pMOS 91-96; 104; 105; 110
.... ; 132; 139
source follower...........57; 188; 190
technologies 6; 51
wires 167; 177; 178
Hot-carrier
degradation.................................21
effects 5; 6; 19; 23; 42; 43; 75
.... ; 107; 111; 134; 180; 185
generation................. 112; 121; 192
injection 8

Hot-electron
channel 7; 43
effects 8
emission 43
substrate 17
Hot-holes
injection 17; 19
Hot-nodes 192
Hysteresis 249; 250

Ideality factor........................ 125; 136
Impact ionisation . 7-15; 23-24; 30-42
........................; 108-115; 133; 172
Impedance........52; 193-206; 212-214
....; 220-228
Implant
diffusion..80
dose........................ 48; 78; 79; 101
energy 79; 86
n-well 85
n^+ 68; 87
p^+ 87
V_t adjust......... 46; 87; 93; 139; 142
Induced barrier lowering.... 6; 44; 110
Inelastic collision 14
Input protection 194
Integrator 231-237
Interface states 7; 16-19
Intermodulation
distortion.......................... 259; 260
Inversion layer......... 38; 54; 130; 174
Inverter 180

Junction
abrupt.................................... 25-30
base-emitter 42; 249; 250
curvature............ 30; 38; 63; 66; 98
cylindrical 32-34
deep 33; 62; 102
diffused 24-38; 104
linearly graded 24-30; 170
metallurgical .. 17; 24; 32-34; 62-63
....; 86; 137
shallow......................... 20; 33; 170
spherical.................. 24; 33; 34; 97

Latchup7; 69; 118; 166; 171-177; 203
Leakage current............. 69; 124; 134
Level shifter
low-standby current .. 188; 206; 246
static 180-206; 220; 246
Lifetime model 20; 21; 22
Linearly graded junction ...24-30; 170
LOCOS 79; 86; 93

Majority carrier
current 171; 172
guards 178
Majority carriers 175
Maximum
channel electric field 19
current density 252
gate current................................ 19
gate voltage 180
output current 180; 209; 216
output power 230
power efficiency.............. 242; 246
substrate current 17
Mean-free-path
re-direction 14
scattering 14
Metal
evaporation....................... 246; 248
wires 171; 228; 252
Metallurgical junction ... 17; 24; 32-34
.............................; 62-63; 86; 137
Minority carrier guards 175; 178
Minority carriers 6; 175; 177
Mirror
current 188; 191; 207; 250
voltage............. 182; 207; 208; 219
Mobility
degradation......................... 17; 157
field dependent 151
Model
lucky electron 13
static device lifetime 21; 22
substrate current 121
Modulation
channel length 144; 145; 157

Modulator
PWM 262
MOSFET
degradation 16
field.......... 118; 166; 173; 177; 178
isolated..52
long-channel 45
short-channel 41
submicron 121
vertical 52
Multiplication factor.. 10; 41; 42; 134

Negative
charge density 16; 19
resistance 41
thermal coefficient 249; 250
Nitride layer........... 78; 79; 86; 92; 93
Noise shaping 229; 230; 231; 264

Pad ring..242
Parasitic
bipolar transistor.... 40; 41; 42; 175
field MOSFET. 166; 173; 177; 178
JFET 56
lateral bipolar tansistor 42
resistance 54; 103; 136
Passivation.....................................80
Poisson's equation. 27; 31; 44; 63; 64
................................; 97; 104; 143
Polysilicon gate................. 58; 79; 93
Potential
barrier 44; 104; 110
distribution... 25; 31; 34; 37; 45; 61
......; 67; 100; 107-113; 120; 152
profiles 25; 27
Power
amplifier 1; 53; 228
applications 131; 265
audio systems...........................236
consumption. 1; 180; 238; 246; 248
dissipation.......... 54; 188; 212; 246
MOSFET 52; 53
spectral density 229
Precautions
layout 117; 176; 177; 203
Production yield.................. 242; 261
Profile
Gaussian 98; 104
Protection
over-current 228; 254; 255
over-temperature 250
transistors 216; 222-225
Punchthrough
breakdown. 5; 58-61; 104-118; 166
current 6; 48; 108; 114; 124

Recovery diode 265
Regulators 1; 52
Relative deviation 183-191
Reliability.....20; 23; 56; 77; 117; 185
....; 192
Repulsion field 14
Resistance
base 171; 172
channel 128
lightly doped drain 124; 127
............................; 136-144; 154
ON state 75; 124; 136; 137; 159
.... ; 160; 197-207; 228; 231
.... ; 242-244; 252-261
parasitic 54; 103; 136
sheet 253
substrate 40; 42
RESURF devices 69; 76
Richardson's expression................. 12
Routing channels................. 171; 178

Safe operating conditions..... 192; 196
....; 222
Sampling frequency234; 241; 261-265
Saturation
current120; 125; 132; 136; 148; 239
mode... 75; 107; 111; 129-158; 183
...........................; 198; 238; 239
region . 8; 9; 42; 128; 138; 149-152
velocity 144
voltage.. 9; 120; 132; 139; 149-158
Scattering mean-free-path 14
Second order Δ-Σ 233-236

Secondary
electrons.......................................7
holes...7
Self-aligned 72; 93
Self-heating 128; 132; 139; 158
Settling time.................................239
Shallow junctions............ 20; 33; 170
Shielding............................ 167; 178
Short-channel 40; 41
Short-circuits 171; 173
Sidewall spacer..............................20
Signal processing..... 1; 2; 6; 178; 179
.... ; 207; 227
Signal-to-noise ratio 229-235; 261-265
Si-SiO_2
interface 7; 8; 13; 17; 43; 64
Slew-rate 192-226; 265
Snapback 5; 7; 23; 40-43
Space charge 23; 42; 64; 65
Spacers...20
Spectral density............................229
Spherical junction........ 24; 33; 34; 97
Spots 43
Stability........ 209; 214; 233; 237; 241
Static device lifetime 21; 22
Static level shifter 180-220; 246
Step approximation 101-105; 146-156
.... ; 167; 170
Stepped field-plate 39; 112
Stress
AC 17; 21; 22
DC 17; 19; 20
Strong inversion..... 45; 210; 212; 215
Submicron........ 6; 121; 130; 143; 144
Substrate current7-21; 40-48; 118-121
.......; 133; 144; 158; 166; 172; 177
Subthreshold slope... 16; 46; 122; 123
.... ; 134
Surface
charge 35; 36
density............2; 29; 228; 242-244
DIBL................................ 122; 134
electric field 17; 56; 64
potential 37; 45; 46
Switched capacitor....... 228; 236; 266
Switching
applications 124
frequency...188; 190; 228-230; 262
losses 246; 261
speed 52

Termination
junction 24; 30; 33; 38
Thermal oxidation 79
Threshold voltage........ 2-7; 16-20; 38
; 46; 53; 63; 75; 79; 121; 122; 134
; 144; 157; 178-183; 190; 191; 239
Transconductance 6; 16; 17; 121; 127
.... ; 138; 157; 162; 163; 200-241
Transfer function................. 229; 231
Transients
output stage ..3; 200; 202; 207; 223
power supply 192-207; 219; 222
Transit time 73
Transition frequency 73
Trapped
electrons 19
holes 18; 19
Traps 19; 22
Tubs 70
Turn
OFF 203; 206; 212; 255
ON..... 75; 171; 222; 224; 248; 251
.... ; 255

Velocity
electrons 133
holes 133
saturation..... 10; 107-113; 139-157
Vertical
bipolar 172
DMOSFET..............53; 54; 56; 69
PNP 172; 175
VLSI 1; 6; 9; 47-49; 121; 143; 178
VMOS 52; 116